AF361752

Physiology and Pathophysiology of Iron Metabolism

Physiology and Pathophysiology of Iron Metabolism

Editor

Dietmar Enko

Basel • Beijing • Wuhan • Barcelona • Belgrade • Novi Sad • Cluj • Manchester

Editor
Dietmar Enko
Medical University of Graz
Graz, Austria

Editorial Office
MDPI
St. Alban-Anlage 66
4052 Basel, Switzerland

This is a reprint of articles from the Special Issue published online in the open access journal *Nutrients* (ISSN 2072-6643) (available at: https://www.mdpi.com/journal/nutrients/special_issues/iron_physiology).

For citation purposes, cite each article independently as indicated on the article page online and as indicated below:

Lastname, A.A.; Lastname, B.B. Article Title. *Journal Name* **Year**, *Volume Number*, Page Range.

ISBN 978-3-0365-9590-0 (Hbk)
ISBN 978-3-0365-9591-7 (PDF)
doi.org/10.3390/books978-3-0365-9591-7

Contents

About the Editor

Dietmar Enko

Docent Dr. Dietmar Enko, medical education completed at the Karl-Franzens University of Graz (Austria) and trained as a general practitioner, is a specialist in medical and chemical laboratory diagnostics and a specialist in transfusion medicine. Habilitation, teaching, and research carried out at the Medical University of Graz (Austria). Since January 2019, he has served as the head of the Institute of Medical and Chemical Laboratory Diagnostics at the Hochsteiermark Hospital, the teaching hospital of the Medical University of Graz (Austria). Research topics: iron metabolism, carbohydrate metabolism, food intolerance, and applied biomarker research.

Preface

This scientific work represents a substantial contribution to society as part of the Third Mission of the Medical University of Graz (Austria). The aim of this book is to make the topic of iron metabolism accessible to a broader audience. The trace element iron is a paradox. On the one hand, it is a vital component of numerous biochemical processes in the human body; on the other hand, this metal is very reactive due to its ease of electron donation and acceptance and can cause great damage due to its high toxic potential.

Dietmar Enko

Editor

Article

Does the Combined Effect of Resistance Training with EPO and Iron Sulfate Improve Iron Metabolism in Older Individuals with End-Stage Renal Disease?

Hugo de Luca Corrêa [1,*], Víctor Manuel Alfaro-Magallanes [2], Sting Ray Gouveia Moura [1], Rodrigo Vanerson Passos Neves [1], Lysleine Alves Deus [1], Fernando Sousa Honorato [1], Victor Lopes Silva [1], Artur Temizio Oppelt Raab [3], Beatriz Carneiro Habbema Maia [3], Isabela Akaishi Padula [3], Lucas Santos de Gusmão Alves [3], Rafaela Araújo Machado [3], Andrea Lucena Reis [1], Jonato Prestes [1], Carlos Ernesto Santos Ferreira [1], Luiz Sinésio da Silva Neto [4], Fernanda Silveira Tavares [3], Rosângela Vieira Andrade [5] and Thiago dos Santos Rosa [1]

[1] Graduate Program of Physical Education, Catholic University of Brasilia, Federal District, Brasilia 71966-700, Brazil; stingraygm@hotmail.com (S.R.G.M.); rpassosneves@yahoo.com.br (R.V.P.N.); lys.deus@gmail.com (L.A.D.); ferhon@gmail.com (F.S.H.); victornutri@yahoo.com.br (V.L.S.); andrealucereis@gmail.com (A.L.R.); jonatop@gmail.com (J.P.); ernestobsb@gmail.com (C.E.S.F.); thiagoacsdkp@yahoo.com.br (T.d.S.R.)
[2] LFE Research Group, Department of Health and Human Performance, Faculty of Physical Activity and Sport Sciences (INEF), Universidad Politécnica de Madrid (UPM), 28040 Madrid, Spain; vm.alfaro@upm.es
[3] Department of Medicine, Catholic University of Brasilia, Federal District, Brasilia 71966-700, Brazil; artur.t.oppelt@gmail.com (A.T.O.R.); beatriz.maia@a.ucb.br (B.C.H.M.); bela.akaishi@gmail.com (I.A.P.); lucasgusmao1204@gmail.com (L.S.d.G.A.); rafaelamacha99@gmail.com (R.A.M.); fernanda.endocrino@gmail.com (F.S.T.)
[4] Faculty of Medicine, Federal University of Tocantins, Para 77402-970, Brazil; luizneto@mail.uft.edu.br
[5] Graduate Program of Genomic Sciences and Biotechnology, Catholic University of Brasilia, Federal District, Brasilia 71966-700, Brazil; rosangelavand@gmail.com
* Correspondence: hugo.efucb@gmail.com

Citation: Corrêa, H.d.L.; Alfaro-Magallanes, V.M.; Moura, S.R.G.; Neves, R.V.P.; Deus, L.A.; Honorato, F.S.; Silva, V.L.; Raab, A.T.O.; Maia, B.C.H.; Padula, I.A.; et al. Does the Combined Effect of Resistance Training with EPO and Iron Sulfate Improve Iron Metabolism in Older Individuals with End-Stage Renal Disease? *Nutrients* **2021**, *13*, 3250. https://doi.org/10.3390/nu13093250

Academic Editor: Dietmar Enko

Received: 11 August 2021
Accepted: 16 September 2021
Published: 18 September 2021

Publisher's Note: MDPI stays neutral with regard to jurisdictional claims in published maps and institutional affiliations.

Abstract: We sought to investigate the effects of resistance training (RT) combined with erythropoietin (EPO) and iron sulfate on the hemoglobin, hepcidin, ferritin, iron status, and inflammatory profile in older individuals with end-stage renal disease (ESRD). ESRD patients (n: 157; age: 66.8 ± 3.6; body mass: 73 ± 15; body mass index: 27 ± 3), were assigned to control (CTL; n: 76) and exercise groups (RT; n: 81). The CTL group was divided according to the iron treatment received: without iron treatment (CTL—none; n = 19), treated only with iron sulfate or EPO (CTL—EPO or IRON; n = 19), and treated with both iron sulfate and EPO (CTL—EPO + IRON; n = 76). The RT group followed the same pattern: (RT—none; n − 20), (RT—EPO or IRON; n = 18), and (RT—EPO + IRON; n − 86). RT consisted of 24 weeks/3 days per week at moderate intensity of full-body resistance exercises prior to the hemodialysis section. The RT group, regardless of the iron treatment, improved iron metabolism in older individuals with ESRD. These results provide some clues on the effects of RT and its combination with EPO and iron sulfate in this population, highlighting RT as an important coadjutant in ESRD-iron deficiency.

Keywords: iron metabolism; chronic kidney disease; hepcidin; erythropoietin; exercise training; anemia; nephrology

1. Introduction

End-stage renal disease (ESRD) patients often present impaired intestinal absorption of dietary iron, blood losses, and low-grade chronic inflammation, which can lead to difficulties in achieving adequate iron status [1,2]. This adverse condition is one of the main causes of hyporesponsiveness to therapy involving erythropoiesis-stimulating agents (ESA), e.g., erythropoietin [2,3]. However, this supplementation, besides being associated with gastrointestinal discomfort, may impair the absorption of other medication, and alter the gut microbiota and systemic metabolome [2–4]. Thus, in the majority of cases, the patient requires an intravenous infusion of iron to treat this condition. Indeed, parenteral administration of iron is safer and more efficient than oral therapy, but not free from adverse events [4]. In this regard, patients with ESRD require oral iron supplementation to manage hemoglobin levels [1–3,5].

This adverse scenario motivates physicians and researchers to seek alternative and non-pharmacological treatments to counteract the adverse events related to ESA therapy and iron supplementation. A recent study from our laboratory outlined the positive effects of resistance training (RT) on iron deficiency in ESRD patients, suggesting this training model as an adjunct treatment for anemia [6]. In our study, 72.61% of the patients underwent EPO treatment (control group (CTL) = 69.73%; resistance training group (RT) = 75.30%) and 54.14% underwent oral iron supplementation (CTL = 55.26%; RT = 53.08%). Given that both treatments may influence multifactorial pathways related to inflammation and iron status, and RT is an important regulator of both aforementioned factors [6–9], two fundamental questions are raised here: (I) Does RT enhance the treatment with EPO and iron sulfate in ESRD older individuals? (II) Is RT an effective treatment on its own to improve iron status?

Attempting to answer this question, we sought to investigate the effects of RT combined with EPO and iron sulfate on the hemoglobin, hepcidin, ferritin, iron status, and inflammatory profile in older individuals with ESRD. We hypothesized that RT would enhance the effects of EPO and iron sulfate in this population. If confirmed, such findings might point to RT as an important adjunct therapy for iron deficiency that may work together with therapy involving erythropoiesis-stimulating agents.

2. Materials and Methods

2.1. Procedures

This study is part of a large trial [6]. Briefly, all participants involved in the study read and agreed with the written informed consent. The experimental protocols were approved by the Local Ethics Committee. All procedures were carried out conforming to the principles outlined in the declaration of Helsinki (1975). The participants were randomized into two groups by simple randomization: the control group (CTL; $n = 76$) and the resistance training group (RT; $n = 81$).

Here, we stratified the patients into six subgroups according to the iron treatment received. The CTL group was divided as follows: without iron treatment (CTL—none; $n = 19$), treated only with iron sulfate or EPO (CTL—EPO or IRON; $n = 19$), and treated with both iron sulfate and EPO (CTL—EPO + IRON; $n = 38$). The RT groups followed the same pattern (RT—none; $n = 20$), (RT—EPO or IRON; $n = 18$), (RT—EPO + IRON; $n = 43$), as described in Figure 1. The protocol of RT occurred before the hemodialysis section and was described in detail by Moura et al. [6].

2.2. Iron and ESA Treatments

The protocols of iron and ESA agents were performed according to the parameters established by the Ministry of Health in Brazil for ESRD [10]. Briefly, patients on hemodialysis are initially treated with one of the following options, later adjusted according to the therapeutic response: -50–100 UI/Kg, subcutaneously, divided into 1 to 3 applications per week; 50–100 UI/Kg, intravenously, divided into 3 applications per week. If after four weeks of treatment, the hemoglobin elevation is less than 0.3 g/dL per week: increase the dose by 25%, respecting the maximum dose limit, which is 300 IU/Kg/week per route subcutaneously and 450 IU/kg/week intravenously. If, after four weeks of treatment, the hemoglobin elevation is in the range of 0.3–0.5 g/dL per week: keep the dose in use. If, after four weeks, the hemoglobin elevation is greater than 0.5 g/dL per week or the hemoglobin level is between 12 and 13 g/dL: reduce the dose by 25% to 50%, respecting the minimum dose limit recommended, which is 50 UI/Kg/week subcutaneously. Temporarily suspend treatment if the hemoglobin level is above 13 g/dL.

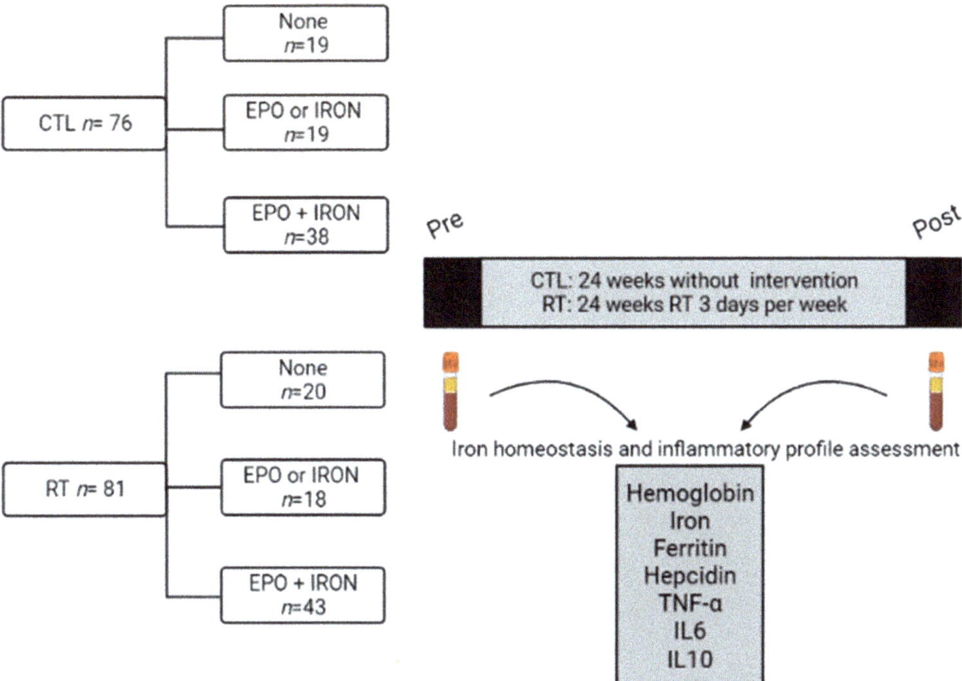

Figure 1. Experimental design. CTL: control group; RT: resistance training group; EPO: Erythropoietin; TNF: tumor necrosis factor; IL: interleukin.

Treatment should be continuous, targeting the hemoglobin at 11 g/dL. Temporary interruption of treatment is recommended if the hemoglobin level is above 13 g/dL, with resumption when hemoglobin levels are below 11 g/dL. Discontinuation should be considered in the event of a serious adverse event.

Iron III hydroxide saccharate is for intravenous use and is presented in 5 ml ampoules containing 100 mg of iron III (20 mg/mL). It must be diluted in 100 mL of saline solution and infused within 15 min, according to the manufacturer. A study demonstrates its safe use in shorter administration times, up to 5 min, without increasing adverse reactions.

2.3. Biochemical Analysis

Venous blood samples were obtained at baseline, and after 24 weeks of training, to measure the iron metabolism status and inflammatory profile. All analyses were described elsewhere [6].

2.4. Statistical Analysis

Initially, normality and homogeneity of data were verified using the Shapiro–Wilk and Levene tests, respectively. A three-way ANOVA $2 \times 2 \times 3$ (Group $\times$ Time $\times$ iron treatment) was performed to compare groups. Deltas (post-pre) were obtained from all groups and compared using a two-way mixed ANOVA 2×3 (group $\times$ iron treatment) followed by Tukey's post-hoc. Results were considered significant at $p < 0.05$. All statistical analyses were performed using SPSS Statistics for Windows, version 22.0 (released 2013) (IBM Corp., Armonk, NY, USA), and GraphPad Prism for Windows, version 8.0.0 (GraphPad Software, San Diego, CA, USA).

3. Results

Subjects that completed the RT protocol did not present adverse effects. Baseline characteristics are displayed in Table 1.

Table 1. Baseline characteristics for each group according to iron treatment.

Variables	CTL			RT			*p* Value	Eta^2
	None	EPO or IRON	EPO + IRON	None	EPO or IRON	EPO + IRON		
Age (years)	66.16 ± 4	66 ± 4.14	66.58 ± 3.77	66.75 ± 3.34	67.83 ± 3.33	67.28 ± 3.19	0.687	0.005
Body mass (kg)	71.47 ± 14.72	72.33 ± 15.46	73.61 ± 14.38	71.38 ± 16.76	77.03 ± 17.31	73.78 ± 16.41	0.740	0.004
Body mass index (kg/m^2)	26.54 ± 2.85	26.78 ± 3.12	26.98 ± 2.88	26.66 ± 3.75	27.95 ± 3.86	27.32 ± 3.78	0.769	0.003
Waist circumference (cm)	94.54 ± 12.84	94.91 ± 12.12	96.27 ± 11.87	93.9 ± 12.24	98.12 ± 12.13	95.47 ± 11.41	0.680	0.005

Data expressed as means and standard deviation. CTL: control; RT: resistance training; EPO: erythropoietin.

Patients from the CTL group did not display changes in iron, hemoglobin, ferritin, and hepcidin after this six-month assessment in all iron treatment groups ($p > 0.05$). RT increased hemoglobin from baseline in all groups. Only the RT group treated with EPO + IRON presented higher hemoglobin in relation to the CTL group treated with EPO + IRON. Serum iron increased in relation to baseline and the CTL groups. Ferritin only decreased from baseline. The RT group presented a decrease in hepcidin in relation to baseline and in relation to the CTL group that received the same iron therapy, as described in Figure 2.

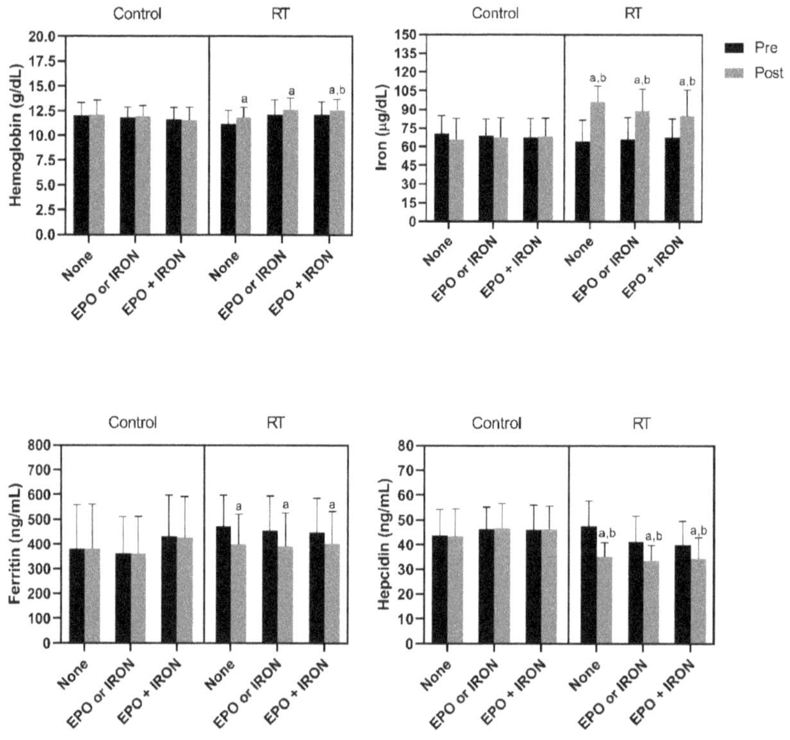

Figure 2. RT modulated iron, hemoglobin, ferritin and hepcidin levels in hemodialysis patients regardless of iron treatment. RT: resistance training; EPO: erythropoietin. Data expressed by mean ± SD. [a] $p < 0.05$ in relation to pre (within group and treatment). [b] $p < 0.05$ in relation to CTL post (between group and within treatment).

Patients from the RT groups displayed an improvement in the inflammatory profile, presenting a decrease in TNF and IL6 and an increase in IL10 when compared to baseline and to the CTL groups, as described in Figure 3.

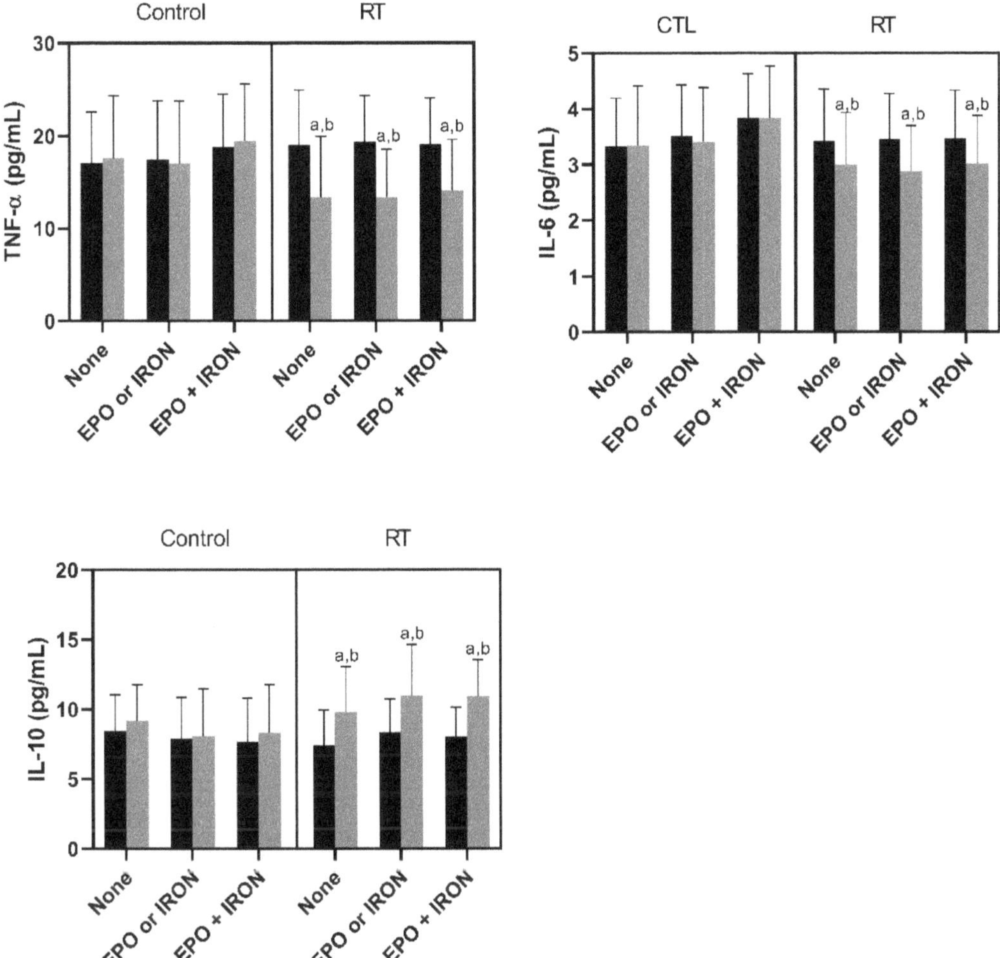

Figure 3. Cytokine modulation following RT in hemodialysis patients. RT: resistance training; EPO: erythropoietin; TNF: tumor necrosis factor; IL: interleukin. Data expressed by mean $\pm$ SD. [a] $p < 0.05$ in relation to pre (within group and treatment). [b] $p < 0.05$ in relation to CTL post (between group and within treatment).

As displayed in Figure 4, regardless of the treatment, the RT group showed modulated hemoglobin, iron, ferritin, hepcidin, and cytokine levels. However, RT—IRON + EPO presented a lower decrease in hepcidin in relation to RT—none.

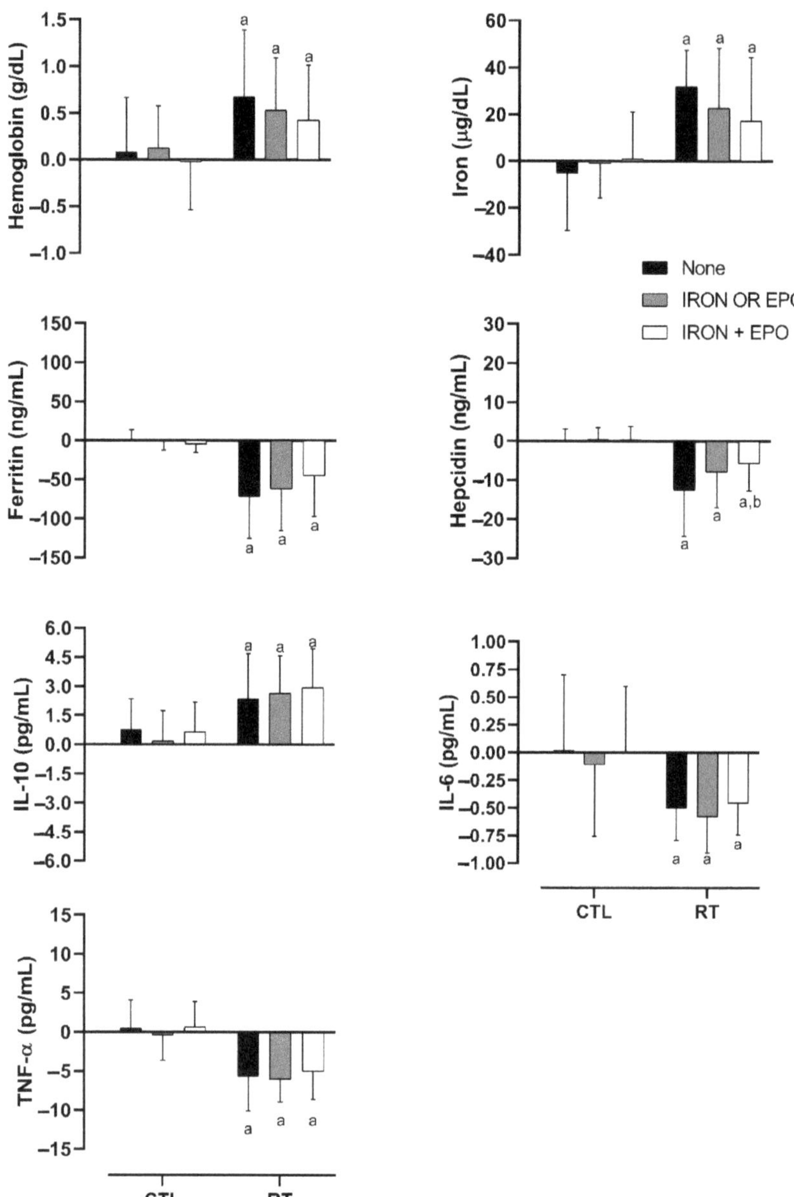

Figure 4. Deltas (post/pre) of hemoglobin, iron, ferritin, hepcidin, and cytokine response to RT following different iron treatments. CTL: control group; RT: resistance training group. [a] $p < 0.05$ in relation to the corresponding treatment in CTL. [b] $p < 0.05$ in relation to RT—none.

4. Discussion

The aim of the present study was to investigate the effects of RT combined with EPO and iron sulfate on the hemoglobin, hepcidin, ferritin, iron status, and inflammatory profile in older individuals with ESRD. Here, we found that regardless of iron treatment, RT appeared to improve serum iron homeostasis parameters in older subjects with chronic kidney disease. These results may point to two important insights: (1) that RT is an effective treatment for improving serum iron parameters in elderly patients with ESRD and (2) that conventional use of ESA + iron in this population is more effective in combination with RT. Interestingly, as RT alone improved serum iron availability, it may provide the opportunity to remove ferrous sulfate and its associated disturbances from the treatment in this population. This would limit iron accumulation and, in theory, further reduce hepcidin, as suggested by the greater reduction in hepcidin in the RT—none group compared to the RT—IRON + EPO group. Furthermore, although the small sample size did not allow us to perform this analysis, the combination of RT + ESA is probably the most promising treatment for lowering hepcidin levels. This combination would reduce the activation of the two main pathways of hepcidin synthesis: the JAK/STAT3 pathway, by lowering inflammation (RT effects), and the BMP/SMAD pathway, by increasing erythropoiesis activity and sequestering erythroferrone BMP2/6 ligands [11,12].

According to the results found by Agarwal et al. [5], non-hemodialysis CKD patients on oral iron therapy have improved hemoglobin, TIBC, transferrin saturation and ferritin. However, although our participants received oral iron supplementation, they were also on hemodialysis treatment, which is known to increase iron losses. It was estimated that CKD patients on hemodialysis lose 1 to 3 g of iron per year, which, coupled with the fact that hemodialysis patients have particularly impaired dietary iron absorption, appears to make oral iron supplementation poorly effective in improving iron markers [13]. In fact, oral iron supplementation was no better than placebo in improving anemia, improving or preventing iron deficiency, or reducing ESA dosage in hemodialysis patients [13–16]. Therefore, the findings from the present study can provide clues on the application of RT in this population to counteract iron-related diseases. To date, no studies have investigated the pooled effects of exercise plus ESA therapy or iron supplementation after 24 weeks.

Anemia and iron deficiency is a common complication of hemodialysis patients [1–3,5]. Because of that, ESA and iron supplementation may be considered to improve health-related parameters in this population [1,2]. RT has appeared as a non-pharmacological therapy to improve iron metabolism in this population [6,8,17]. The possibility of using RT as part of the treatment of iron deficiency may lead to relevant management of anemia biomarkers (Figures 2 and 3). Moreover, it could lead to a cost reduction in the treatment of anemia in ESRD patients, due to the possibility of reducing or removing drugs administered for this purpose. Nonetheless, although this study may point to important insights for future research, the present findings should be interpreted cautiously, because we did not control and randomize the sample according to the iron treatment.

Anemia leads to reduced quality of life, fatigue, dyspnea, and impaired cognitive capacity. Furthermore, it is associated with a greater risk of adverse cardiovascular events and mortality. Therefore, these conditions usually end up leading to a greater number of hospitalizations, with increased costs for the health system. In Brazil, it is estimated that 133,464 patients were on dialysis in 2018, and the use of EPO was part of the treatment of more than 80% of these patients. The annual cost of treatment for anemia in patients with chronic kidney disease can reach US $3241.65 dollars, while the minimum monthly wage for Brazilian citizens is around US $222 dollars, leading to an economic burden for the treatment of anemia [18]. Therefore, a key finding of the present study was the improvement in iron, hemoglobin, ferritin and hepcidin markers, regardless of the use of EPO, suggesting that the application of RT in hemodialysis clinics would be more cost-effective.

The present manuscript had an important limitation: the study did not control for the time of iron supplementation or EPO, which might influence the dependent variables. We recommend further studies to control for this condition. As this study is an additional analysis of a larger trial [6], it was not initially designed for this subgroup analysis, which is why there is a low sample size for each iron treatment group. The lack of analysis related to nutritional markers and electrolytes also limits our study since it could influence iron metabolism. However, to date, this is the first study to demonstrate that RT is capable of inducing changes in iron metabolism regardless of iron treatment. Moreover, these additional findings may open perspectives about the combined effect of RT and pharmacological iron treatments and encourage further studies to be designed to answer this question in ESRD patients with iron deficiency.

5. Conclusions

We conclude that regardless of the iron treatment, RT could improve hemoglobin, iron, ferritin and hepcidin in older individuals with ESRD. These novel findings provide some clues on the combined effect of RT plus EPO and iron sulfate in this population. Moreover, RT alone may also be an effective strategy to improve iron metabolism in hemodialysis patients. Therefore, it is rational to infer that the application of RT programs should be strongly recommended in dialysis care. This would improve the prognosis of several ESRD patients, especially those with iron deficiency and anemia. Further studies are needed to determine whether the treatment with EPO and iron sulfate is more effective in treating iron deficiency and anemia when combined with exercise training in patients with ESRD.

Author Contributions: Conceptualization, H.d.L.C. and T.d.S.R.; methodology, H.d.L.C., V.M.A.-M., S.R.G.M., R.V.P.N. and T.d.S.R.; software, H.d.L.C.; validation, H.d.L.C., L.A.D., F.S.H., V.L.S., A.T.O.R. and T.d.S.R.; formal analysis, H.d.L.C.; investigation, H.d.L.C., B.C.H.M., I.A.P., L.S.d.G.A. and T.d.S.R.; resources, T.d.S.R.; data curation, R.A.M., A.L.R., J.P., C.E.S.F., L.S.d.S.N., F.S.T., R.V.A. and T.d.S.R.; writing—original draft preparation, all authors; writing—review and editing, all authors; visualization, all authors; supervision, T.d.S.R.; project administration, H.d.L.C. and T.d.S.R.; funding acquisition, H.d.L.C. and T.d.S.R. All authors have read and agreed to the published version of the manuscript.

Funding: H.d.L.C. and L.A.D. were supported by a grant provided by the Coordenação de Aperfeiçoamento de Pessoal de Nível Superior—Brazil (CAPES)—Finance Code 001. V.M.A.-M. was supported by a grant provided by the Universidad Politécnica de Madrid.

Institutional Review Board Statement: Written informed consent was obtained from all participants involved in the study. All experimental protocols were approved by the Local Ethics Committee, under the number: 23007319.0.0000.0029. All procedures were carried out conforming to the principles outlined in the declaration of Helsinki (1975).

Informed Consent Statement: Informed consent was obtained from all subjects involved in the study.

Data Availability Statement: Data ara available upon reasonable request to the corresponding author.

Conflicts of Interest: The authors declare no conflict of interest.

References

1. Collister, D.; Rigatto, C.; Tangri, N. Anemia management in chronic kidney disease and dialysis: A narrative review. *Curr. Opin. Nephrol. Hypertens.* **2017**, *26*, 214–218. [CrossRef] [PubMed]
2. Hörl, W.H. Clinical aspects of iron use in the anemia of kidney disease. *J. Am. Soc. Nephrol.* **2007**, *18*, 382–393. [CrossRef] [PubMed]
3. Besarab, A.; Coyne, D.W. Iron supplementation to treat anemia in patients with chronic kidney disease. *Nat. Rev. Nephrol.* **2010**, *6*, 699–710. [CrossRef] [PubMed]
4. Babitt, J.L.; Eisenga, M.F.; Haase, V.H.; Kshirsagar, A.V.; Levin, A.; Locatelli, F.; Małyszko, J.; Swinkels, D.W.; Tarng, D.C.; Cheung, M.; et al. Controversies in optimal anemia management: Conclusions from a Kidney Disease: Improving Global Outcomes (KDIGO) Conference. *Kidney Int.* **2021**, *99*, 1280–1295. [CrossRef] [PubMed]
5. Agarwal, R.; Kusek, J.W.; Pappas, M.K. A randomized trial of intravenous and oral iron in chronic kidney disease. *Kidney Int.* **2015**, *88*, 905–914. [CrossRef] [PubMed]
6. Moura, S.R.G.; Corrêa, H.L.; Neves, R.V.P.; Santos, C.A.R.; Neto, L.S.S.; Silva, V.L.; Souza, M.K.; Deus, L.A.; Reis, A.L.; Simões, H.G.; et al. Effects of resistance training on hepcidin levels and iron bioavailability in older individuals with end-stage renal disease: A randomized controlled trial. *Exp. Gerontol.* **2020**, *139*, 111017. [CrossRef] [PubMed]
7. Corrêa, H.L.; Moura, S.R.G.; Neves, R.V.P.; Tzanno-Martins, C.; Souza, M.K.; Haro, A.S.; Costa, F.; Silva, J.A.B.; Stone, W.; Honorato, F.S.; et al. Resistance training improves sleep quality, redox balance and inflammatory profile in maintenance hemodialysis patients: A randomized controlled trial. *Sci. Rep.* **2020**, *10*, 11708. [CrossRef] [PubMed]
8. da Silva, V.; Corrêa, H.; Neves, R.; Deus, L.; Reis, A.; Souza, M.; dos Santos, C.; de Castro, D.; Honorato, F.; Simões, H.; et al. Impact of Low Hemoglobin on Body Composition, Strength, and Redox Status of Older Hemodialysis Patients Following Resistance Training. *Front. Physiol.* **2021**, *12*, 619054. [CrossRef] [PubMed]

9. Neves, R.V.P.; Corrêa, H.L.; Deus, L.A.; Reis, A.L.; Souza, M.K.; Simões, H.G.; Navalta, J.W.; Moraes, M.R.; Prestes, J.; Rosa, T.S. Dynamic not isometric training blunts osteo-renal disease and improves the sclerostin/FGF23/Klotho axis in maintenance hemodialysis patients: A randomized clinical trial. *J. Appl. Physiol.* **2021**, *130*, 508–516. [CrossRef] [PubMed]
10. Beltrame, A. *Portaria N° 224, de 10 de Maio de 2010*; Ministério da Saúde: Rio de Janeiro, Brazil, 2010.
11. Bay, M.L.; Pedersen, B.K. Muscle-Organ Crosstalk: Focus on Immunometabolism. *Front. Physiol.* **2020**, *11*, 567881. [CrossRef] [PubMed]
12. Xiao, X.; Alfaro-Magallanes, V.M.; Babitt, J.L. Bone morphogenic proteins in iron homeostasis. *Bone* **2020**, *138*, 115495. [CrossRef] [PubMed]
13. Babitt, J.L.; Lin, H.Y. Mechanisms of anemia in CKD. *J. Am. Soc. Nephrol.* **2012**, *23*, 1631–1634. [CrossRef] [PubMed]
14. Fudin, R.; Jaichenko, J.; Shostak, A.; Bennett, M.; Gotloib, L. Correction of uremic iron deficiency anemia in hemodialyzed patients: A prospective study. *Nephron* **1998**, *79*, 299–305. [CrossRef] [PubMed]
15. Macdougall, I.C.; Tucker, B.; Thompson, J.; Tomson, C.R.; Baker, L.R.; Raine, A.E. A randomized controlled study of iron supplementation in patients treated with erythropoietin. *Kidney Int.* **1996**, *50*, 1694–1699. [CrossRef] [PubMed]
16. Markowitz, G.S.; Kahn, G.A.; Feingold, R.E.; Coco, M.; Lynn, R.I. An evaluation of the effectiveness of oral iron therapy in hemodialysis patients receiving recombinant human erythropoietin. *Clin. Nephrol.* **1997**, *48*, 34–40. [PubMed]
17. Gadelha, A.B.; Cesari, M.; Corrêa, H.L.; Neves, R.V.P.; Sousa, C.V.; Deus, L.A.; Souza, M.K.; Reis, A.L.; Moraes, M.R.; Prestes, J.; et al. Effects of pre-dialysis resistance training on sarcopenia, inflammatory profile, and anemia biomarkers in older community-dwelling patients with chronic kidney disease: A randomized controlled trial. *Int. Urol. Nephrol.* **2021**, in press. [CrossRef] [PubMed]
18. Neves, P.; Sesso, R.C.C.; Thomé, F.S.; Lugon, J.R.; Nasicmento, M.M. Brazilian Dialysis Census: Analysis of data from the 2009–2018 decade. *J. Bras. Nefrol.* **2020**, *42*, 191–200. [CrossRef] [PubMed]

 *nutrients*

Article

Investigating the Links between Lower Iron Status in Pregnancy and Respiratory Disease in Offspring Using Murine Models

Henry M. Gomez [1,†], Amber L. Pillar [1,†], Alexandra C. Brown [1], Richard Y. Kim [1,2], Md Khadem Ali [1], Ama-Tawiah Essilfie [3], Rebecca L. Vanders [1], David M. Frazer [3,4], Gregory J. Anderson [3,5], Philip M. Hansbro [1,6], Adam M. Collison [7], Megan E. Jensen [7], Vanessa E. Murphy [7], Daniel M. Johnstone [8], David Reid [3], Elizabeth A. Milward [1], Chantal Donovan [1,2,†] and Jay C. Horvat [1,*,†]

[1] School of Biomedical Sciences and Pharmacy, College of Health, Medicine and Wellbeing, and Priority Research Centre for Healthy Lungs, University of Newcastle and Hunter Medical Research Institute, Callaghan, NSW 2308, Australia; henry.gomez@newcastle.edu.au (H.M.G.); Amber.pillar@uon.edu.au (A.L.P.); alexandra.brown@newcastle.edu.au (A.C.B.); richard.kim@uts.edu.au (R.Y.K.); mdali@stanford.edu (M.K.A.); rebecca.vanders@newcastle.edu.au (R.L.V.); philip.hansbro@uts.edu.au (P.M.H.); liz.milward@newcastle.edu.au (E.A.M.); chantal.donovan@uts.edu.au (C.D.)
[2] Faculty of Science, School of Life Sciences, University of Technology Sydney, Sydney, NSW 2007, Australia
[3] QIMR Berghofer Medical Research Institute, Herston, QLD 4006, Australia; Ama-Tawiah.Essilfie@qimrberghofer.edu.au (A.-T.E.); David.Frazer@qimrberghofer.edu.au (D.M.F.); Greg.Anderson@qimrberghofer.edu.au (G.J.A.); David.Reid@qimrberghofer.edu.au (D.R.)
[4] School of Biomedical Sciences, The University of Queensland, St Lucia, QLD 4067, Australia
[5] School of Chemistry and Molecular Bioscience, The University of Queensland, St Lucia, QLD 4067, Australia
[6] Centre for Inflammation, School of Life Sciences, Faculty of Science, Centenary Institute and University of Technology Sydney, Sydney, NSW 2007, Australia
[7] School of Medicine and Public Health, College of Health, Medicine and Wellbeing, and Priority Research Centre for GrowUpWell, The University of Newcastle and Hunter Medical Research Institute, Callaghan, NSW 2308, Australia; Adam.collison@newcastle.edu.au (A.M.C.); megan.jensen@newcastle.edu.au (M.E.J.); vanessa.murphy@newcastle.edu.au (V.E.M.)
[8] School of Medical Sciences, University of Sydney, Camperdown, NSW 2050, Australia; daniel.johnstone@sydney.edu.au
* Correspondence: jay.horvat@newcastle.edu.au; Tel.: +612-4042-0220
† These authors contributed equally.

Citation: Gomez, H.M.; Pillar, A.L.; Brown, A.C.; Kim, R.Y.; Ali, M.K.; Essilfie, A.-T.; Vanders, R.L.; Frazer, D.M.; Anderson, G.J.; Hansbro, P.M.; et al. Investigating the Links between Lower Iron Status in Pregnancy and Respiratory Disease in Offspring Using Murine Models. *Nutrients* **2021**, *13*, 4461. https://doi.org/10.3390/nu13124461

Academic Editor: Dietmar Enko

Received: 24 November 2021
Accepted: 8 December 2021
Published: 14 December 2021

Publisher's Note: MDPI stays neutral with regard to jurisdictional claims in published maps and institutional affiliations.

Abstract: Maternal iron deficiency occurs in 40–50% of all pregnancies and is associated with an increased risk of respiratory disease and asthma in children. We used murine models to examine the effects of lower iron status during pregnancy on lung function, inflammation and structure, as well as its contribution to increased severity of asthma in the offspring. A low iron diet during pregnancy impairs lung function, increases airway inflammation, and alters lung structure in the absence and presence of experimental asthma. A low iron diet during pregnancy further increases these major disease features in offspring with experimental asthma. Importantly, a low iron diet increases neutrophilic inflammation, which is indicative of more severe disease, in asthma. Together, our data demonstrate that lower dietary iron and systemic deficiency during pregnancy can lead to physiological, immunological and anatomical changes in the lungs and airways of offspring that predispose to greater susceptibility to respiratory disease. These findings suggest that correcting iron deficiency in pregnancy using iron supplements may play an important role in preventing or reducing the severity of respiratory disease in offspring. They also highlight the utility of experimental models for understanding how iron status in pregnancy affects disease outcomes in offspring and provide a means for testing the efficacy of different iron supplements for preventing disease.

Keywords: iron deficiency; pregnancy; respiratory disease; asthma; offspring

1. Introduction

Iron deficiency is present in 50% of all pregnancies [1] and nutrient deficiency may be an important aetiological factor. Other aetiologies of iron deficiency include malabsorption diseases such as celiac disease and atrophic gastritis, and blood loss through gastroesophageal reflux and oesophagitis, gastritis, peptic ulcers, inflammatory bowel diseases and menorrhagia [2,3]. Significantly, clinical evidence associates maternal iron deficiency with increased risk of respiratory disease, including asthma, in children [4,5]. Children born to asthmatic mothers with anaemia during pregnancy are more likely to have recurrent wheeze in the first year of life and experience asthma by six years of age [4].

In humans, maternal iron deficiency and increased risk of respiratory disease in offspring implicates iron as a potentially key micronutrient in lung development. A lower maternal haemoglobin concentration during pregnancy is associated with elevated IgE and increased risk of allergic sensitisation in the offspring [6]. Furthermore, maternal anaemia limits foetal intrauterine iron supply [1,7], which may have implications for foetal lung development. Post-partum, infant growth velocity is significantly correlated with iron deficiency anaemia in the first two years of life [8]. However, beyond these studies that associate maternal iron deficiency and increased risk of respiratory disease, no studies have demonstrated that low maternal systemic iron status during pregnancy alters lung development and function, whilst increasing the risk of asthma and disease severity, in offspring.

Given the high proportion of women affected by iron deficiency in pregnancy globally and the links with respiratory disease, it is critical to enhance our understanding of how lower maternal iron status during pregnancy affects respiratory disease in the offspring and determine whether these changes can be prevented and/or mitigated. To address this, we placed mice on a low iron diet (LID) prior to, and maintained throughout, pregnancy to induce iron deficiency [9], then studied house dust mite (HDM)-induced asthma in the offspring to investigate how lower iron status during pregnancy impacts lung function, structure and inflammatory responses.

2. Materials and Methods

2.1. Animal Ethics Statement

Animal procedures were performed in accordance with the Australian Code of Practice for the Care and Use of Animals for Scientific Purposes as issued by the National Health and Medical Research Council (NHMRC) Australia and approved by The University of Newcastle Animal Care and Ethics Committee (A-2019-941, approved 6 December 2019).

2.2. Mice

Wild-type BALB/c mice were sourced from the Central Animal House of The University of Newcastle (Callaghan, NSW, Australia). Animals were housed under specific pathogen free (SPF), PC2 conditions in individually ventilated cages (Bioresources facility, HMRI Building, New Lambton Heights, NSW, Australia). All mice were provided with food and water *ad libitum*.

2.3. Low Systemic Iron Diet during Pregnancy and Time Mating Protocol (F0 Generation)

Six to eight-week-old female, wild-type BALB/c mice, received either a LID (~2.5 mg Fe/kg, SF01-017 diet based on AIN-93G, <5 mg Iron/Kg, Specialty Feeds, Glen Forrest, Australia) to induce systemic iron deficiency [9] or a control chow diet (CC, SF09-091, 75 mg Iron/Kg, Specialty Feeds, Glen Forrest, Australia) commencing 5 weeks (Figure 1A) prior to mating. Timed mating periods were conducted by firstly synchronising the induction of oestrous cycles at week five of the protocol. Males (8–10 weeks old) were housed individually for seven days before being removed to new, clean cages. Female mice were introduced to the dirty cages previously occupied by the males (two females per male cage) for three days before being removed and introduced to the new, clean cages occupied by the males (two females per male). Mating occurred over three days and presence of mucus plugs were checked daily. Females with plugs were transferred into individual housing in clean cages for the duration of pregnancy and the birth and weaning of offspring. The status of pregnant females was monitored by regular weighing, and births occurred during weeks eight of the protocol.

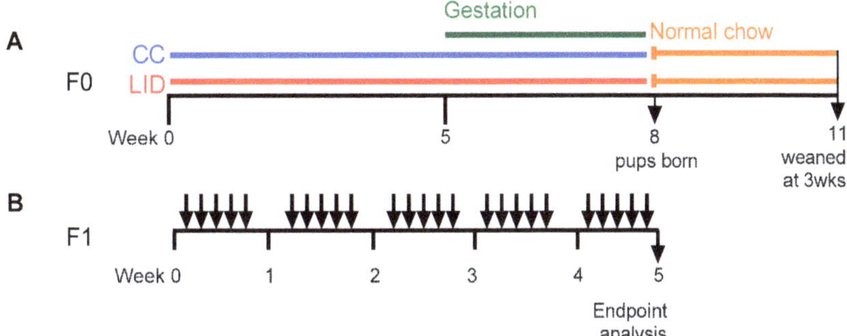

Figure 1. Protocols. (**A**) Six to eight-week-old, female wild type BALB/c mice (F0) received control chow (CC) or low iron diet (LID) from week 0. Timed mating occurred at week five and offspring weaned at three weeks of age. (**B**) Eight-week-old female offspring (F1) treated with house dust mite (HDM; 25 µg/50 µL Saline) or saline (Sal; 50 µL) five days/week for five weeks.

Post-partum, all mice were placed onto a normal mouse chow (meat free mouse and rat diet, 200 mg Iron/Kg Specialty Feeds, Glen Forrest, Australia) and offspring (F1) were weaned at three weeks of age.

2.4. Establishment of HDM-Induced Experimental Asthma (F1 Generation)

HDM-induced experimental asthma was induced in the F1 generation at eight weeks of age via intranasal administration of house dust mite (HDM) extract (25 µg, *Dermatophagoides pternyssinus* extract FD, Citeq biologics, Groningen, Netherlands) in medical grade sterile saline (50 µL), once daily, five times per week, for five weeks under isoflurane anaesthesia (Figure 1B). Due to a female-biased offspring sex ratio in the F1 generation, the experiments that examined the effects of maternal LID on asthma severity were conducted in female F1 offspring only.

2.5. Assessment of Airway Inflammation

Airway inflammation was measured in the bronchoalveolar lavage fluid (BALF) as previously described [9,10]. Briefly, the right multi-lobes of the lung were tied off, and the left lung was flushed with Hank's Buffered Salt Solution (HBSS, 2 × 500 µL). Samples were centrifuged (5 min, 132× g, 4 °C), and the cell pellets were re-suspended in red cell lysis buffer (200 µL, 5 min on ice), spun down and cell pellets were re-suspended in HBSS (200 µL). The proportion of viable cells was assessed using the trypan blue exclusion method by counting using a haemocytometer. Cell suspensions (90 µL) were spun onto slides using cytocentrifugation (5 min, 300 rpm, room temperature). Differential leukocyte populations were determined under a light microscope at 400× magnification, following May Grunwald-Giemsa stain as per the manufacturer's instruction (Sigma-Aldrich, Castle Hill, Australia).

2.6. Lung Function Analyses

Lung function tests were administered as previously described [9,10]. Briefly, mice were anaesthetised by intraperitoneal administration of ketamine (400 mg/kg; Parnell) and xylazine (10 mg/kg; Troy Laboratories, Glendenning, Australia) in saline (200 µL final volume) prior to tracheotomy and cannulation. Mice were ventilated at 450 breaths/min, and lung function parameters were assessed using the FlexiVent apparatus (FX1 System;

SCIREQ^{TM}, Montreal, Canada) with increasing doses of nebulised methacholine (0, 0.1, 1, 3 and 10 mg/mL) as previously described [9,10]. A minimum of three measurements per dose was performed for each parameter. Data were analysed using Flexiware software v7.6 (SCIREQ^{TM}, Montreal, Canada), Microsoft Excel v10 (Microsoft, Redmond, WA, USA) and GraphPad v8 (Graphpad Software Inc., San Diego, CA, USA).

2.7. Histological Analyses

Following collection of BALF, the left lobes of the lungs were perfused with 0.9% saline and inflated with, and drop-fixed in, 10% neutral buffered formalin (up to 500 μL; Sigma-Aldrich, Castle Hill, Australia), paraffin embedded, sectioned and stained (HMRI histology services, New Lambton Heights, Australia).

Alveolar diameter was measured using Haematoxylin and Eosin-stained slides; The number of mucus secreting cells (MSCs)/μm of basement membrane (BM), using Alcian Blue Periodic Acid-Schiff (AB PAS) stain; Small airway collagen deposition, using Sirius Red/Fast Green stain; The number of eosinophils/100 μm of the basement membrane using Congo-Red (CR); as previously described [9–11]. Images were captured using a Zeiss AxioImager.M2 microscope (Zeiss Australia, North Ryde, NSW, Australia) and Zeiss ZEN software (v3.2, Zeiss Australia, North Ryde, Australia) and analysed using custom scripts in ImageJ v1.5 (NIH, Bethesda, MD, USA) as previously described [9–11].

2.8. Statistical Analyses

All data are presented as mean ± S.E.M. Comparisons between two groups were made using an unpaired students *t*-test. Comparisons between multiple groups were performed using a one-way analysis of variance (ANOVA) with a post-hoc Fisher's least significant difference (LSD) test. Airway hyperresponsiveness (AHR) data were analysed using two-way repeated measures ANOVA with a post-hoc Tukey test. All statistical analyses were performed using GraphPad Prism software v8 (Graphpad Software Inc., San Diego, CA, USA).

3. Results

3.1. Low Iron Diet (LID) during Pregnancy Impairs Lung Function, Increases Airway Inflammation, and Increases Small Airway Collagen Deposition in Offspring

To determine whether lower iron status during pregnancy alters lung function in the offspring, mice were fed an LID or CC [9], time mated at week five, the offspring weaned at three weeks of age, and treated with saline five days/week for five weeks (Figure 1A,B). Baseline lung function measurements and response to methacholine provocation were assessed. An LID during pregnancy increases baseline central airway resistance (Rn; $p = 0.058$; Figure 2A), tissue damping (Figure 2B), tissue elastance (Figure 2C), transpulmonary resistance (Rrs; Figure 2D) and transpulmonary elastance (Ers; Figure 2E) and decreases compliance (Crs; Figure 2F) in saline-treated offspring (LID/Sal), compared to saline-treated offspring from mothers on a CC (CC/Sal). We next measured the response to methacholine provocation to assess AHR in the offspring (Figure 2G–L). An LID during pregnancy increases AHR in terms of Rn, Rrs, Ers and decreases Crs in Saline-treated offspring (LID/Sal), compared to those from mothers on a CC (CC/Sal; Figure 2G,J–L). An LID during pregnancy had no effect on tissue damping or elastance (Figure 2H,I).

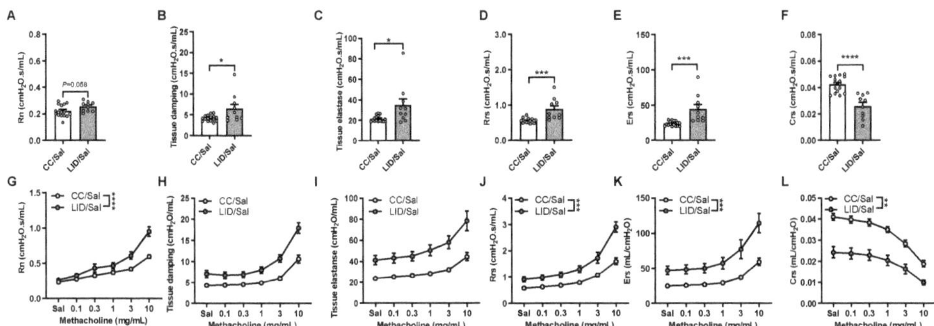

Figure 2. Low iron diet (LID) during pregnancy increases airway hyperresponsiveness (AHR), indicative of impaired lung function, in male and female offspring. Baseline lung function was assessed in terms of (**A**) central airway resistance (Rn), (**B**) tissue damping, (**C**) tissue elastance, (**D**) transpulmonary resistance (Rrs), (**E**) elastance (Ers), and (**F**) compliance (Crs). The response to methacholine provocation was assessed in terms of (**G**) Rn, (**H**) tissue damping, (**I**) tissue elastance, (**J**) Rrs, (**K**) Ers, (**L**) Crs. (**A–F**) Analysed by unpaired student *t*-test, (**G–L**) analysed by 2-way ANOVA and statistics at maximal dose from AHR curves presented. n = 10–16 mice per group. Data are presented as mean $\pm$ SEM (* $p < 0.05$; ** $p < 0.01$; *** $p < 0.001$; **** $p < 0.0001$).

To determine whether an LID during pregnancy altered airway inflammation in male and female offspring, BALF was assessed for total leukocytes, macrophages, eosinophils, neutrophils and lymphocytes (Figure 3A–E). An LID during pregnancy significantly increases total leukocyte, macrophage, and lymphocyte numbers in the airways of saline-treated offspring (LID/Sal), but not neutrophils or eosinophils, compared to mothers on a CC (CC/Sal; Figure 3A–E). To assess lung structure in the offspring, alveolar diameter, mucus-secreting cells, collagen and airway-associated eosinophils were assessed. An LID during pregnancy does not affect alveolar diameter (Figure 3F) or mucus-secreting cells (Figure 3G) but increases small airway collagen deposition (Figure 3H). Furthermore, there are no changes in tissue eosinophil numbers around the small airways (Figure 3I). Together, these data show that LID during pregnancy impairs lung function, increases airway inflammation and increases collagen deposition around the small airways in the offspring.

3.2. Low Iron Diet (LID) during Pregnancy Impairs Lung Function, Increases Airway Inflammation, and Increases Small Airway Collagen Deposition in Female Offspring Treated with HDM

To assess whether an LID during pregnancy affects HDM-induced experimental asthma in female offspring (Figure 1A,B), lung function, airway inflammation and lung structure were assessed.

HDM-treated female offspring from mothers on a CC (CC/HDM) have similar baseline lung function to saline-treated offspring from mothers on an LID (LID/Sal; Figure 4A–F). HDM-treated offspring from mothers on an LID (LID/HDM) have increased tissue damping (Figure 4B), but no other parameters are affected, compared to saline-treated offspring from mothers on an LID (LID/Sal). We next measured the response to methacholine provocation to assess AHR in the offspring (Figure 4G–L). An LID during pregnancy increases AHR in terms of Rn, Rrs, Ers and decreased Crs in saline-treated offspring (LID/Sal), compared to those from mothers on a CC (CC/Sal; Figure 4G,J–L). A LID during pregnancy has no effect on tissue damping or elastance (Figure 4H,I). HDM-treated offspring from mothers on a CC (CC/HDM) have increased AHR compared to saline-treated offspring from mothers on a CC (CC/Sal; Figure 4G–L). Interestingly, HDM-treated offspring from mothers on a LID (LID/HDM) exhibited further increases in AHR in terms of Rn, elastance, Rrs and Ers, but no change in tissue damping or Crs, compared to HDM-treated offspring from mothers on a CC (CC/HDM; Figure 4G–L). Together these data show that saline-treated female offspring from mothers on an LID have decreased baseline lung function and increased AHR compared to female offspring from mothers on a CC. Furthermore, HDM-treated female offspring from mothers on an LID have further impairment of baseline lung function and increased magnitude of AHR compared to HDM-treated female offspring from mothers on a CC (Figure 4).

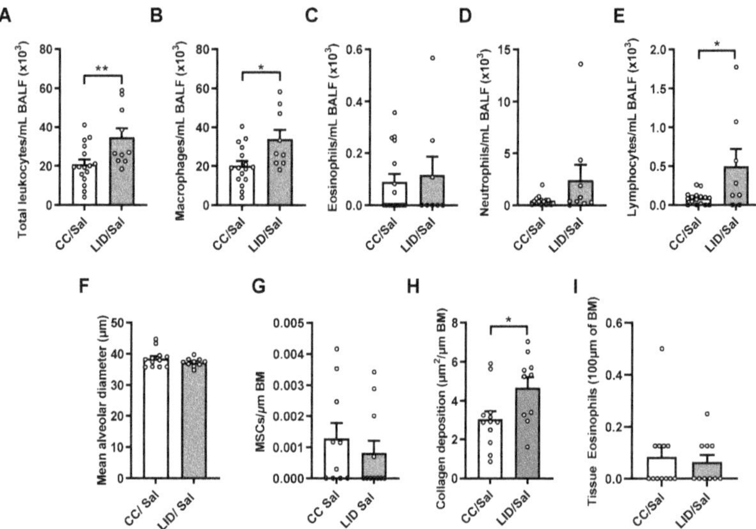

Figure 3. Low iron diet (LID) during pregnancy increases airway inflammation and collagen deposition in male and female offspring. (**A**) Total leukocytes, (**B**) macrophages, (**C**) eosinophils (**D**) neutrophils and (**E**) lymphocytes were enumerated in bronchoalveolar lavage fluid (BALF) at five weeks. Histopathology was assessed in terms of (**F**) mean alveolar diameter, (**G**) mucus secreting cells (MSCs) per μm of basement membrane (BM), (**H**) small airway collagen deposition and (**I**) airway-associated eosinophils. n = 10–16 mice per group. Analysed by unpaired student *t*-test. Data are presented as mean ± SEM (* $p < 0.05$; ** $p < 0.01$).

To determine whether an LID during pregnancy alters airway inflammation in control and HDM-treated female offspring, BALF was assessed for total leukocytes, macrophages, neutrophils, eosinophils, and lymphocytes (Figure 5A–E). An LID during pregnancy significantly increases total leukocyte, macrophage, and lymphocyte numbers in the airways of saline-treated offspring (LID/Sal), but not eosinophils or neutrophils, compared to saline-treated controls (CC/Sal; Figure 5A–E). HDM-treated female offspring from mothers on a CC (CC/HDM) have increased total leukocytes, eosinophils, neutrophils and lymphocytes, but not macrophages, compared to saline-treated female offspring from mothers on a CC (CC/Sal; Figure 5A–E). Interestingly, HDM-treated offspring from mothers on an LID (LID/HDM) had a further increase in total leukocytes, macrophages and neutrophils, and no change in eosinophils or lymphocytes, compared to HDM-treated offspring from mothers on a CC (CC/HDM; Figure 5A–E). Together, these data show that an LID during pregnancy increases airway inflammation in offspring, and that the inflammation associated with HDM-induced experimental asthma is further increased in offspring from mothers on an LID.

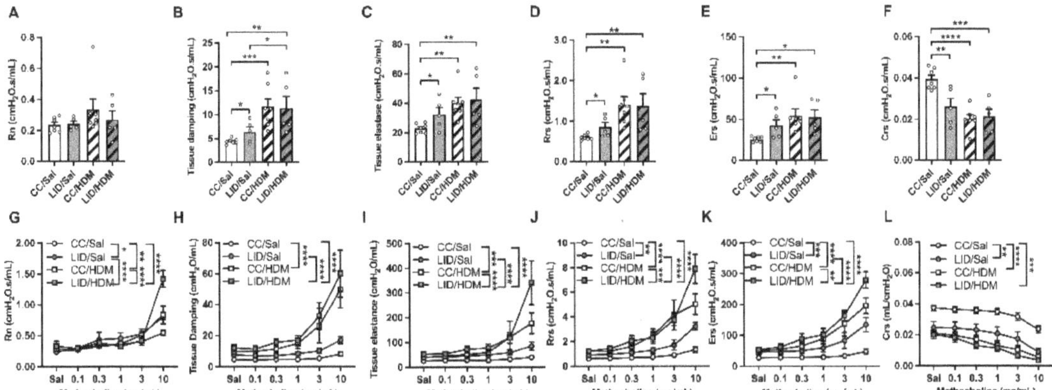

Figure 4. Low iron diet (LID) during pregnancy impairs lung function and increases AHR in female offspring in the absence and presence of HDM. Baseline lung function was assessed in terms of (**A**) central airway resistance (Rn), (**B**) tissue damping, (**C**) tissue elastance, (**D**) transpulmonary resistance (Rrs), (**E**) elastance (Ers), (**F**) compliance (Crs). Response to methacholine provocation was assessed in terms of (**G**) Rn, (**H**) tissue damping, (**I**) tissue elastance, (**J**) Rrs, (**K**) Ers, (**L**) Crs. (**A–F**) Analysed by one-way ANOVA with Fishers LSD and unpaired student *t*-test, (**G–L**) analysed by 2-way ANOVA and statistics at maximal dose from AHR curves presented. n = 5–7 mice per group. Data are presented as mean ± SEM (* $p < 0.05$; ** $p < 0.01$; *** $p < 0.001$; **** $p < 0.0001$).

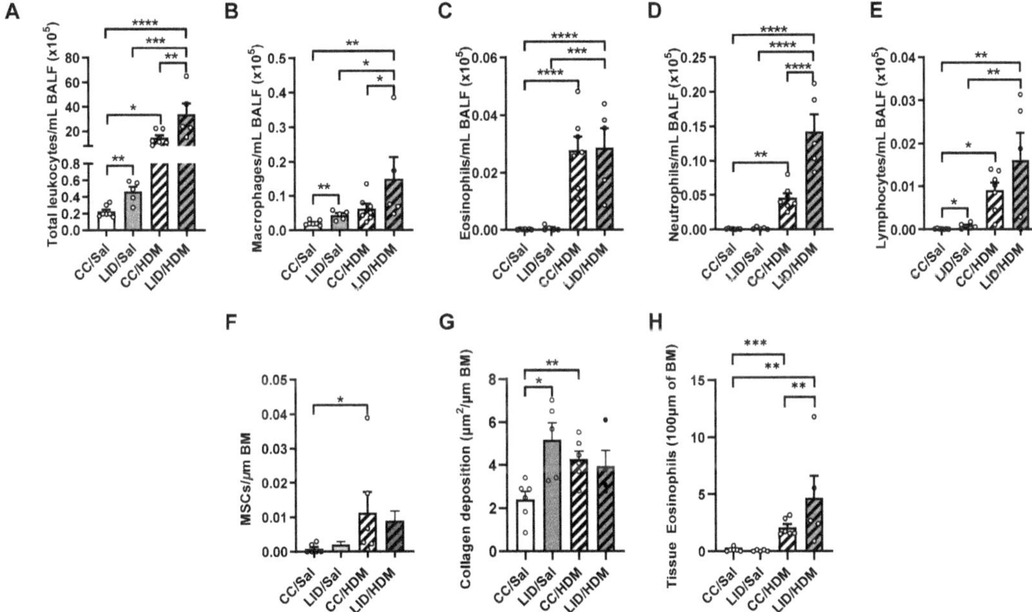

Figure 5. Low iron diet (LID) during pregnancy increases airway inflammation and collagen deposition in female offspring in the absence and presence of HDM. (**A**) Total leukocytes, (**B**) macrophages, (**C**) eosinophils (**D**) neutrophils and (**E**) lymphocytes were enumerated in bronchoalveolar lavage fluid (BALF) at five weeks. Histopathology was assessed in terms of (**F**) mucus secreting cells (MSCs) per μm of basement membrane (BM), (**G**) small airway collagen deposition and (**H**) airway-associated eosinophils. *n* = 5–7 mice per group. Data analysed by one-way ANOVA with Fishers LSD and unpaired student *t*-test Data are presented as mean ± SEM (* $p < 0.05$; ** $p < 0.01$; *** $p < 0.001$; **** $p < 0.0001$).

We next assessed whether an LID during pregnancy altered lung structure in control and HDM-treated female offspring. An LID during pregnancy had no effect on mucus-secreting cells (Figure 5F), however, mucus-secreting cells are increased in HDM-treated female offspring from mothers on a CC (CC/HDM) compared to saline-treated female offspring from mothers on a CC (CC/Sal; Figure 5F). An LID during pregnancy significantly increased small airway collagen deposition in saline-treated female offspring (LID/Sal; Figure 5G) to levels similar to those observed in HDM-treated female offspring from mothers on a CC (CC/HDM; Figure 5G). Furthermore, there are no changes in tissue eosinophil numbers around the small airways (Figure 5H). Together, these data show that an LID during pregnancy and HDM-treatment in offspring result in increased small airway collagen deposition.

4. Discussion

Iron dysregulation is strongly implicated in the development and/or exacerbation of many diseases. We and others [9,10,12–17] have previously shown that lung iron levels and their regulation play important roles in the pathogenesis and severity of lung infection and respiratory diseases such as asthma, chronic obstructive pulmonary disease, cystic fibrosis and idiopathic pulmonary fibrosis [9,10]. Furthermore, increasing clinical evidence suggests that maternal anaemia or an LID during pregnancy are linked with asthma and wheezing in offspring in later life [4]. However, no studies have demonstrated that low maternal iron status during pregnancy results in altered lung function, inflammatory response and structure, and whether this results in increased severity of asthma in offspring. Using our mouse models, we show for the first time that LID during pregnancy impairs baseline lung function, increases AHR, increases airway inflammation and promotes small airway collagen deposition in offspring. Furthermore, HDM-induced experimental asthma in the offspring of mothers on a LID during pregnancy was characterised by exaggerated features of disease, indicating that iron dysregulation during pregnancy imparts functional and structural changes in the lung of offspring in the presence or absence of experimental asthma.

It is known that maternal insults such as hypoxia that cause intrauterine growth restriction, lead to profound deleterious effects on lung and diaphragm function in offspring independent of sex and changes in lung structure [18,19]. Our findings now strongly suggest that LID in pregnancy is another significant maternal insult that can have profound deleterious effects on lung function in the offspring. A previously published study demonstrated that low foetal iron is linked to increased susceptibility to eosinophilia in infancy [20]. Here we show increased airway tissue eosinophils in asthmatic offspring from mothers on an LID diet. We also observed increases in small airway collagen deposition in offspring born to mothers on an LID, and it is known that collagen deposition is associated with AHR and airway remodelling in asthma [21–23]. Furthermore, we observed increases in macrophage and lymphocytes that may have also contributed to AHR in these offspring. Further studies that interrogate the mechanistic interplay between these features and how they contribute to increased AHR in offspring from mothers on an LID diet are warranted.

Our findings demonstrate that lung structure is altered by a low iron status *in utero*. This concept is supported by a study showing that desferrioxamine-mediated iron chelation in ex vivo lung buds from mouse embryos restricted the development of vascular networks and reduced epithelial branching [24]. Mothers in our study were on a maternal diet containing normal iron levels post-partum, demonstrating that the changes induced in the offspring in utero could not be reversed by a normal diet post-partum. This contrasts with a study that showed reversal of desferrioxamine-mediated iron chelation following iron exposure in mouse embryos [24]. Thus, in our model, post-partum maternal exposure to a diet containing normal iron levels did not correct the disease-causing effects of LID in utero on the offspring.

A longitudinal study in humans demonstrated that lower iron status in the umbilical cord was linked to increased risk of atopy in children in later life [25]. Our observations that HDM-treated offspring from mothers on an LID had increased disease features, including impaired lung function and airway inflammation are consistent with these clinical observations. Interestingly, our observation of increased AHR was not associated with further increases in mucus-secreting cells or collagen deposition around the small airways, but in this context the increase in AHR may be explained by the increases in total leukocytes and neutrophils observed in the airways of HDM-treated female offspring from mothers on a LID compared to mothers on a CC diet. Our findings demonstrate that an LID during pregnancy promotes a shift from classically eosinophil-dominated allergic airway inflammation to a mixed eosinophilic/neutrophilic inflammatory profile in HDM-induced allergic airway disease. We and others have shown that increased neutrophils in the airways in both clinical and experimental asthma is associated with more severe disease [26–30], indicating that responses to specific environmental exposures in adulthood play important roles in the development of severe disease. Here, we show that lower iron status in pregnancy results in increased neutrophilic inflammation during asthma in offspring, which is reminiscent of the increased neutrophilic responses that are observed in severe asthma [26–28]. Significantly, our study suggests that in utero responses to low iron status may be considered as another environmental exposure that has profound effects on the respiratory milieu of offspring, and that correcting low iron status in pregnancy may be beneficial for reducing the severity of asthma in the offspring. These findings suggest that studies that assess the links between low iron status in pregnancy and the prevalence and severity of asthma in the offspring may be warranted.

The limitations of our study include that LID during pregnancy was administered before and during pregnancy but not specifically during pregnancy and/or postpartum which may have additional impacts on disease in the offspring. Whilst our study examines the effects of maternal LID intake, further insights into the effects of lower iron status during pregnancy on disease development in offspring may be gained by testing the effects of diets with different concentrations of iron during pregnancy. It is known the iron deficiency can contribute to neurological changes through epigenetic mechanisms, however no studies to date have assessed the epigenetic alternations in the lungs of offspring born to mothers with LID during pregnancy. An examination of this relationship would be informative and warrants further investigation.

In addition, there is evidence to suggest that low iron can affect the composition of the gut microbiome and it is established that vertical transmission of the microbiome from mothers to offspring can occur [31–33]. These data present the possibility that maternal LID may alter the maternal gut microbiome and, through vertical transmission, may contribute to changes in offspring gut microbiome and lung changes.

Whilst such studies are beyond the scope of the current manuscript, future studies that investigate the role of epigenetics or microbiomes in driving the effects of low iron status in pregnancy on lung structure, function, and inflammatory responses in the lungs of offspring are warranted.

In our HDM-treated offspring from mothers on an LID, we used female mice; however, it would be informative to compare male vs. female in this context. Importantly, our observations in this study suggest that correcting iron deficiency during pregnancy with iron supplements and/or nutraceuticals may be beneficial for lung function and structure in the offspring, and this could be tested in intervention studies using our model. One such approach may be to boost iron intake through a measured nutraceutical approach involving dietary supplementation of iron, and/or point-of-use fortification, which has been proposed to be beneficial in chronic inflammatory lung diseases [34].

5. Conclusions

Our observations show that low iron status in pregnancy can have profound effects on lung function and increases the severity of disease features in offspring. Importantly, our study reconciles clinical observations that have linked lower iron levels in pregnancy and asthma risk in offspring by demonstrating a causal link. Our findings demonstrate how lower iron status in pregnancy affects respiratory disease and further support the importance of treating low iron during pregnancy with iron supplements. The mouse model of LID in pregnancy that we have developed recapitulates the effects of low iron in pregnancy on human respiratory disease, and it therefore provides not only an ideal platform for studying the disease-causing mechanisms of lower iron status in pregnancy but will also allow pre-clinical testing of the efficacy of iron-correcting interventions to be conducted.

Author Contributions: Conceptualization, H.M.G., A.L.P., A.C.B., R.Y.K., A.M.C., M.E.J., V.E.M., D.M.J., D.R., E.A.M., J.C.H.; methodology, H.M.G., A.L.P., A.C.B., M.K.A., A.-T.E.; formal analysis, H.M.G., A.L.P., C.D., J.C.H.; resources, D.M.F., G.J.A., D.M.J., D.R., E.A.M., C.D., J.C.H.; data curation, H.M.G., A.L.P., A.C.B., R.Y.K., D.M.F., G.J.A., C.D., J.C.H.; writing—original draft preparation, H.M.G., A.L.P., R.Y.K., C.D., J.C.H.; writing—review and editing, H.M.G., A.L.P., A.C.B., R.Y.K., M.K.A., A.-T.E., R.L.V., D.M.F., G.J.A., P.M.H., A.M.C., M.E.J., V.E.M., D.M.J., D.R., E.A.M., C.D., J.C.H.; supervision, H.M.G., A.C.B., R.L.V., P.M.H., J.C.H.; project administration, J.C.H. All authors have read and agreed to the published version of the manuscript.

Funding: This research was supported by funding to J.C.H: Gladys M. Brawn Career Development fellowship and funding from the College of Health and Wellbeing, University of Newcastle. A.L.P. is supported by an Australian Government Research Training Program Scholarship. P.M.H is funded by a fellowship and grants from the National Health and Medical Research Council (NHMRC) of Australia (1175134) and by UTS.

Institutional Review Board Statement: This study was conducted according to the guidelines of the Declaration of Helsinki and approved by the Animal Care and Ethics Committee of The University of Newcastle (protocol: A-2019-941 and approved 6 of December 2019).

Informed Consent Statement: Not applicable.

Acknowledgments: The authors thank Tegan Hunter for technical assistance.

Conflicts of Interest: The authors declare no conflict of interest.

References

1. Abu-Ouf, N.M.; Jan, M.M. The impact of maternal iron deficiency and iron deficiency anemia on child's health. *Saudi Med. J.* **2015**, *36*, 146–149. [CrossRef] [PubMed]
2. Annibale, B.; Capurso, G.; Chistolini, A.; D'Ambra, G.; DiGiulio, E.; Monarca, B.; DelleFave, G. Gastrointestinal causes of refractory iron deficiency anemia in patients without gastrointestinal symptoms. *Am. J. Med.* **2001**, *111*, 439–445. [CrossRef]
3. Johnson-Wimbley, T.D.; Graham, D.Y. Diagnosis and management of iron deficiency anemia in the 21st century. *Therap. Adv. Gastroenterol.* **2011**, *4*, 177–184. [CrossRef] [PubMed]
4. Triche, E.W.; Lundsberg, L.S.; Wickner, P.G.; Belanger, K.; Leaderer, B.P.; Bracken, M.B. Association of maternal anemia with increased wheeze and asthma in children. *Ann. Allergy Asthma Immunol.* **2011**, *106*, 131–139. [CrossRef]
5. Bedard, A.; Lewis, S.J.; Burgess, S.; Henderson, A.J.; Shaheen, S.O. Maternal iron status during pregnancy and respiratory and atopic outcomes in the offspring: A Mendelian randomisation study. *BMJ Open Respir. Res.* **2018**, *5*, e000275. [CrossRef]
6. Shaheen, S.O.; Macdonald-Wallis, C.; Lawlor, D.A.; Henderson, A.J. Haemoglobin concentrations in pregnancy and respiratory and allergic outcomes in childhood: Birth cohort study. *Clin. Exp. Allergy* **2017**, *47*, 1615–1624. [CrossRef] [PubMed]
7. Singla, P.N.; Tyagi, M.; Shankar, R.; Dash, D.; Kumar, A. Fetal iron status in maternal anemia. *Acta Paediatr.* **1996**, *85*, 1327–1330. [CrossRef] [PubMed]
8. Soliman, A.T.; Al Dabbagh, M.M.; Habboub, A.H.; Adel, A.; Humaidy, N.A.; Abushahin, A. Linear growth in children with iron deficiency anemia before and after treatment. *J. Trop. Pediatr.* **2009**, *55*, 324–327. [CrossRef] [PubMed]
9. Ali, M.K.; Kim, R.Y.; Brown, A.C.; Mayall, J.R.; Karim, R.; Pinkerton, J.W.; Liu, G.; Martin, K.L.; Starkey, M.R.; Pillar, A.L.; et al. Crucial role for lung iron level and regulation in the pathogenesis and severity of asthma. *Eur. Respir. J.* **2020**, *55*, 1901340. [CrossRef]
10. Ali, M.K.; Kim, R.Y.; Brown, A.C.; Donovan, C.; Vanka, K.S.; Mayall, J.R.; Liu, G.; Pillar, A.L.; Jones-Freeman, B.; Xenaki, D.; et al. Critical role for iron accumulation in the pathogenesis of fibrotic lung disease. *J. Pathol.* **2020**, *251*, 49–62. [CrossRef] [PubMed]
11. Horvat, J.C.; Starkey, M.R.; Kim, R.Y.; Phipps, S.; Gibson, P.G.; Beagley, K.W.; Foster, P.S.; Hansbro, P.M. Early-life chlamydial lung infection enhances allergic airways disease through age-dependent differences in immunopathology. *J. Allergy Clin. Immunol.* **2010**, *125*, 617–625. [CrossRef]
12. Ali, M.K.; Kim, R.Y.; Karim, R.; Mayall, J.R.; Martin, K.L.; Shahandeh, A.; Abbasian, F.; Starkey, M.R.; Loustaud-Ratti, V.; Johnstone, D.; et al. Role of iron in the pathogenesis of respiratory disease. *Int. J. Biochem. Cell Biol.* **2017**, *88*, 181–195. [CrossRef] [PubMed]

13. Brown, A.C.; Horvat, J.C. Casting iron in the pathogenesis of fibrotic lung disease. *Am. J. Respir. Cell Mol. Biol.* **2021**, *65*, 130–131. [CrossRef] [PubMed]
14. Zhu, Y.; Chang, J.; Tan, K.; Huang, S.K.; Liu, X.; Wang, X.; Cao, M.; Zhang, H.; Li, S.; Duan, X.; et al. Clioquinol attenuates pulmonary fibrosis through inactivation of fibroblasts via iron chelation. *Am. J. Respir. Cell Mol. Biol.* **2021**, *65*, 189–200. [CrossRef]
15. Neves, J.; Leitz, D.; Kraut, S.; Brandenberger, C.; Agrawal, R.; Weissmann, N.; Muhlfeld, C.; Mall, M.A.; Altamura, S.; Muckenthaler, M.U. Disruption of the hepcidin/ferroportin regulatory system causes pulmonary iron overload and restrictive lung disease. *eBioMedicine* **2017**, *20*, 230–239. [CrossRef]
16. Janssen-Heininger, Y.M.; Aesif, S.W.; van der Velden, J.; Guala, A.S.; Reiss, J.N.; Roberson, E.C.; Budd, R.C.; Reynaert, N.L.; Anathy, V. Regulation of apoptosis through cysteine oxidation: Implications for fibrotic lung disease. *Ann. N. Y. Acad. Sci.* **2010**, *1203*, 23–28. [CrossRef] [PubMed]
17. Zhang, W.Z.; Butler, J.J.; Cloonan, S.M. Smoking-induced iron dysregulation in the lung. *Free Radic. Biol. Med.* **2019**, *133*, 238–247. [CrossRef] [PubMed]
18. Francis, M.R.; Pinniger, G.J.; Noble, P.B.; Wang, K.C.W. Intrauterine growth restriction affects diaphragm function in adult female and male mice. *Pediatr. Pulmonol.* **2020**, *55*, 229–235. [CrossRef] [PubMed]
19. Kalotas, J.O.; Wang, C.J.; Noble, P.B.; Wang, K.C.W. Intrauterine growth restriction promotes postnatal airway hyperresponsiveness independent of allergic disease. *Front. Med.* **2021**, *8*, 674324. [CrossRef]
20. Weigert, R.; Dosch, N.C.; Bacsik-Campbell, M.E.; Guilbert, T.W.; Coe, C.L.; Kling, P.J. Maternal pregnancy weight gain and cord blood iron status are associated with eosinophilia in infancy. *J. Perinatol.* **2015**, *35*, 621–626. [CrossRef] [PubMed]
21. Donovan, C.; Royce, S.G.; Esposito, J.; Tran, J.; Ibrahim, Z.A.; Tang, M.L.; Bailey, S.; Bourke, J.E. Differential effects of allergen challenge on large and small airway reactivity in mice. *PLoS ONE* **2013**, *8*, e74101. [CrossRef]
22. Ward, C.; Johns, D.P.; Bish, R.; Pais, M.; Reid, D.W.; Ingram, C.; Feltis, B.; Walters, E.H. Reduced airway distensibility, fixed airflow limitation, and airway wall remodeling in asthma. *Am. J. Respir. Crit. Care Med.* **2001**, *164*, 1718–1721. [CrossRef] [PubMed]
23. Ward, C.; Pais, M.; Bish, R.; Reid, D.; Feltis, B.; Johns, D.; Walters, E.H. Airway inflammation, basement membrane thickening and bronchial hyperresponsiveness in asthma. *Thorax* **2002**, *57*, 309–316. [CrossRef]
24. Groenman, F.A.; Rutter, M.; Wang, J.; Caniggia, I.; Tibboel, D.; Post, M. Effect of chemical stabilizers of hypoxia-inducible factors on early lung development. *Am. J. Physiol. Lung Cell Mol. Physiol.* **2007**, *293*, L557–L567. [CrossRef] [PubMed]
25. Shaheen, S.O.; Newson, R.B.; Henderson, A.J.; Emmett, P.M.; Sherriff, A.; Cooke, M.; Team, A.S. Umbilical cord trace elements and minerals and risk of early childhood wheezing and eczema. *Eur. Respir. J.* **2004**, *24*, 292–297. [CrossRef]
26. Kim, R.Y.; Horvat, J.C.; Pinkerton, J.W.; Starkey, M.R.; Essilfie, A.T.; Mayall, J.R.; Nair, P.M.; Hansbro, N.G.; Jones, B.; Haw, T.J.; et al. MicroRNA-21 drives severe, steroid-insensitive experimental asthma by amplifying phosphoinositide 3-kinase-mediated suppression of histone deacetylase 2. *J. Allergy Clin. Immunol.* **2017**, *139*, 519–532. [CrossRef] [PubMed]
27. Essilfie, A.T.; Horvat, J.C.; Kim, R.Y.; Mayall, J.R.; Pinkerton, J.W.; Beckett, E.L.; Starkey, M.R.; Simpson, J.L.; Foster, P.S.; Gibson, P.G.; et al. Macrolide therapy suppresses key features of experimental steroid-sensitive and steroid-insensitive asthma. *Thorax* **2015**, *70*, 458–467. [CrossRef]
28. Kim, R.Y.; Pinkerton, J.W.; Essilfie, A.T.; Robertson, A.A.B.; Baines, K.J.; Brown, A.C.; Mayall, J.R.; Ali, M.K.; Starkey, M.R.; Hansbro, N.G.; et al. Role for NLRP3 inflammasome-mediated, IL-1beta-dependent responses in severe, steroid-resistant asthma. *Am. J. Respir. Crit. Care Med.* **2017**, *196*, 283–297. [CrossRef]
29. Pinkerton, J.W.; Kim, R.Y.; Brown, A.C.; Rae, B.E.; Donovan, C.; Mayall, J.R.; Carroll, O.R.; Khadem Ali, M.; Scott, H.A.; Berthon, B.S.; et al. Relationship between type 2 cytokine and inflammasome responses in obesity-associated asthma. *J. Allergy Clin. Immunol.* **2021**. [CrossRef] [PubMed]
30. Rossios, C.; Pavlidis, S.; Hoda, U.; Kuo, C.H.; Wiegman, C.; Russell, K.; Sun, K.; Loza, M.J.; Baribaud, F.; Durham, A.L.; et al. Sputum transcriptomics reveal upregulation of IL-1 receptor family members in patients with severe asthma. *J. Allergy Clin. Immunol.* **2018**, *141*, 560–570. [CrossRef] [PubMed]
31. Coe, G.L.; Pinkham, N.V.; Celis, A.I.; Johnson, C.; DuBois, J.L.; Walk, S.T. Dynamic gut microbiome changes in response to low-iron challenge. *Appl. Environ. Microbiol.* **2021**, *87*, e02307–e02320. [CrossRef] [PubMed]
32. Li, W.; Tapiainen, T.; Brinkac, L.; Lorenzi, H.A.; Moncera, K.; Tejesvi, M.V.; Salo, J.; Nelson, K.E. Vertical transmission of gut microbiome and antimicrobial resistance genes in infants exposed to antibiotics at birth. *J. Infect. Dis.* **2021**, *224*, 1236–1246. [CrossRef] [PubMed]
33. Torres, J.; Hu, J.; Seki, A.; Eisele, C.; Nair, N.; Huang, R.; Tarassishin, L.; Jharap, B.; Cote-Daigneault, J.; Mao, Q.; et al. Infants born to mothers with IBD present with altered gut microbiome that transfers abnormalities of the adaptive immune system to germ-free mice. *Gut* **2020**, *69*, 42–51. [CrossRef] [PubMed]
34. Chan, Y.; Raju Allam, V.S.R.; Paudel, K.R.; Singh, S.K.; Gulati, M.; Dhanasekaran, M.; Gupta, P.K.; Jha, N.K.; Devkota, H.P.; Gupta, G.; et al. Nutraceuticals: Unlocking newer paradigms in the mitigation of inflammatory lung diseases. *Crit. Rev. Food Sci. Nutr.* **2021**, 1–31. [CrossRef] [PubMed]

Article

Iron Deficiency in Cystic Fibrosis: A Cross-Sectional Single-Centre Study in a Referral Adult Centre

Hervé Lobbes [1,2,*], Stéphane Durupt [3], Sabine Mainbourg [3,4], Bruno Pereira [5], Raphaele Nove-Josserand [3], Isabelle Durieu [3,6] and Quitterie Reynaud [3,6]

[1] Service de Médecine Interne, Hôpital Estaing, CHU de Clermont-Ferrand, F-63000 Clermont-Ferrand, France
[2] SIGMA Clermont, Institut Pascal, CHU Clermont-Ferrand, Université Clermont Auvergne, CNRS, F-63000 Clermont-Ferrand, France
[3] Département de Médecine Interne et Centre de Référence Mucoviscidose, Centre Hospitalier Lyon Sud, Hospices Civils de Lyon, F-69310 Pierre-Bénite, France; durupt@chu-lyon.fr (S.D.); sabine.mainbourg@chu-lyon.fr (S.M.); raphaele.nove-josserand@chu-lyon.fr (R.N.-J.); isabelle.durieu@chu-lyon.fr (I.D.); quitterie.reynaud@chu-lyon.fr (Q.R.)
[4] Equipe Evaluation et Modélisation des Effets Thérapeutiques, UMR 5558, Laboratoire de Biométrie et Biologie Evolutive, CNRS, Claude Bernard University Lyon 1, F-69622 Villeurbanne, France
[5] Biostatistics Unit, Centre Hospitalier Universitaire de Clermont-Ferrand, F-63000 Clermont-Ferrand, France; bpereira@chu-clermontferrand.fr
[6] Research on Healthcare Performance (REHSAPE), INSERM U1290, Université Claude Bernard Lyon 1, F-69373 Lyon, France
* Correspondence: hlobbes@chu-clermontferrand.fr; Tel.: +33-4-73-750-085; Fax: +33-4-73-750-361

Abstract: Iron deficiency (ID) diagnosis in cystic fibrosis (CF) is challenging because of frequent systemic inflammation. We aimed to determine the prevalence and risk factors of ID in adult patients with CF. We conducted a single-centre prospective study in a referral centre. ID was defined by transferrin saturation $\leq$16% or ferritin $\leq$20 (women) or 30 (men) µg/L, or $\leq$100 µg/L in the case of systemic inflammation. Apparent exacerbation was an exclusion criterion. We included 165 patients (78 women), mean age—31.1 $\pm$ 8.9 years. ID prevalence was 44.2%. ID was significantly associated with female gender (58.9% vs. 38%), lower age (29.4 $\pm$ 8.5 vs. 32.5 $\pm$ 9.1), lower body mass index (20.5 $\pm$ 2.2 vs. 21.3 $\pm$ 2.5), and *Pseudomonas aeruginosa* colonization (70.8% vs. 55.1%). Diabetes mellitus, antiacid drug use and low pulmonary function were more frequent in patients with ID with no statistical significance. The use of CFTR correctors was not associated with ID. In the multivariate analysis, ID was associated with female gender (OR 2.64, CI95% 1.31–5.31), age < 30 years (OR 2.30, CI95% 1.16–4.56), and *P. aeruginosa* (OR 2.09, CI95% 1.04–4.19).

Keywords: iron deficiency; cystic fibrosis; anaemia; ferritin

Citation: Lobbes, H.; Durupt, S.; Mainbourg, S.; Pereira, B.; Nove-Josserand, R.; Durieu, I.; Reynaud, Q. Iron Deficiency in Cystic Fibrosis: A Cross-Sectional Single-Centre Study in a Referral Adult Centre. *Nutrients* **2022**, *14*, 673. https://doi.org/10.3390/nu14030673

Academic Editor: Dietmar Enko

Received: 14 January 2022
Accepted: 2 February 2022
Published: 5 February 2022

Publisher's Note: MDPI stays neutral with regard to jurisdictional claims in published maps and institutional affiliations.

1. Introduction

Iron is an essential micronutrient ensuring several vital body functions [1]. Iron deficiency (ID) has been frequently reported in cystic fibrosis (CF) and is attributed to a combination of chronic inflammation, impaired dietary iron absorption, malnutrition, and increased iron loss via sputum [2]. A specific concern regarding iron homeostasis in cystic fibrosis is its association with colonization by *Pseudomonas aeruginosa*, which is a well-known compounding factor of patients' respiratory conditions [3]. In addition, there is a fear that iron supplementation may promote *P. aeruginosa*-related infections, as iron has been shown to increase bacterial growth in vitro [4].

Most of the data on the prevalence of iron deficiency in CF come from previous studies, before advances were made in the treatment of CF exacerbation, and especially in the improved nutritional management of patients with CF (pwCF) [2,5–9]. A recent study investigated the iron store status in an adult cohort of pwCF, revealing a prevalence of 41.8%, which was associated with anaemia and poor lung function [10].

In a general setting, assessing the serum ferritin level is the most sensitive and specific test for ID diagnosis, with a threshold of 20 µg/L (women) or 30 µg/L (men), providing 92% sensitivity and 98% specificity. Other biological markers can occasionally be used, such as transferrin saturation (TSAT, threshold < 16%) [11]. The main challenge in ID diagnosis for pwCF is the effect of systemic inflammation on biological iron markers. Inflammation leads to an increase in ferritin levels due to the enhanced hepcidin production through the effects of interleukin-6 [12]. In pwCF studies, very heterogeneous biological definitions of iron deficiency coexist, probably related to the concurrent recruitment of adult and paediatric populations.

Therefore, we aimed to identify the prevalence of ID in a recent homogeneous cohort of adult pwCF, using the most common biological definitions. The secondary objective of our study was to identify the risk factors for iron deficiency.

2. Materials and Methods

We conducted an observational cross-sectional single-centre study. Patients were screened for participation during their annual medical systematic visit to the Referral Cystic Fibrosis Centre in Lyon, Hospices Civils, France.

2.1. Patients

The inclusion criteria were: (i) genetically proven CF, (ii) age ≥ 18 years, and (iii) serum iron parameter assessment. The exclusion criteria were: (i) current or past use of dietary iron supplement or iron supplement drugs (oral or intravenous) in the previous year, and (ii) clinically apparent exacerbation of CF between the previous 7 days and the day of the medical consultation, defined by one or more of the following symptoms (Fuchs criteria [13]:

- New or increased cough, sputum production or chest congestion.
- Decreased exercise tolerance, increased dyspnoea.
- Increased fatigue, decreased appetite.
- Increase respiratory rate or dyspnoea at rest.
- Change in sputum appearance.
- Fever.

2.2. Data Collection

Baseline data were extracted from medical records: gender, CFTR mutations, history of diabetes mellitus, solid organ transplantation, oxygen requirement, current treatment, including CFTR modulators, antiacids (alginate, anti-H2 and proton pump inhibitors: PPI), pancreatic enzymes, and vitamin supplement (A, D, K and E).

The following parameters were prospectively collected the day of the annual medical check-up: (i) clinical data: age, weight, height, and body mass index (BMI); (ii) biological data: bronchial colonization, iron stores (ferritin, TSAT), C reactive protein (CRP), liver function tests, creatinine, complete blood count, blood glucose, albumin, vitamin level (A, D, E, K), and (iii) pulmonary function test (PFT), including the measurement of forced expiratory volume in one second (FEV1) and forced vital capacity (FVC).

Sputum samples were collected the day of inclusion through physiotherapy manoeuvres (oscillatory positive expiratory pressure and forced expiration with acceleration of expiratory flow) and sent to the microbiology department for culture to search for *P. aeruginosa* colonization.

The main laboratory assays used during the study were: quantitative immunoturbidimetric (Architect, Abbott) for C reactive protein (normal < 5 mg/L) and transferrin (normal ranges: 1.74–3.64 g/L), chemiluminescent microparticle immunoassay (Architect, Abbott) for ferritin (normal ranges: 22–275 µg/L), and direct FERENE photometry (without deproteinization—Architect, Abbott) for serum iron (12.0–31.0 µmol/L).

2.3. Anaemia and Micronutrient Deficiency Definitions

Anaemia was defined using the WHO criteria [14]: haemoglobin < 120 (women) or 130 (men) g/L.

Two definitions of ID were studied in our protocol:

(i) The international recommended biological definition of ID [11] was used as our primary endpoint: ferritin ≤20 (women) or 30 (men) µg/L, or ≤100 µg/L in the case of systemic inflammation (CRP ≥ 10 mg/L) or TSAT ≤ 16%.

(ii) The historical paediatric CF definition of ID [5,15,16] was used as a secondary endpoint: ferritin ≤ 12 µg/L or TSAT ≤ 16%.

We also studied a population with mildly depleted iron stores, defined by ferritin ≤ 50 µg/L, as several studies showed an improvement of life quality parameters after intravenous iron supplementation in this population [17].

Fat-soluble vitamin deficiencies were defined as follows: vitamin K < 1000 ng/L, vitamin D < 30 ng/mL [18], vitamin E < 12 µmol/L [19], and vitamin A < 0.52 µmol/L [20].

2.4. Statistics

Categorical data were described with numbers (percentages). Continuous data were expressed as mean and standard deviation or median and interquartile range, according to the statistical distribution. The assumption normality of the data was assessed using the Shapiro-Wilk test. The comparisons between independent groups according to ID were performed using the Student's *t*-test or the Mann–Whitney test when the assumptions of the *t*-test were not met. Homoscedasticity was assessed using the Fisher–Snedecor test. Categorical data were compared between groups (ID yes/no) using chi-squared or Fisher's exact tests.

Then, a multivariable analysis was conducted to determine the factors associated with ID. Generalized linear modelling (logistic for binary dependent outcome: ID yes/no) was performed with covariates fixed according to the univariate results and to the clinical relevance, with particular attention paid to multicollinearity. Furthermore, age and BMI, which did not follow a Gaussian distribution, were categorized according to their statistical distribution. The results were expressed using odds ratios (OR) and 95% confidence intervals.

To ensure the robustness of our results, the final model was validated by a two-step bootstrapping process. In each step, 1000 bootstrap samples with replacements were created from the training set. In the first one, using the stepwise procedure, we determined the percentage of models, including each of the initial variables. In the second step, we independently estimated the logistic model parameters of the final model. The bootstrap estimates of each covariate coefficient and standard errors were averaged from these replicates. Statistical analyses were performed using Stata 15 (StataCorp, College Station, TX, USA). All statistical tests were two sided, with a type I error set at 5%.

2.5. Ethics

The study was conducted in accordance with the Declaration of Helsinki and registered on https://clinicaltrials.gov/ct2/show/NCT04584489 (accessed on 14 October 2020). According to French regulations, patient consent was waived. The study protocol was approved by the local International Review Board (69HCL20_0793).

3. Results

We assessed 226 PwCF from 6 October 2020 to 23 February 2021 for eligibility. Figure 1 provides a STROBE-compliant flow diagram [21].

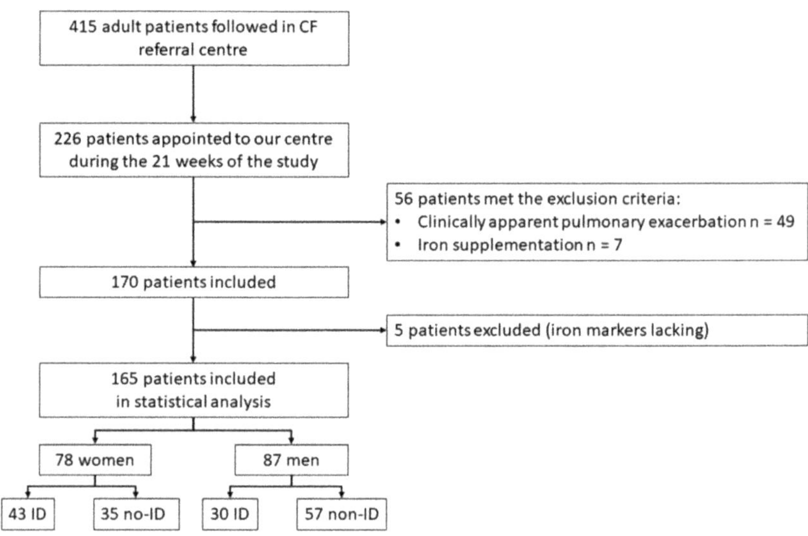

Figure 1. Flowchart. CF: cystic fibrosis. No-ID: no iron deficiency.

3.1. Baseline Characteristics

Table 1 summarizes the characteristics of the 165 patients (78 women, 47%) included in the final analysis. The mean age was 31.1 ± 8.9 years. The types of CFTR mutations are available as Supplementary Materials (Supplementary Table S1). The most frequent CF complication was diabetes mellitus (27/165, 16.3%), which was significantly more frequent among women. Regarding solid organ transplantation, 14/165 (8.5%) had lung transplants (median delay—76 months), two patients had liver transplants, and one patient had a kidney transplant. In total, 57/165 (34.5%) patients were treated by CFTR correctors: ivacaftor (n = 3), ivacaftor/lumacaftor (n = 42), ivacaftor/tezacaftor (n = 1), and ivacaftor/tezacaftor/elexacaftor (n = 11).

Table 1. Baseline characteristics of the population.

	Men (n = 87)	Women (n = 78)	p
Age (years)	30.2 ± 7.7	32.1 ± 10.1	0.37
Genotype			
p.PheF508del heterozygote (n, %)	32 (36.8%)	29 (37.2%)	
p.PheF508del homozygote (n, %)	45 (51.7%)	36 (46.1%)	0.59
other genotypes (n, %)	10 (11.5%)	13 (16.7%)	
BMI (kg·m^{-2})	21.4 ± 2.5	20.3 ± 2.1	0.005
Diabetes (n, %)	8 (9.2%)	19 (24.4%)	0.009
CF-related liver disease (n, %)	6 (6.9%)	11 (14.1%)	0.12
P. aeruginosa colonization (n, %)	54/85 (63.5%)	46/76 (60.5%)	0.69
Anaemia (n, %)	2/80 (2.5%)	7/67 (10.4%)	0.08
Ferritin (µg/L)	87.4 ± 69.2	49.7 ± 64.8	<0.001
TSAT (%)	22.8 ± 8.8	17.5 ± 6.1	<0.001
CRP (mg/L)	5.8 ± 8.2	8 ± 14.3	0.78
FEV1 (n, %)	n = 86	n = 76	
>79%	50 (58.1%)	32 (42.1%)	
50–79%	17 (19.8%)	27 (35.5%)	
30–49%	19 (22.1%)	16 (21.1%)	0.05
<30%	0	1 (1.3%)	

Data are presented as mean ± standard deviation or as number and percentage (%). CF: cystic fibrosis. CRP: C reactive protein. FEV: Forced expiratory volume in one second. n: number. TSAT: transferrin saturation.

The mean CRP was 6.8 ± 11.4 mg/L. Thirty-two patients had significantly increased CRP > 10 mg/L, showing systemic inflammatory conditions. The prevalence of vitamin A, D, E and K deficiency was 3/142 (2%), 19/137 (13%), 11/142 (7.7%), and 33/141 (23%), respectively.

3.2. Iron Deficiency Prevalence

A total of 73 of the 165 patients (44.2%) pwCF had ID, according to the definition retained for our primary endpoint. Among them, only 9/73 were classified as ID because the ferritin level was ≤100 μg/L with CRP ≥ 10 mg/L.

Using the historical definition of ID in CF, 53/165 (32.1%) pwCF had ID. As illustrated in Figure 2, each patient with ID defined through the historical definition of ID was included in the primary endpoint definition.

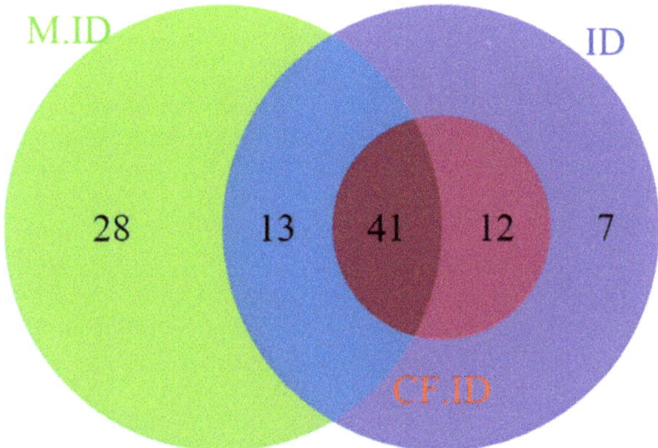

Figure 2. Venn diagram of iron-deficient patients, according to the different biological definitions. M.ID: patients with mild iron depletion (ferritin ≤ 50 μg/L). CF.ID: historical paediatric definition of iron deficiency in cystic fibrosis (ferritin ≤ 12 μg/L or TSAT ≤ 16%). ID: primary endpoint definition of iron deficiency, according to international criteria (ferritin ≤ 20 (women) or 30 (men) μg/L or ≤100 μg/L in the case of systemic inflammation (C reactive protein ≥ 10 mg/L) or transferrin saturation ≤ 16%).

Finally, 82 patients (49.7%) of our cohort had mildly depleted iron stores (ferritin ≤ 50 μg/L), including 54/73 patients with ID, according to the primary endpoint definition.

Nine patients were anaemic: seven women (four with ID) and two men (one with ID). The mean corpuscular volume (MCV) was significantly lower in iron-deficient patients (85.8 ± 5.2 vs. 88.5 ± 4.7 fL, $p < 0.001$). MCV was significantly lower in ID patients irrespective of gender (men: 85.2 ± 3.3 vs. 88.1 ± 3.6 fL, $p < 0.01$, women: 86.2 ± 6.2 vs. 89.1 ± 6.0 fL, $p = 0.01$).

3.3. Iron Deficiency Risk Factors

3.3.1. Univariate Analysis

Table 2 shows the association of ID and risk factors. ID was significantly associated with female gender (58.9% vs. 38%, $p = 0.008$), lower age (29.4 ± 8.5 vs. 32.5 ± 9.1 years, $p = 0.02$), lower BMI (20.5 ± 2.2 vs. 21.3 ± 2.5, $p = 0.05$), and *P. aeruginosa* colonization (70.8% vs. 55.1%, $p = 0.04$).

Diabetes mellitus and the use of antiacid drugs or pump proton inhibitors were more frequent in the ID group, but the difference did not reach statistical significance ($p = 0.08$ and $p = 0.06$, respectively). The prevalence of ID tended to be correlated with worsened PFT, illustrated by the decrease in FEV1 ($p = 0.07$). The proportion of patients treated with

CFTR correctors was similar in the ID and no-ID groups (p = 0.60). Fat-soluble vitamin deficiency was not significantly associated with ID.

Table 2. Risk factors and 95% CI for iron deficiency in univariate analysis.

	No-ID (*n* = 92)	ID (*n* = 73)	*p*
Age (year)	32.5 ± 9.1	29.4 ± 8.5	0.02
Female gender (%)	38%	58.9%	0.008
BMI (kg/m^2)	21.3 ± 2.5	20.5 ± 2.2	0.05
P. aeruginosa (%)	55.1%	70.8%	0.04
Diabetes mellitus (*n*)	11	16	0.08
CF-related liver disease (*n*)	9	8	0.80
Antiacid drugs/PPI (%)	25%	38.4%	0.06
CFTR corrector drugs (%)	32.6%	37%	0.60
FEV1			
<30%	0	1	
30–49%	16	19	0.07
50–79%	35	33	
≥80%	39	19	

BMI: body mass index; CF: cystic fibrosis; CFTR: cystic fibrosis transmembrane regulators; ID: iron-deficient; FEV1: forced expiratory volume in one second; PPI: pump proton inhibitors. NID: No iron deficiency.

3.3.2. Multivariate Analysis

In the multivariate analysis (Figure 3), female gender (OR: 2.64, CI95%: 1.31–5.31, p = 0.006), age below 30 years (OR: 2.30, CI95%: 1.16–4.56, p = 0.02), and *P. aeruginosa* (OR: 2.09, CI95%: 1.04–4.19, p = 0.04) were significantly associated with ID, whereas diabetes mellitus and BMI were not associated with ID.

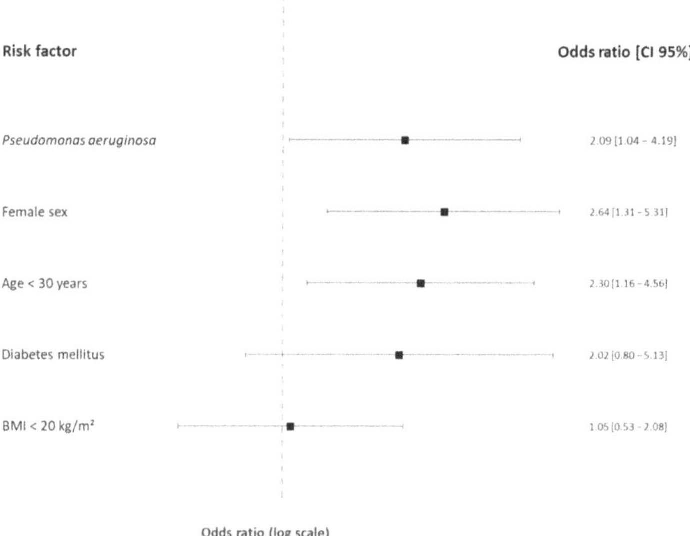

Figure 3. Odds ratios and 95% confidence interval of iron deficiency risk factors in cystic fibrosis in the multivariate analysis. BMI: body mass index (kg/m^2).

Noticeably, we repeated the multivariate analysis including only the 64 patients defined as ID with low ferritin or TSAT level and excluding those with CRP ≤ 100 µg/L in the case of systemic inflammation (CRP ≥ 10 mg/L) or TSAT ≤ 16%. Similar results were found: ID was significantly associated with *P. aeruginosa* colonization (OR: 2.15, CI95%: 1.03–4.48, p = 0.04), female sex (OR: 3.38, CI95%: 1.65–6.93, p = 0.001), and age < 30 years (OR: 2.57,

CI95%: 1.26–5.27, p = 0.009), whereas diabetes mellitus (OR: 1.84, CI95%: 0.72–4.69, p = 0.20) and BMI < 20 kg/m^2 (OR: 0.88, CI95%: 0.43–1.78, p = 0.71) were not associated with ID.

4. Discussion

In this prospective study, we found a high prevalence of ID in adult pwCF, similar to the previous most recent study but a low prevalence of anaemia [10]. To circumvent the non-specific elevation of ferritin during systemic inflammation, we recruited stable patients during their annual evaluation and used the most recent guidelines for ID diagnosis, which recommend increasing the threshold of ferritin in the case of high CRP levels [11]. The use of serum iron concentration or soluble transferrin receptor (sTfR) has been suggested in such situations, but serum iron concentration is reduced in the case of systemic inflammation and the sTfR can be affected by enhanced erythropoiesis and has a lower sensitivity and specificity than ferritin [22]. Moreover, the use of sTfR is limited by the lack of standardization of sTfR measurement [23].

Very few data about the risk factors of ID in adult pwCF are available at present. Previous studies focused on children [7] or mixed populations (both adult and child patients [8] and were impaired by their sample size, missing data, or confounding factors [16]. Table 3 presents the main previously published studies during the past 40 years (PubMed literature review using the following mesh terms—"Iron", "Iron deficiency" and "Cystic Fibrosis"), showing the great heterogeneity of biological definitions used to define ID in pwCF.

Table 3. PubMed literature systematic review of studies reporting iron deficiency among patients with cystic fibrosis.

	n	Patients	ID Biological Definition	ID Prevalence	Exacerbation
Gettle, 2020 [10]	67	A	ferritin < 12 µg/L and/or TSAT < 16%	41.8%	PEx+
Kałużna-Czyż, 2018 [24]	46	P	ferritin < 12 µg/L (<5 yo) ferritin < 15 µg/L (>5 yo)	39%	PEx+ and PEx−
Yadav, 2014 [25]	27	P	SI < 4 µmol/L	48.1%	PEx+ and PEx−
Gifford, 2012 [26]	12	A	SI < 12 µmol/L	83%	PEx+
Gifford, 2011 [27]	39	A	SI < 12 µmol/L	76.9%	PEx+ and PEx−
von Drygalski, 2008 [8]	26	A + P	SI ≤ 40 µg/dL or TSAT ≤ 20% or ferritin ≤ 35 µg/L	61% * 87.5% #	NA
Khalid, 2007 [16]	127	A	ferritin < 12 µg/L (women) and 20 µg/L (men) or SI < 12 µmol/L or TSAT ≤ 15% or sTfR < 1.74 mg/L	18.9% (ferritin) 42.5% (TSAT)15% (sTfR)	PEx+ and PEx−
Reid, 2002 [6]	30	A	SI < 12 µmol/L or TSAT < 16%	74%	PEx−
Jaffe, 2002 [28]	144	P	NA	58%	NA
Keevil, 2000 [5]	70	A	ferritin < 12 µg/L (women) ferritin < 20 µg/L (men) SI < 12 µmol/L TSAT < 16% sTfR < 1.74 mg/L	11% (ferritin) 69% (TSAT) 29% (sTfR)	NA
Pond, 1996 [29]	71	A	TSAT < 16%	62%	NA
Zempsky, 1989 [30]	13	A	ferritin ≤ 25 µg/L	38.4%	PEx−
Ehrhardt, 1987 [15]	127	A + P	ferritin < 12 µg/L	32.3%	PEx+ and PEx−
Ater, 1983 [31]	39	A + P	ferritin < 12 µg/L SI < 40 µg/dL TSAT < 16%	33% (ferritin) 25% (SI) 28% (TSAT)	PEx+ and Pex−

* among anaemic patients; # among non-anaemic patients. A: adult population; n: sample size; NA: not available; P: paediatric population; Pex+: included patients with clinically apparent pulmonary exacerbation; PEx−: included patients without apparent pulmonary exacerbation; SI: serum iron; sTfR: soluble transferrin receptor; TSAT: transferrin saturation.

Unlike the study of Gettle et al. [10], antiacid drug use was not associated with ID in our cohort, whereas a lower BMI, lower age, and female gender appeared as risk factors of ID in the multivariate analysis. These discrepancies might be due to a younger population and a lower mean BMI in our work or to the systematic inflammation assessment that allowed the reclassification of nine patients who were misdiagnosed as non-iron deficient. In previous studies [6], FEV1 was correlated with ID. A similar trend was present in our results, without statistical significance. In our study, patients were less frequently colonized by *P. aeruginosa* and mean FEV1 was higher than in previous cohorts, reflecting the improved medical management of CF. We expected a lower prevalence of ID in lung transplant patients, related to decreased systemic inflammation linked to the disappearance of *P. aeruginosa* colonization. However, the difference was not statistically significant, probably because of the small number of lung transplant recipients with ID (2/14).

ID aetiology in CF remained uncertain. The high prevalence in men compared to the general population (2% according to the Centre for Disease Control [32]) highlighted that ID in CF is not supported by the same blood loss mechanism. The treatment of CF exacerbation has been shown to be linked to an increase in iron stores, supporting the hypothesis that systemic inflammation through enhanced hepcidin secretion is a central cause of ID [26]. As such, the increasing use of CFTR modulators, which are known to reduce the systemic inflammation, appears as a promising option for preventing ID [33]. One-third of our patients received these treatments with no difference between ID and no-ID groups. A lack of power due to the sample size could explain these results: a prospective assessment of iron status before and after highly effective CFTR modulator use appears necessary. Indeed, triple combination therapy is expected to provide better control of pulmonary inflammation, which could thus help to prevent ID and anaemia [34].

The treatment of ID is controversial and no recommendations are available [35]. Iron supplementation has been suspected to enhance bacterial growth. In a small case series, the use of ferric carboxymaltose was associated with the worsening of pulmonary symptoms and FEV1 within a few days following infusion [36]. However, in a blinded trial, weekly intravenous ferrous sulphate administration in anaemic ID adult pwCF did not increase CF exacerbation [28,37]. Future studies should focus not only on the iron supplementation tolerance profile but also on the improvement of health quality and fatigue related to the correction of ID [38,39]

Some limitations of our study can be raised. First, our study is a single-centre design and took place in a referral centre: thus, our results might not be generalizable to the whole pwCF population. Furthermore, our results are only applicable to adult patients and specific studies should be conducted in the paediatric population. Second, although we included 40% of our cohort, the sample size may limit multiple-comparison and subgroup analyses. Third, the cross-sectional nature of our study does not allow us to study the effect of treatments, such as antibiotics or CFTR modulators, on the incidence of ID. To the best of our knowledge, our study is the largest published to date, representing a first-step analysis with moderate statistical power, providing a basis and hypothesis for future prospective studies to confirm these results in larger samples.

5. Conclusions

ID is highly prevalent in adult patients with CF. The current guidelines for pwCF recommend reviewing iron status annually [40], but a recent survey among CF clinicians showed that few centres are routinely screening for ID. To improve ID screening in pwCF, our results suggest that special attention should be directed to young women with low BMI and *P. aeruginosa* colonization. However, male patients would also benefit from ID screening as a much higher prevalence was identified compared to young men in the general population. The therapeutic revolution of CFTR modulators could dramatically improve the prevalence of ID in CF, but further prospective studies are required to determine the effect of inflammatory syndrome normalization on iron metabolism.

Supplementary Materials: The following supporting information can be downloaded at: https://www.mdpi.com/article/10.3390/nu14030673/s1, Table S1: genotypes of our cohort.

Author Contributions: Conceptualization, H.L., S.D. and I.D.; methodology, H.L., S.D. and I.D.; formal analysis, H.L., S.M. and B.P.; investigation, H.L., S.D., R.N.-J., I.D. and Q.R.; data curation, H.L.; writing—original draft preparation, H.L.; writing—review and editing, H.L., S.D., S.M., B.P., R.N.-J., I.D. and Q.R.; visualization, H.L., S.D., S.M., I.D. and Q.R.; supervision, I.D. and Q.R.; project administration, Q.R. All authors have read and agreed to the published version of the manuscript.

Funding: This research received no external funding.

Institutional Review Board Statement: The study was conducted in accordance with the Declaration of Helsinki and approved by the Institutional Review Board (69HCL20_0793).

Informed Consent Statement: According to French regulations, patient consent was waived.

Data Availability Statement: The data presented in this study are available on request from the corresponding author.

Acknowledgments: We are grateful to Marie Roux-Perceval for her valuable support in collecting the data for this study.

Conflicts of Interest: Isabelle Durieu reports a relationship outside of this work with the French Ministry of Health and with non-profit organization "Vaincre la mucoviscidose" that includes: funding grants; with Zambon outside of this work that includes: travel reimbursement. Isabelle Durieu reports a relationship with Vertex that includes: board membership, outside of this work. The other authors report no conflict of interest.

References

1. GBD. 2017 Disease and Injury Incidence and Prevalence Collaborators Global, Regional, and National Incidence, Prevalence, and Years Lived with Disability for 354 Diseases and Injuries for 195 Countries and Territories, 1990-2017: A Systematic Analysis for the Global Burden of Disease Study 2017. *Lancet* **2018**, *392*, 1789–1858. [CrossRef]
2. Fischer, R.; Simmerlein, R.; Huber, R.M.; Schiffl, H.; Lang, S.M. Lung Disease Severity, Chronic Inflammation, Iron Deficiency, and Erythropoietin Response in Adults with Cystic Fibrosis. *Pediatric Pulmonol.* **2007**, *42*, 1193–1197. [CrossRef] [PubMed]
3. Staudinger, B.J.; Muller, J.F.; Halldórsson, S.; Boles, B.; Angermeyer, A.; Nguyen, D.; Rosen, H.; Baldursson, Ó.; Gottfreðsson, M.; Guðmundsson, G.H.; et al. Conditions Associated with the Cystic Fibrosis Defect Promote Chronic Pseudomonas Aeruginosa Infection. *Am. J. Respir. Crit. Care Med.* **2014**, *189*, 812–824. [CrossRef]
4. Smith, D.J.; Lamont, I.L.; Anderson, G.J.; Reid, D.W. Targeting Iron Uptake to Control Pseudomonas Aeruginosa Infections in Cystic Fibrosis. *Eur. Respir. J.* **2013**, *42*, 1723–1736. [CrossRef] [PubMed]
5. Keevil, B.; Rowlands, D.; Burton, I.; Webb, A.K. Assessment of Iron Status in Cystic Fibrosis Patients. *Ann. Clin. Biochem.* **2000**, *37 Pt 5*, 662–665. [CrossRef]
6. Reid, D.W.; Withers, N.J.; Francis, L.; Wilson, J.W.; Kotsimbos, T.C. Iron Deficiency in Cystic Fibrosis: Relationship to Lung Disease Severity and Chronic Pseudomonas Aeruginosa Infection. *Chest* **2002**, *121*, 48–54. [CrossRef] [PubMed]
7. Uijterschout, L.; Nuijsink, M.; Hendriks, D.; Vos, R.; Brus, F. Iron Deficiency Occurs Frequently in Children with Cystic Fibrosis. *Pediatric Pulmonol.* **2014**, *49*, 458–462. [CrossRef]
8. von Drygalski, A.; Biller, J. Anemia in Cystic Fibrosis: Incidence, Mechanisms, and Association with Pulmonary Function and Vitamin Deficiency. *Nutr. Clin. Pract.* **2008**, *23*, 557–563. [CrossRef]
9. Smith, D.J.; Anderson, G.J.; Lamont, I.L.; Masel, P.; Bell, S.C.; Reid, D.W. Accurate Assessment of Systemic Iron Status in Cystic Fibrosis Will Avoid the Hazards of Inappropriate Iron Supplementation. *J. Cyst. Fibros.* **2013**, *12*, 303–304. [CrossRef] [PubMed]
10. Gettle, L.S.; Harden, A.; Bridges, M.; Albon, D. Prevalence and Risk Factors for Iron Deficiency in Adults With Cystic Fibrosis. *Nutr. Clin. Pract.* **2020**, *35*, 1101–1109. [CrossRef] [PubMed]
11. Camaschella, C. Iron Deficiency. *Blood* **2019**, *133*, 30–39. [CrossRef] [PubMed]
12. Nemeth, E.; Ganz, T. Hepcidin-Ferroportin Interaction Controls Systemic Iron Homeostasis. *Int. J. Mol. Sci.* **2021**, *22*, 6493. [CrossRef] [PubMed]
13. Fuchs, H.J.; Borowitz, D.S.; Christiansen, D.H.; Morris, E.M.; Nash, M.L.; Ramsey, B.W.; Rosenstein, B.J.; Smith, A.L.; Wohl, M.E. Effect of Aerosolized Recombinant Human DNase on Exacerbations of Respiratory Symptoms and on Pulmonary Function in Patients with Cystic Fibrosis. The Pulmozyme Study Group. *N. Engl. J. Med.* **1994**, *331*, 637–642. [CrossRef] [PubMed]
14. World Health Organization. *Haemoglobin Concentrations for the Diagnosis of Anaemia and Assessment of Severity*; World Health Organization: Geneva, Switzerland, 2011.
15. Ehrhardt, P.; Miller, M.G.; Littlewood, J.M. Iron Deficiency in Cystic Fibrosis. *Arch. Dis. Child.* **1987**, *62*, 185–187. [CrossRef] [PubMed]

16. Khalid, S.; McGrowder, D.; Kemp, M.; Johnson, P. The Use of Soluble Transferin Receptor to Assess Iron Deficiency in Adults with Cystic Fibrosis. *Clin. Chim. Acta* **2007**, *378*, 194–200. [CrossRef]

17. Favrat, B.; Balck, K.; Breymann, C.; Hedenus, M.; Keller, T.; Mezzacasa, A.; Gasche, C. Evaluation of a Single Dose of Ferric Carboxymaltose in Fatigued, Iron-Deficient Women—PREFER a Randomized, Placebo-Controlled Study. *PLoS ONE* **2014**, *9*, e94217. [CrossRef]

18. Holick, M.F. Vitamin D Deficiency. *N. Engl. J. Med.* **2007**, *357*, 266–281. [CrossRef]

19. Raederstorff, D.; Wyss, A.; Calder, P.C.; Weber, P.; Eggersdorfer, M. Vitamin E Function and Requirements in Relation to PUFA. *Br. J. Nutr.* **2015**, *114*, 1113–1122. [CrossRef]

20. Carazo, A.; Macáková, K.; Matoušová, K.; Krčmová, L.K.; Protti, M.; Mladěnka, P. Vitamin A Update: Forms, Sources, Kinetics, Detection, Function, Deficiency, Therapeutic Use and Toxicity. *Nutrients* **2021**, *13*, 1703. [CrossRef]

21. von Elm, E.; Altman, D.G.; Egger, M.; Pocock, S.J.; Gøtzsche, P.C.; Vandenbroucke, J.P. STROBE Initiative The Strengthening the Reporting of Observational Studies in Epidemiology (STROBE) Statement: Guidelines for Reporting Observational Studies. *Ann. Intern. Med.* **2007**, *147*, 573–577. [CrossRef] [PubMed]

22. Wish, J.B. Assessing Iron Status: Beyond Serum Ferritin and Transferrin Saturation. *Clin. J. Am. Soc. Nephrol.* **2006**, *1* (Suppl. 1), S4–S8. [CrossRef]

23. Dignass, A.; Farrag, K.; Stein, J. Limitations of Serum Ferritin in Diagnosing Iron Deficiency in Inflammatory Conditions. *Int. J. Chronic. Dis.* **2018**, *2018*, 9394060. [CrossRef]

24. Kałużna-Czyż, M.; Grzybowska-Chlebowczyk, U.; Woś, H.; Więcek, S. Serum Hepcidin Level as a Marker of Iron Status in Children with Cystic Fibrosis. *Mediat. Inflamm.* **2018**, *2018*, 3040346. [CrossRef] [PubMed]

25. Yadav, K.; Singh, M.; Angurana, S.K.; Attri, S.V.; Sharma, G.; Tageja, M.; Bhalla, A.K. Evaluation of Micronutrient Profile of North Indian Children with Cystic Fibrosis: A Case–Control Study. *Pediatr. Res.* **2014**, *75*, 762–766. [CrossRef] [PubMed]

26. Gifford, A.H.; Moulton, L.A.; Dorman, D.B.; Olbina, G.; Westerman, M.; Parker, H.W.; Stanton, B.A.; O'Toole, G.A. Iron Homeostasis during Cystic Fibrosis Pulmonary Exacerbation. *Clin. Transl. Sci.* **2012**, *5*, 368–373. [CrossRef]

27. Gifford, A.H.; Miller, S.D.; Jackson, B.P.; Hampton, T.H.; O'Toole, G.A.; Stanton, B.A.; Parker, H.W. Iron and CF-Related Anemia: Expanding Clinical and Biochemical Relationships. *Pediatr. Pulmonol.* **2011**, *46*, 160–165. [CrossRef] [PubMed]

28. Jaffé, A.; Buchdahl, R.; Bush, A.; Balfour-Lynn, I.M. Are Annual Blood Tests in Preschool Cystic Fibrosis Patients Worthwhile? *Arch. Dis. Child.* **2002**, *87*, 518–520. [CrossRef]

29. Pond, M.N.; Morton, A.M.; Conway, S.P. Functional Iron Deficiency in Adults with Cystic Fibrosis. *Respir. Med.* **1996**, *90*, 409–413. [CrossRef]

30. Zempsky, W.T.; Rosenstein, B.J.; Carroll, J.A.; Oski, F.A. Effect of Pancreatic Enzyme Supplements on Iron Absorption. *Am. J. Dis. Child.* **1989**, *143*, 969–972. [CrossRef]

31. Ater, J.L.; Herbst, J.J.; Landaw, S.A.; O'Brien, R.T. Relative Anemia and Iron Deficiency in Cystic Fibrosis. *Pediatrics* **1983**, *71*, 810–814. [CrossRef] [PubMed]

32. Centers for Disease Control and Prevention (CDC) Iron Deficiency–United States, 1999-2000. *MMWR Morb. Mortal. Wkly. Rep.* **2002**, *51*, 897–899.

33. Gillan, J.L.; Davidson, D.J.; Gray, R.D. Targeting Cystic Fibrosis Inflammation in the Age of CFTR Modulators: Focus on Macrophages. *Eur. Respir. J.* **2021**, *57*, 2003502. [CrossRef]

34. Gifford, A.H.; Heltshe, S.L.; Goss, C.H. CFTR Modulator Use Is Associated with Higher Hemoglobin Levels in Individuals with Cystic Fibrosis. *Ann. Am. Thorac. Soc.* **2019**, *16*, 331–340. [CrossRef]

35. Garlow, G.M.; Gettle, L.S.; Felicetti, N.J.; Polineni, D.; Gifford, A.H. Perspectives on Anemia and Iron Deficiency from the Cystic Fibrosis Care Community. *Pediatr. Pulmonol.* **2019**, *54*, 939–940. [CrossRef]

36. Hoo, Z.H.; Wildman, M.J. Intravenous Iron among Cystic Fibrosis Patients. *J. Cyst. Fibros.* **2012**, *11*, 560–562. [CrossRef]

37. Gifford, A.H.; Alexandru, D.M.; Li, Z.; Dorman, D.B.; Moulton, L.A.; Price, K.E.; Hampton, T.H.; Sogin, M.L.; Zuckerman, J.B.; Parker, H.W.; et al. Iron Supplementation Does Not Worsen Respiratory Health or Alter the Sputum Microbiome in Cystic Fibrosis. *J. Cyst. Fibros.* **2014**, *13*, 311–318. [CrossRef]

38. Sharma, R.; Stanek, J.R.; Koch, T.L.; Grooms, L.; O'Brien, S.H. Intravenous Iron Therapy in Non-Anemic Iron-Deficient Menstruating Adolescent Females with Fatigue. *Am. J. Hematol.* **2016**, *91*, 973–977. [CrossRef] [PubMed]

39. Jankowska, E.A.; Kirwan, B.-A.; Kosiborod, M.; Butler, J.; Anker, S.D.; McDonagh, T.; Dorobantu, M.; Drozdz, J.; Filippatos, G.; Keren, A.; et al. The Effect of Intravenous Ferric Carboxymaltose on Health-Related Quality of Life in Iron-Deficient Patients with Acute Heart Failure: The Results of the AFFIRM-AHF Study. *Eur. Heart J.* **2021**, ehab234. [CrossRef] [PubMed]

40. Turck, D.; Braegger, C.P.; Colombo, C.; Declercq, D.; Morton, A.; Pancheva, R.; Robberecht, E.; Stern, M.; Strandvik, B.; Wolfe, S.; et al. ESPEN-ESPGHAN-ECFS Guidelines on Nutrition Care for Infants, Children, and Adults with Cystic Fibrosis. *Clin. Nutr.* **2016**, *35*, 557–577. [CrossRef]

Article

Heat-Killed *Lactococcus lactis* subsp. *cremoris* H61 Altered the Iron Status of Young Women: A Randomized, Double-Blinded, Placebo-Controlled, Parallel-Group Comparative Study

Mizuki Takaragawa, Keishoku Sakuraba and Yoshio Suzuki *

Graduate School of Health and Sports Science, Juntendo University, Inzai 270-1695, Japan; mizuki854@gmail.com (M.T.); sakuraba@kf6.so-net.ne.jp (K.S.)
* Correspondence: yssuzuki@juntendo.ac.jp

Abstract: Women are prone to iron deficiency because of increased iron excretion associated with menstruation. This is often treated by oral iron supplementation, although this treatment can cause side effects, such as stomach pain and nausea, with low absorption of ingested iron. Previously, a significant increase in serum iron was observed in association with the consumption of foods containing *Lactococcus lactis* subsp. *cremoris* H61 (H61). However, the causal relationship between H61 ingestion and elevated serum iron is still unclear. Therefore, in this study, we aimed to determine the effects of H61 ingestion on the iron status of young women. Healthy young Japanese women (18–25 years of age) ingested either heat-killed H61 or placebo for 4 weeks. Serum iron, transferrin saturation, and ferritin were significantly elevated in the H61 group but remained unchanged in the placebo group. Compared to before the intervention, iron intake remained unchanged during the intervention period, so the change in the iron status of the H61 group was not due to increased iron intake. These results suggest that heat-killed H61 may elevate iron status by enhancing iron absorption.

Keywords: ferritin; probiotics; serum iron; transferrin saturation; unsaturated iron-binding capacity

Citation: Takaragawa, M.; Sakuraba, K.; Suzuki, Y. Heat-Killed *Lactococcus lactis* subsp. *cremoris* H61 Altered the Iron Status of Young Women: A Randomized, Double-Blinded, Placebo-Controlled, Parallel-Group Comparative Study. *Nutrients* **2022**, *14*, 3144. https://doi.org/10.3390/nu14153144

Academic Editor: James H. Swain

Received: 2 July 2022
Accepted: 26 July 2022
Published: 30 July 2022

Publisher's Note: MDPI stays neutral with regard to jurisdictional claims in published maps and institutional affiliations.

1. Introduction

Iron is an important nutrient because it is involved in the electron transport system and oxygen transport in the human body. However, due to its low absorption from food [1,2], iron deficiency is among the most common nutritional problems worldwide [3]. Although there is no physiological pathway that actively excretes iron, approximately 1–2 mg/day of iron is inevitably lost due to epithelial sloughing, sweating, and bleeding [4,5]. In addition, menstruating women are prone to iron deficiency anemia due to high iron discharge associated with menstruation [6].

Iron is absorbed from food in the intestine and enters circulation. Then, the iron released from enterocytes into the circulatory system is inhibited by the hepatic hormone hepcidin [7–9]. As serum iron concentration is tightly regulated by hepcidin, its oversecretion causes anemia, whereas its undersecretion causes iron overload [7]. Exercise [10,11] and excessive iron intake [12] enhance hepcidin secretion, so physically active women are more susceptible to iron deficiency anemia, even with adequate iron intake.

Iron deficiency is generally treated with oral iron supplementation. Due to the rapid increase in iron exposure in the gastrointestinal tract, oral supplementation may cause side effects, such as stomach pain and nausea [13]. An increase in serum iron was observed in people consuming either yogurt fermented with *Lactococcus lactis* subsp. *Cremoris* H61 (H61) [14] or a commercial dietary supplement containing heat-killed H61 [15]. However, in previous studies, participants were not instructed to refrain from iron supplementation during the intervention period [14], and the dietary supplement used contained other probiotics and vitamins [15]. Therefore, the causal relationship between H61 ingestion and elevated serum iron is not yet clarified.

Thus, in this study, a double-blind, randomized, placebo-controlled trial was conducted to determine the effect of H61 on the iron status of healthy young women.

2. Materials and Methods

2.1. Participants

A total of 50 young women were recruited. The inclusion criteria were: (1) healthy women aged 18–25 years (2) in good physical condition (3) willing to voluntarily participated in the study and provide written consent. The exclusion criteria were as follows: (1) a history of serious cardiovascular, hepatic, renal, respiratory, endocrine, or metabolic disorders; (2) a history of chest pain or syncope; (3) those at risk of developing allergies related to test supplements; (4) those who had 200 mL of blood drawn within 1 month or 400 mL within 3 months prior to the start of this study, (e.g., blood donation); (5) smokers; and (6) those who were otherwise deemed unsuitable by the study investigator. The purpose and methods of this study were explained to participants orally and in writing, and written consent was obtained.

The participants were randomly allocated to the test (H61) or placebo group. One participant in the placebo group withdrew prior to the intervention. Another participant in the placebo group was excluded because her initial serum iron concentration was not normal. During the intervention, six participants in the H61 group dropped out because they did not appear in the second blood sampling (n = 5) and lost the test supplement (n = 1). Forty-two participants completed the intervention. Among them, 13 participants were excluded for common cold (n = 4), iron use (n = 6), compliance of <80% (n = 3), or missing data (n = 1). The remaining 29 participants were analyzed (Figure 1, Table 1).

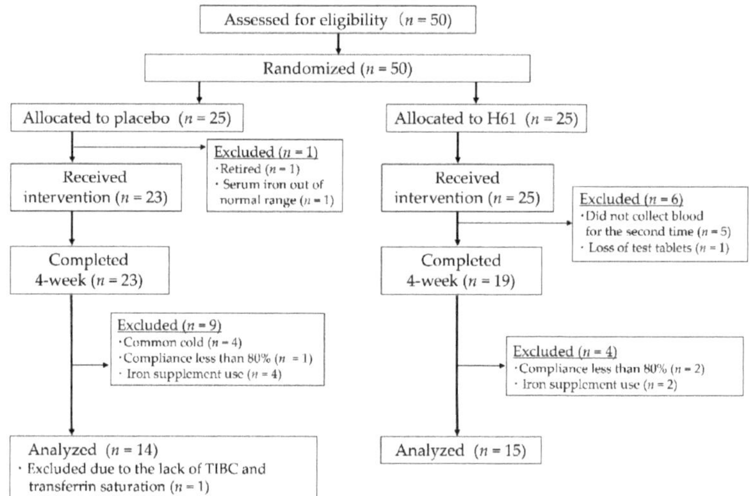

Figure 1. Flow diagram of the participants.

Table 1. Characteristics of the participants.

	Placebo (n = 15)	H61 (n = 14)	p
Age (years)	19.9 ± 1.5	19.9 ± 1.1	0.772
Height (cm)	163.0 ± 6.4	161.7 ± 6.7	0.603
Body weight (kg)	58.5 ± 5.9	56.6 ± 5.5	0.381

Data are expressed as means ± SD.

This study was conducted according to the Declaration of Helsinki (approved in 1964, amended in 2013), with the approval of the ethics committee of Juntendo University

Graduate School of Health and Sports Science (approve No. 29-162). The study was also registered in UMIN-CTR (UMIN000030815) prior to its initiation.

2.2. Experimental Design

A randomized, double-blinded, placebo-controlled, parallel-group comparative trial was conducted. The participants took H61 (60 mg/day) or placebo at the same time daily for 4 weeks. During the intervention, the participants were instructed not to change their dietary habits. Blood was drawn before (pre) and after (post) the intervention to assess iron status. A dietary survey was conducted using a brief-type, self-administered diet history questionnaire (BDHQ) within 3 days before and after blood sampling. The participants recorded their test supplement intake, physical condition, medications/dietary supplement use, and menstrual status in a daily logbook during the intervention.

2.3. Test Supplement

The H61 supplement contained heat-killed H61 (Supplementary Figure S1). One tablet contained 30.0 mg of heat-killed H61, 167.5 mg dextrin, 50.0 mg crystalline cellulose, and 2.5 mg calcium stearate. The placebo tablet contained 197.5 mg dextrin, 50.0 mg crystalline cellulose, and 2.5 mg calcium stearate. The appearance of the H61 and placebo supplements was indistinguishable.

The test supplements were provided by Toa Biopharma (Tokyo, Japan). These were packaged in aluminum pouches for each individual identified by a unique key code. The key codes were kept by Toa Biopharma and opened after the intervention was completed and all data were fixed.

2.4. Blood Collection and Measurements

Blood was collected from the cubital vein between 12:00 and 13:00. Blood and biochemical analyses were performed in a certified clinical laboratory (SRL, Tokyo, Japan). Briefly, red blood cell count, hemoglobin (Hgb), and hematocrit were assessed using a Sysmex XE-2100 automated hematology analyzer (Sysmex Corporation, Hyogo, Japan). Serum ferritin, serum iron, and total iron-binding capacity (TIBC) were evaluated using latex agglutination turbidimetry, direct colorimetry, and 2-nitroso-5-(N-propyl-N-sulfopropylamino) phenol (nitroso-PSAP) methods, respectively, using a JCA-BM8060 automatic analyzer (JEOL Ltd., Tokyo, Japan). Transferrin saturation (TSAT) and unsaturated iron-binding capacity (UIBC) were calculated as serum iron/TIBC × 100 and TIBC−serum iron, respectively.

Serum hepcidin concentrations were measured in duplicate using a hepcidin-25 extraction-free ELISA (Cosmo Bio, Tokyo, Japan) according to the protocol provided by the manufacturer. If the difference was <10% of the mean, the mean value was used as the concentration; otherwise, the measurement was repeated until the difference was <10%.

2.5. Dietary Survey

A dietary survey was conducted using BDHQ, a self-administered questionnaire developed for the Japanese population that was previously validated [16]. BDHQ was used to measure nutrient intake for one month from the time of the survey. The daily intake of iron and vitamin C was estimated according to the density of each nutrient (mg/1000 kcal) and estimated energy requirements (2200 kcal/day). The estimated energy requirement was based on physical activity level category III (high) for women aged 18–29 as defined in the Dietary Reference Intakes for Japanese (2015 edition), as all participants were active and belonged to a collegiate athletic club (e.g., basketball and soccer).

2.6. Menstrual Cycle

The menstrual cycle on the day of blood collection was classified into two phases according to the diary kept by participant: follicular or luteal [17]. The luteal phase was defined as up to 14 days from the first day of bleeding, whereas the follicular phase was defined as any other day.

2.7. Statistical Analyses

The changes in hematological parameters were analyzed using a generalized estimated equation of a generalized linear model controlled for menstrual cycle. The model included a subject ID as a subject variable, intervention (H61, placebo), measure point (Pre, Post), and menstrual cycle as within-subject variables and interactions; intervention × measure point was fixed, whereas intervention × menstrual cycle and measure point × menstrual cycle were included in the model if they reduced the quasi-information criterion. The model with the smallest criterion was adopted.

Statistical significance was set at $p < 0.05$. Statistical analyses were conducted using SPSS Statistics ver. 24 (IBM Japan, Tokyo, Japan).

3. Results

3.1. Iron Status

After 4 weeks of intervention, serum iron ($p < 0.05$), TSAT ($p < 0.05$), and ferritin ($p < 0.001$) increased significantly, whereas UIBC decreased significantly in the H61 group ($p < 0.001$). In the placebo group, MHC ($p < 0.001$) and hepcidin ($p < 0.01$) levels decreased significantly. A significant decrease in MCV was also observed in both groups ($p < 0.05$).

After the intervention (Post), the H61 group had significantly higher serum iron ($p < 0.05$), TSAT ($p < 0.05$), and ferritin ($p < 0.05$) and significantly lower UIBC ($p < 0.01$) than the placebo group. However, before the intervention (Pre), the H61 group also had significantly higher serum iron ($p < 0.05$) and TSAT ($p < 0.01$) and significantly lower UIBC ($p < 0.01$) than the placebo group (Figure 2, Supplementary Table S1).

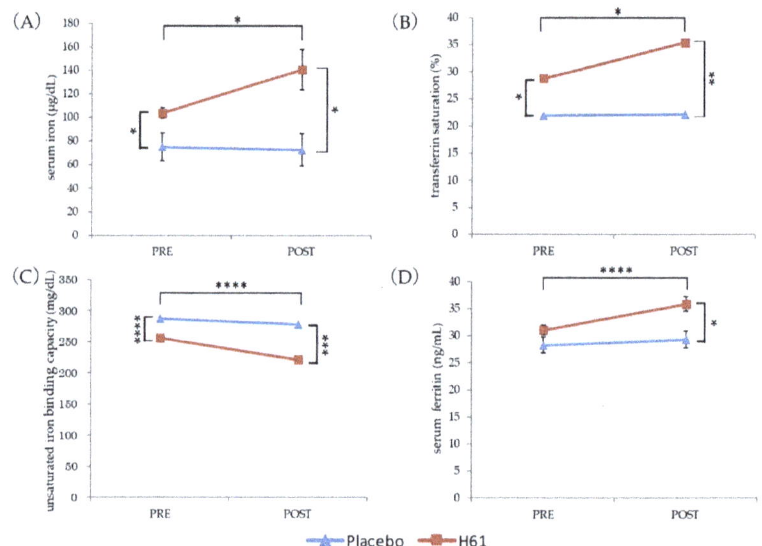

Figure 2. Changes in hematological parameters. (**A**) Serum iron. (**B**) Transferrin saturation. (**C**) Unsaturated iron-binding capacity. (**D**) Serum ferritin. * $p < 0.05$, ** $p < 0.01$, *** $p < 0.005$, **** $p < 0.001$.

Iron status, depending on menstrual cycle (follicular phase vs. luteal phase), was also examined. Ferritin was significantly higher ($p < 0.05$) in the follicular phase (EMM 33.9, SE 1.0 ng/mL) than in the luteal phase (EMM, 28.5; SE, 1.2 ng/mL). There were no significant differences in other parameters, such as serum iron and TSAT (Supplementary Table S2).

Throughout the study, the highest ferritin and TSAT were 103.0 ng/mL and 47.8%, respectively; therefore, no subjects were suspected of having iron overload.

3.2. Iron and Vitamin C Intake

During the intervention, the participants' iron and vitamin C intake did not change relative to pre-intervention. The mean iron intake was lower than the estimated average requirement (EAR) in both groups, with 64% (9/14) of the H61 group and 67% (10/15) of the placebo group below the EAR during the intervention. The mean vitamin C intake was greater than the EAR in both groups, with 29% (4/14) of the H61 group and 40% (6/15) of the placebo group below the EAR during the intervention (Table 2).

Table 2. Changes in the estimated daily intake of iron and vitamin C.

	EAR	Group	PRE			POST			p [b]
			Mean	SD	p [a]	Mean	SD	p [a]	
Iron	8.5 [#]	H61	8.2	1.6	0.215	7.8	2.4	0.766	0.361
(mg/day)		placebo	7.5	1.3		7.6	1.9		0.900
Vitamin C	85	H61	126.5	44.4	0.234	110.8	51.8	0.779	0.121
(mg/day)		placebo	107.6	38.5		105.9	40.6		0.891

EAR, estimated average requirement; [#], women with menstruation aged 18–29; p [a], H61 vs. placebo; p [b], PRE vs. POST.

Table 2 summarizes the intergroup differences. The estimated daily intake of iron and vitamin C did not differ significantly between the H61 and placebo groups pre- and post-intervention.

4. Discussion

The 4 weeks of heat-killed H61 consumption significantly increased serum iron, ferritin, and TSAT levels and significantly decreased UIBC in healthy women. Iron and vitamin C were not supplemented during the intervention. The estimated daily iron and vitamin C intake during the intervention did not differ from pre-intervention levels.

TSAT was calculated as Fe/TIBC, whereas UIBC was calculated as TIBC−Fe. Because serum iron increased, whereas TIBC did not change in the H61 group, the increase in serum iron may account for the increase in TSAT and decrease in UIBC.

When multiple measurements are taken, if a population with extreme initial values is selected, the next measurement will be closer to the overall mean. This is a statistical phenomenon called regression toward the mean, and examples in clinical parameters include blood pressure and serum cholesterol [18]. In this study, the participants were randomly allocated before the intervention, and pre-intervention serum iron was significantly higher in the H61 group (EMM, 103.8 µg/dL; SE, 4.4 µg/dL) than in the placebo group (EMM, 75.2 µg/dL; SE, 11.8 µg/dL). According to the National Health and Nutrition Survey, the mean serum iron level of Japanese women aged 20−29 years is 75.2 µg/dL (SD, 39.6 µg/dL) [19], suggesting that participants with high serum iron levels were disproportionately allocated to the H61 group. Therefore, if the intervention had no effect on serum iron, the post-intervention serum iron in the H61 group should have been lower than the pre-intervention level because of regression toward the mean. However, after the intervention, serum iron (EMM, 140.8 µg/dL; SE, 17.1 µg/dL) was significantly elevated in the H61 group, whereas it was unchanged in the placebo group. Thus, the four weeks of administration of heat-killed H61 may have increased the serum iron levels of healthy young women, even with a higher serum iron population.

Previously, an increase in serum iron was observed in young women consuming 300 mL/day of yogurt fermented with H61 for 4 weeks [14]. However, in this study, the mean serum iron was also elevated in the control group. Furthermore, a dietary survey was not included, and iron supplement use could not be excluded because the participants were female track and field athletes. Thus, the change in serum iron levels was not discussed as the influence of H61. In another study, serum iron increased when men took a commercially available dietary supplement containing H61 (1.6×10^8 cells before being heat-killed/day) for 30 days [15]. However, the dietary supplement also contained another probiotic,

Lactobacillus sporegenes, and vitamins (vitamin C, vitamin E, niacin, calcium pantothenate, vitamin B1, vitamin B6, vitamin B2, vitamin A, folic acid, vitamin D, and vitamin B12). Therefore, H61 could not be concluded as responsible for the increase in serum iron. In this double-blind, randomized, placebo-controlled study, the effect of heat-killed H61 was examined, and an increase in serum iron was observed.

The National Health and Nutrition Survey reported a mean daily iron intake of 7.1 mg (SD, 3.1 mg) for Japanese women aged 20–29 years and a mean vitamin C intake of 73 mg (SD, 52 mg) [19], which is involved in iron absorption [20]. Although a direct comparison of intakes cannot be made because of the difference in survey methods, the iron and vitamin C intakes of the participants in this study, although suboptimal, were not significantly different from those of average Japanese women aged 20–29 years. In addition, iron or vitamin C supplementation was not provided in this study, and the estimated dietary iron and vitamin C intakes during the intervention did not differ from pre-intervention levels. Thus, the increase in serum iron observed in the H61 group was not due to increased iron or vitamin C intake.

It is noteworthy that H61 increased the serum iron concentrations in healthy women with suboptimal dietary iron intake. Orally ingested iron is absorbed in the duodenum and upper small intestine. Heme iron is absorbed as-is, whereas non-heme iron (Fe^{3+}) is reduced to Fe^{2+} by ferric reductase in the apical membrane [21] and is absorbed by divalent metal transporter 1 (DMT1) and human copper transporter 1 [22]. Therefore, the absorption of non-heme iron competes with absorption of zinc and copper, which are also absorbed by DMT1 [23]. Iron absorbed in enterocytes is converted to Fe^{3+} by transmembrane copper-dependent ferroxidase hephaestin and is released into the blood by the basal membrane iron transporter ferroportin (SLC40A1) and binds to transferrin in the serum [7,20]. The hepatic hormone hepcidin inhibits iron transport by ferroportin. [24]. In this study, serum iron was significantly increased in the H61 group. Therefore, H61 may have an influence at some point in this process.

Dietary iron absorption is estimated to be 14% with a Swedish diet, 16% with a French diet, and 16.6% with a U.S. diet [25]. Dietary non-heme iron (Fe^{3+}) must be reduced to divalent iron (Fe^{2+}) in the lumen before it can be absorbed. Vitamin C promotes iron absorption because it can reduce iron [20,26]. Lactate also enhances iron absorption [26,27]. The ingestion of heat-killed H61 was reported to increase the abundance of intestinal *Lactobacillales* [28]. *Lactobacillales*, commonly called lactic acid bacteria, ferment carbohydrates to produce lactate. Therefore, the ingestion of heat-killed H61 may have increased the abundance of *Lactobacillales* to increase the concentration of lactate in the intestines, enhancing iron absorption. However, the intestinal microbiota was not examined in this study. As the composition of the microbiota is differs considerably between humans and animals, the changes in the intestinal microbiota in rats resulting from heat-killed H61 ingestion cannot be directly applied to humans. Meanwhile, a recent systematic review showed that *Lactobacillus plantarum* 299v increases iron absorption [29]. Thereby, it seems possible that H16 may also increase iron absorption. Therefore, the mechanism to increase serum iron should be clarified.

Serum ferritin reflects the amount of stored iron [30]. In this study, the H61 group had an increased serum ferritin and serum iron after the intervention. Therefore, stored iron also seemed to increase in this group.

Iron status fluctuates with the menstrual cycle. Serum iron, TSAT, and hepcidin decrease before menstruation; recover from the onset to the end of menstrual bleeding; and peak during the follicular phase [31]. In female cyclists, Hgb and serum ferritin levels increase from the menstrual phase to the follicular phase and decrease from the follicular phase to the luteal phase [32]. In this study, serum ferritin was higher in the follicular phase than in the luteal phase, similar to a previous report [32], although TSAT and Hgb did not differ according to menstrual cycle. Thus, the menstrual cycle should be incorporated when assessing a woman's iron status.

In a previous report, menstrual cycle was not considered, although an increase in serum iron was observed in young women who consumed yogurt fermented with H61 [14]. In this study, generalized estimating equations were used to control for menstrual cycle. Therefore, the results of this study may demonstrate the effects of heat-killed H61 without the influence of the menstrual cycle.

The treatment of iron deficiency anemia typically involves oral iron supplementation. However, such treatment has adverse effects, such as stomach pain and nausea, which can lead to poor compliance [13]. In contrast, the increase in serum iron observed in this study was not due to increased iron intake. None of the participants complained of gastrointestinal symptoms during the intervention. Therefore, instead of oral iron supplementation, H61 can be used to improve iron deficiency anemia without causing gastrointestinal symptoms or physical discomfort.

This study is subject to some limitations. We did not determine the amount of menstrual bleeding and individual differences in this parameter. This may have impacted the participants' iron status. The initial serum iron level significantly differed between the participants in the H61 and placebo groups; this was an accidental bias resulting from random allocation upon entry. However, the possible difference in the participants' iron bioavailability or genetic predispositions cannot be ruled as having influenced the results. Therefore, the effects of H61 should be re-examined in a population with uniform iron metabolism by measuring the amount of iron loss due to menstrual bleeding. In this study, the 4-week consumption of heat-killed H61 improved iron status with no signs of iron overload. However, it is necessary to determine the risk of iron overload with larger doses and/or longer duration.

5. Conclusions

Four-week consumption of heat-killed *Lactococcus lactis* subsp. *cremoris* H61 was demonstrated to increase serum iron, TSAT, and ferritin levels and decrease UIBC. As the improvement in iron status was not attributed to the amount of iron intake, dietary iron availability was suggested to be enhanced by heat-killed H61.

Supplementary Materials: The following supporting information can be downloaded at: https://www.mdpi.com/article/10.3390/nu14153144/s1, Table S1: Changes in iron status of the women analyzed by generalized estimating equation, Table S2: Iron status and menstrual cycle analyzed by generalized estimating equation. Figure S1: Microscopic view of the heat-killed *Lactococcus lactis* subsp. *cremoris* H61.

Author Contributions: Formal analysis, M.T. and Y.S.; investigation, M.T., K.S. and Y.S.; resources, Y.S.; data curation, Y.S.; writing—original draft preparation, M.T.; writing—review and editing, Y.S.; visualization, M.T. and Y.S.; supervision, Y.S.; project administration, Y.S.; funding acquisition, Y.S. All authors have read and agreed to the published version of the manuscript.

Funding: This study was partly funded by the Cross-Ministerial Strategic Innovation Promotion Program of the Cabinet Office, Government of Japan (grant no.: 14532924). English editing and APC were supported by Toa Biopharma Co., Ltd. (Tokyo, Japan).

Institutional Review Board Statement: The study was conducted according to the guidelines of the Declaration of Helsinki and approved by the Ethics Committee of Juntendo University Graduate School of Health and Sports Science (Approval number: 29-162).

Informed Consent Statement: Informed consent was obtained from all subjects involved in the study. Written informed consent was obtained from the patients to publish this paper.

Data Availability Statement: The data presented in this study are available upon request from the corresponding author. The data are not publicly available due to ethical restrictions.

Acknowledgments: The authors thank all participants. The authors also thank Toa Biopharma Co., Ltd. (Tokyo, Japan) for providing test supplements and hepcidin ELISA kits.

Conflicts of Interest: The authors declare that they have no competing interest.

References

1. Bjorn-Rasmussen, E.; Hallberg, L.; Isaksson, B.; Arvidsson, B. Food iron absorption in man. Applications of the two-pool extrinsic tag method to measure heme and nonheme iron absorption from the whole diet. *J. Clin. Investig.* **1974**, *53*, 247–255. [CrossRef]
2. Hurrell, R. How to ensure adequate iron absorption from iron-fortified food. *Nutr. Rev.* **2002**, *60*, S7–S15, discussion S43. [CrossRef] [PubMed]
3. Stoltzfus, R.J.; Dreyfuss, M.L. Guidelines for the Use of Iron Supplements to Prevent and Treat Iron Deficiency Anaemia. World Health Organization (WHO). 1998. Available online: https://motherchildnutrition.org/nutrition-protection-promotion/pdf/mcn-guidelines-for-iron-supplementation.pdf (accessed on 12 March 2022).
4. Hentze, M.W.; Muckenthaler, M.U.; Andrews, N.C. Balancing acts: Molecular control of mammalian iron metabolism. *Cell* **2004**, *117*, 285–297. [CrossRef]
5. Andrews, N.C. Disorders of iron metabolism. *N. Engl. J. Med.* **1999**, *341*, 1986–1995. [CrossRef]
6. Zimmermann, M.B.; Hurrell, R.F. Nutritional iron deficiency. *Lancet* **2007**, *370*, 511–520. [CrossRef]
7. Hentze, M.W.; Muckenthaler, M.U.; Galy, B.; Camaschella, C. Two to tango: Regulation of Mammalian iron metabolism. *Cell* **2010**, *142*, 24–38. [CrossRef] [PubMed]
8. Core, A.B.; Canali, S.; Babitt, J.L. Hemojuvelin and bone morphogenetic protein (BMP) signaling in iron homeostasis. *Front. Pharmacol.* **2014**, *5*, 104. [CrossRef] [PubMed]
9. Ganz, T. Systemic iron homeostasis. *Physiol. Rev.* **2013**, *93*, 1721–1741. [CrossRef] [PubMed]
10. Ziemann, E.; Kasprowicz, K.; Kasperska, A.; Zembron-Lacny, A.; Antosiewicz, J.; Laskowski, R. Do high blood hepcidin concentrations contribute to low ferritin levels in young tennis players at the end of tournament season? *J. Sports Sci. Med.* **2013**, *12*, 249–258. [PubMed]
11. Ishibashi, A.; Maeda, N.; Sumi, D.; Goto, K. Elevated Serum Hepcidin Levels during an Intensified Training Period in Well-Trained Female Long-Distance Runners. *Nutrients* **2017**, *9*, 277. [CrossRef]
12. Tomosugi, N.; Kawabata, H.; Wakatabe, R.; Higuchi, M.; Yamaya, H.; Umehara, H.; Ishikawa, I. Detection of serum hepcidin in renal failure and inflammation by using ProteinChip System. *Blood* **2006**, *108*, 1381–1387. [CrossRef] [PubMed]
13. Rimon, E.; Kagansky, N.; Kagansky, M.; Mechnick, L.; Mashiah, T.; Namir, M.; Levy, S. Are we giving too much iron? Low-dose iron therapy is effective in octogenarians. *Am. J. Med.* **2005**, *118*, 1142–1147. [CrossRef] [PubMed]
14. Kimoto-Nira, H.; Nagakura, Y.; Kodama, C.; Shimizu, T.; Okuta, M.; Sasaki, K.; Koikawa, N.; Sakuraba, K.; Suzuki, C.; Suzuki, Y. Effects of ingesting milk fermented by Lactococcus lactis H61 on skin health in young women: A randomized double-blind study. *J. Dairy Sci.* **2014**, *97*, 5898–5903. [CrossRef] [PubMed]
15. Suzuki, Y.; Takaragawa, M.; Miyahara, T.; Sakuraba, K.; Nagato, S.; Matsumoto, N.; Misawa, Y.; Minami, S.; Morio, K. Lactococcus lactis subsp. cremoris H61 improved iron status in male distance runners. *Int. J. Anal. Bio-Sci.* **2022**, *10*, 33–41.
16. Kobayashi, S.; Murakami, K.; Sasaki, S.; Okubo, H.; Hirota, N.; Notsu, A.; Fukui, M.; Date, C. Comparison of relative validity of food group intakes estimated by comprehensive and brief-type self-administered diet history questionnaires against 16 d dietary records in Japanese adults. *Public Health Nutr.* **2011**, *14*, 1200–1211. [CrossRef] [PubMed]
17. Reed, B.G.; Carr, B.R. The Normal Menstrual Cycle and the Control of Ovulation. In *Endotext*; Feingold, K.R., Anawalt, B., Boyce, A., Chrousos, G., de Herder, W.W., Dhatariya, K., Dungan, K., Hershman, J.M., Hofland, J., Kalra, S., et al., Eds.; MDText. com, Inc.: South Dartmouth, MA, USA, 2000.
18. Bland, J.M.; Altman, D.G. Some examples of regression towards the mean. *BMJ* **1994**, *309*, 780. [CrossRef] [PubMed]
19. Ministry of Health, Labour and Welfare, Japan. The National Health and Nutrition Survey in Japan, 2017. 2018. Available online: https://www.mhlw.go.jp/stf/seisakunitsuite/bunya/kenkou_iryou/kenkou/eiyou/h30-houkoku_00001.html (accessed on 22 May 2022).
20. Lane, D.J.; Richardson, D.R. The active role of vitamin C in mammalian iron metabolism: Much more than just enhanced iron absorption! *Free. Radic. Biol. Med.* **2014**, *75*, 69–83. [CrossRef]
21. McKie, A.T.; Barrow, D.; Latunde-Dada, G.O.; Rolfs, A.; Sager, G.; Mudaly, E.; Mudaly, M.; Richardson, C.; Barlow, D.; Bomford, A.; et al. An iron-regulated ferric reductase associated with the absorption of dietary iron. *Science* **2001**, *291*, 1755–1759. [CrossRef] [PubMed]
22. Espinoza, A.; Le Blanc, S.; Olivares, M.; Pizarro, F.; Ruz, M.; Arredondo, M. Iron, copper, and zinc transport: Inhibition of divalent metal transporter 1 (DMT1) and human copper transporter 1 (hCTR1) by shRNA. *Biol. Trace Elem. Res.* **2012**, *146*, 281–286. [CrossRef]
23. Arredondo, M.; Martinez, R.; Nunez, M.T.; Ruz, M.; Olivares, M. Inhibition of iron and copper uptake by iron, copper and zinc. *Biol. Res.* **2006**, *39*, 95–102. [CrossRef] [PubMed]
24. Ganz, T.; Nemeth, E. Hepcidin and iron homeostasis. *Biochim. Et Biophys. Acta* **2012**, *1823*, 1434–1443. [CrossRef] [PubMed]
25. Hallberg, L.; Rossander-Hulten, L. Iron requirements in menstruating women. *Am. J. Clin. Nutr.* **1991**, *54*, 1047–1058. [CrossRef] [PubMed]
26. Teucher, B.; Olivares, M.; Cori, H. Enhancers of iron absorption: Ascorbic acid and other organic acids. *Int. J. Vitam. Nutr. Res.* **2004**, *74*, 403–419. [CrossRef]
27. Gillooly, M.; Bothwell, T.H.; Torrance, J.D.; MacPhail, A.P.; Derman, D.P.; Bezwoda, W.R.; Mills, W.; Charlton, R.W.; Mayet, F. The effects of organic acids, phytates and polyphenols on the absorption of iron from vegetables. *Br. J. Nutr.* **1983**, *49*, 331–342. [CrossRef] [PubMed]

28. Oike, H.; Aoki-Yoshida, A.; Kimoto-Nira, H.; Yamagishi, N.; Tomita, S.; Sekiyama, Y.; Wakagi, M.; Sakurai, M.; Ippoushi, K.; Suzuki, C.; et al. Dietary intake of heat-killed Lactococcus lactis H61 delays age-related hearing loss in C57BL/6J mice. *Sci. Rep.* **2016**, *6*, 23556. [CrossRef]
29. Vonderheid, S.C.; Tussing-Humphreys, L.; Park, C.; Pauls, H.; OjiNjideka Hemphill, N.; LaBomascus, B.; McLeod, A.; Koenig, M.D. A Systematic Review and Meta-Analysis on the Effects of Probiotic Species on Iron Absorption and Iron Status. *Nutrients* **2019**, *11*, 2938. [CrossRef]
30. Knovich, M.A.; Storey, J.A.; Coffman, L.G.; Torti, S.V.; Torti, F.M. Ferritin for the clinician. *Blood Rev.* **2009**, *23*, 95–104. [CrossRef] [PubMed]
31. Laine, F.; Angeli, A.; Ropert, M.; Jezequel, C.; Bardou-Jacquet, E.; Deugnier, Y.; Gissot, V.; Lacut, K.; Sacher-Huvelin, S.; Lavenu, A.; et al. Variations of hepcidin and iron-status parameters during the menstrual cycle in healthy women. *Br. J. Haematol.* **2016**, *175*, 980–982. [CrossRef] [PubMed]
32. Suzuki, Y.; Sakuraba, K.; Sunohara, M.; Takaragawa, M. Variations in iron status linked to menstrual cycles among Japanese female athletes. *Int. J. Anal. Bio-Sci.* **2018**, *6*, 45–50.

Article

Liposomal Mineral Absorption: A Randomized Crossover Trial

Grant M. Tinsley *, Patrick S. Harty, Matthew T. Stratton, Madelin R. Siedler and Christian Rodriguez

Energy Balance & Body Composition Laboratory, Department of Kinesiology & Sport Management, Texas Tech University, Lubbock, TX 79409, USA
* Correspondence: grant.tinsley@ttu.edu

Abstract: Multivitamin/mineral (MVM) supplements are one of the most popular dietary supplement categories. The purpose of this analysis was to determine if a novel liposomal delivery mechanism improves mineral absorption from an MVM product. In a randomized crossover trial, 25 healthy participants (12 females, 13 males) completed two testing sessions in which blood samples were collected at baseline and 2, 4, and 6 h following the ingestion of either a liposomal MVM or a nutrient-matched standard MVM. Analysis of MVM products indicated an elemental iron content of 9.4 and 10.1 mg (~50% U.S. FDA Daily Value) and an elemental magnesium content of 22.0 and 23.3 mg (~5% U.S. FDA Daily Value) in the liposomal and standard MVM products, respectively. Blood samples were analyzed for concentrations of iron and magnesium using colorimetric assays. Changes in mineral concentrations were analyzed using linear mixed models, and pharmacokinetic parameters were compared between conditions. For iron, statistically significant condition $\times$ time interactions were observed for percent change from baseline ($p = 0.002$), rank of percent change from baseline ($p = 0.01$), and raw concentrations ($p = 0.02$). Follow-up testing indicated that the liposomal condition exhibited larger changes from baseline than the standard MVM condition at 4 ($p = 0.0001$; +14.3 ± 18.5% vs. −6.0 ± 13.1%) and 6 h ($p = 0.0002$; +1.0 ± 20.9% vs. −21.0 ± 15.3%) following MVM ingestion. These changes were further supported by a 50% greater mean incremental area under the curve in the liposomal condition (33.2 ± 30.9 vs. 19.8 ± 19.8 mcg/dL × 6 h; $p = 0.02$, Cohen's d effect size = 0.52). In contrast, no differential effects for magnesium absorption were observed. In conclusion, iron absorption from an MVM product is enhanced by a liposomal delivery mechanism.

Keywords: iron; magnesium; multivitamin; liposomes; micronutrients; bioavailability; absorption

Citation: Tinsley, G.M.; Harty, P.S.; Stratton, M.T.; Siedler, M.R.; Rodriguez, C. Liposomal Mineral Absorption: A Randomized Crossover Trial. *Nutrients* **2022**, *14*, 3321. https://doi.org/10.3390/nu14163321

Academic Editor: Dietmar Enko

Received: 20 July 2022
Accepted: 12 August 2022
Published: 13 August 2022

Publisher's Note: MDPI stays neutral with regard to jurisdictional claims in published maps and institutional affiliations.

1. Introduction

Multivitamin/mineral (MVM) supplements have been consumed since the early 1940s and remain popular today [1,2]. National Health and Nutrition Examination Survey (NHANES) data indicate that MVM products are consumed more frequently than any other type of dietary supplement, and that the proportion of the population consuming MVM products increases with age. Based on NHANES 2017–2018 data, it was estimated that MVM products are consumed by 24% of adults aged 20–39 years, 30% of adults aged 40–59 years, and 39% of adults aged 60 years and older [3]. It is also estimated that, in the United States, 14% of all dietary supplement purchases and 38% of all vitamin and mineral sales are attributable to MVM products.

Several investigations have supported the utility of MVM products for increasing nutrient intake and improving nutrient status [4–8]. However, it is noteworthy that a standardized definition of MVM is not currently available, and a variety of classifications have been used in research, monitoring, and commercial contexts [1,9]. Additionally, distinct subcategories of MVM, such as basic (broad spectrum), high potency, and specialized (condition specific) have been described [1]. These considerations, as well as the multitude of specific formulations used in extant research, preclude clear conclusions regarding the utility of MVM supplements as a broad category [1,10]. As such, the attributes of specific, commercially available MVM products should be evaluated. One relevant consideration is the bioavailability and absorption profiles of individual nutrients contained within a product, factors which are affected by the product characteristics (e.g., formulation, excipients, fillers, coatings, etc.) and dissolution ability [9]. The need to establish accurate composition and bioavailability data across micronutrient products, including MVM, has been noted as a priority for future research [10]. One major challenge to establishing the bioavailability of MVM products is variance between label claims and actual contents. In this regard, production and examination of a certificate of analysis (COA) for each investigated product is warranted to aid in objective verification of nutrient content. Furthermore, it has been noted that minimal bioavailability data are available for MVM products, in contrast to individual nutrient supplements or foods, and it cannot be assumed that the same nutrients within an MVM will be absorbed similarly to a single-ingredient preparation [10]. The unique matrix of nutrients, corresponding nutrient–nutrient interactions, and the physical form of the product (i.e., capsule, tablet, liquid, etc.) may influence the bioavailability of nutrients within an MVM, but limited information is available in this regard [10].

A particularly notable development influencing the physical form of products and potentially altering the bioavailability of nutrients within MVMs is the introduction of liposomal delivery mechanisms, in which nutrients are packaged in liposomes to promote enhanced absorption and bioavailability [11]. Liposomes are spherical vesicles composed of one or more phospholipid bilayers (Figure 1) [12]. This structure allows for packaging of both water- and fat-soluble compounds. Hydrophobic compounds are encapsulated within the interior of the sphere, adjacent to the hydrophilic phospholipid heads, whereas hydrophobic compounds can be accommodated among the hydrophobic fatty acid tails [11]. In some cases, liposomal packaging allows for protection of the contents from a hostile gastrointestinal environment, and also provides the potential for improved cellular uptake due to interactions between the liposomal membrane and the cell membrane. Liposomes have several additional advantages as a compound delivery mode, including increased stability, biocompatibility, and flexibility as compared to conventional methods [12,13].

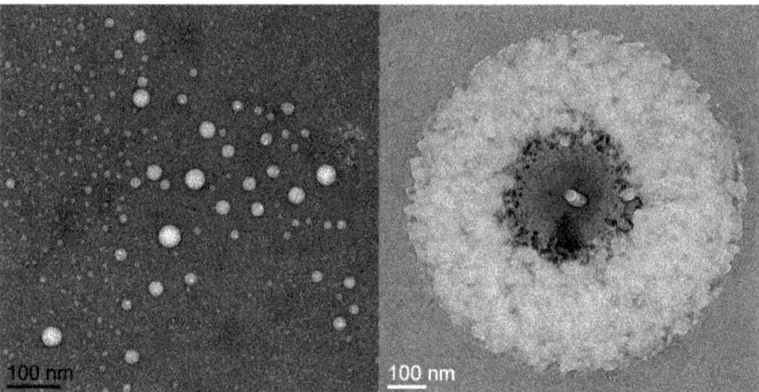

Figure 1. Liposomes as contained in the liposomal multivitamin/mineral product in the present investigation. Images were captured using cryogenic electron microscopy. Images courtesy of Dr. David Belnap, EM Core Lab, University of Utah.

Several investigations support the ability of liposomal packaging to improve vitamin absorption [14–17]. However, few studies have examined the influence of liposomal packaging on mineral absorption, particularly within the context of an MVM product. This is notable due to well-documented impacts of inadequate intakes of specific minerals, including iron, magnesium, and others [18,19]. The World Health Organization (WHO) estimated that one in three non-pregnant women, corresponding to ~500 million individuals, were anemic in 2011, and that iron deficiency likely contributed to at least half of these cases [20]. Improved understanding of optimal supplementation strategies may help inform interventions to combat these global concerns. In this regard, previous studies have reported the absorption characteristics of non-liposomal forms of iron and magnesium. For example, increases in serum and plasma iron concentrations 2 h following ingestion of iron as iron sulfate or ferroglycin sulphate have been reported [21,22], and the change in iron concentration 2 h following ingestion has been posited as an appropriate time interval for low-dose oral iron absorption tests [22]. Additionally, multiple investigations have reported increased serum magnesium concentrations 4 h after ingestion of supplements containing magnesium citrate or magnesium malate [23,24]. While these investigations do not provide information regarding absorption of minerals from liposomal products, they indicate the potential to examine differential absorption of iron and magnesium several hours after ingestion.

Due to the promise of liposomal technology for compound delivery, additional research is needed to clarify the impact of this delivery mechanism on the pharmacokinetic properties of individual nutrients, such as iron, contained within MVM products. Accordingly, the purpose of the present investigation was to determine if a novel liposomal delivery mechanism improves absorption of iron and magnesium contained in an MVM product. It was hypothesized that superior nutrient absorption would be observed with the liposomal MVM as compared to a nutrient-matched standard MVM product.

2. Materials and Methods

2.1. Overview

This study was a randomized crossover trial examining the pharmacokinetic profiles of mineral absorption from traditional and liposomal MVM formulations in healthy adults. Each participant completed two research visits, which were identical except for which MVM product was consumed. At each visit, participants reported to the laboratory after an overnight fast. After a baseline blood sample was collected, an MVM product was consumed alongside a standardized breakfast. At 2, 4, and 6 h post-ingestion, additional blood samples were collected. Concentrations of iron and magnesium were quantified, and the pharmacokinetic profiles of each nutrient were examined. The study was conducted according to the guidelines of the Declaration of Helsinki and approved by the Institutional Review Board of Texas Tech University (protocol code 2021-527; date of approval: 22 July 2021). The study was registered on clinicaltrials.gov (identifier: NCT05060367; first posted: 29 September 2021). While the study was originally designed to additionally examine vitamin concentrations, analytical complications at the partner laboratory prevented use of these data.

2.2. Participants

Healthy adult participants were recruited for participation. Inclusion criteria were age of 18 to 65 years, body mass of $\geq$50 kg (due to blood draws), and anticipated ability to comply with study procedures and scheduling requirements. Exclusion criteria were presence of a disease or medical condition—such as cardiovascular disease, cancer, respiratory disease, gastrointestinal disease, or metabolic disease—or current use of medication that could reasonably influence study outcomes or make participation inadvisable; inability to abstain from medication, supplement, or substance ingestion during the overnight fast and duration of the visit; anticipated inability to provide blood samples; current pregnancy or breastfeeding; and allergy that would prevent safe consumption of the standardized breakfast or MVM products. Written informed consent was obtained from all subjects involved in the study. Participants were asked to follow their normal lifestyle practices— including their typical diet—throughout the entire study involvement, with the exception of the pre-visit restrictions described below. At the first testing visit, body composition was estimated using multi-frequency bioelectrical impedance analysis (Seca mBCA 515/514, Seca, Hamburg, Germany). Participant characteristics are displayed in Table 1.

Table 1. Participant Characteristics [1].

	All (*n* = 25)	**Males (*n* = 13)**	**Females (*n* = 12)**
Age (y)	26.0 ± 3.4	26.2 ± 3.6	25.9 ± 3.5
Height (cm)	169.8 ± 9.0	175.9 ± 6.6	163.3 ± 6.1
Weight (kg)	74.7 ± 15.6	83.0 ± 10.9	65.8 ± 15.3
BMI (kg/m^2)	25.7 ± 3.9	26.8 ± 2.8	24.5 ± 4.6
Body fat (%)	25.4 ± 8.3	21.4 ± 7.2	29.7 ± 7.5

[1] Values displayed as mean ± SD.

2.3. Testing Visits

For each visit, participants reported to the laboratory after an overnight fast ($\geq$12 h) from food, dietary supplements, medications, and intake of all substances except water. Additionally, participants were asked to abstain from any dietary supplement consumption for the three days prior to each testing visit. A baseline blood draw was collected using standard phlebotomy procedures. Blood was collected into serum separation tubes (SST; BD Vacutainer). Processing procedures were based on manufacturer recommendations. Upon collection, the SST were gently inverted five times, then allowed to clot for 30 min in the upright position. Thereafter, the SST were centrifuged at room temperature (22 °C) at a speed of 1000 RCF (g) for 10 min in a swinging bucket centrifuge. After centrifugation, serum samples were aliquoted into microcentrifuge tubes and frozen at −80 °C until shipment to the partner laboratory for analysis.

Following the baseline blood draw, participants were provided with the standardized breakfast and MVM product. The standardized breakfast consisted of two packages (4 total bars) of Nature Valley Oats n' Honey Crunchy granola bars. Based on product labeling, this breakfast provided 380 kcal, 14 g fat, 58 g carbohydrate, 6 g protein, 280 mg sodium, and 2 mg iron. Ingredients in this product were whole grain oats, sugar, canola and/or sunflower oil, rice flour, honey, salt, brown sugar syrup, baking soda, soy lecithin, and natural flavor. This product was selected due to its relative lack of micronutrient fortification, unlike most packaged breakfast products. The first package (i.e., two bars) of the breakfast product was consumed, followed by MVM ingestion (liposomal or standard) and consumption of the second package. In a double-blind fashion, each participant ingested one serving (2 capsules) of the specified MVM product. The order of MVM ingestion for each participant was randomized using the *randomizR* software package [25] for R (date of randomization: 8 September 2021). Both MVM products were manufactured by Nutraceutical International Corporation using CELLg8® delivery technology, and certificates of analysis (COAs) were provided for the liposomal multivitamin (Lot #: SRC256641; date: 31 August 2021) and the standard multivitamin (Lot #: SRC258623; date: 2 September 2021). Elemental iron content (from ferrous glycinate) of the liposomal and standard MVM was 9.4 and 10.1 mg, respectively (~50% of U.S. FDA Daily Value). Elemental magnesium content (from magnesium glycinate) of the liposomal and standard MVM was 22.0 and 23.3 mg, respectively (~5% of U.S. FDA Daily Value). The full nutritional content of the liposomal and standard MVM products as obtained from the COAs is displayed in Table 2. The exact time of MVM ingestion was noted, and all subsequent blood draws were based on this time point. Bottled water (Purified Drinking Water, Great Value) was provided to all participants and was allowed ad libitum during the first testing visit, with a matched amount of bottled water provided at the second visit. Water consumption during the two visits was (mean ± SD) 1.8 ± 0.9 L and 2.0 ± 1.0 L, respectively.

At 2, 4, and 6 h after MVM ingestion, additional blood samples were collected and processed using the aforementioned procedures. These time intervals were selected for two reasons: (1) based on similarity to previous research examining absorption of iron and magnesium [21–24], and (2) due to practical constraints related to performing the study in a university laboratory rather than an inpatient facility. Upon completion of the first study visit, each participant began a washout period of at least seven days before returning to the laboratory to complete the second testing visit based on their scheduling availability (mean $\pm$ SD duration of washout period: 7.7 $\pm$ 2.0 days). Participants were instructed to continue maintaining their usual lifestyle practices, including diet, during the washout period.

Table 2. Nutrient content of multivitamin/mineral products.

Raw Material [1]	Nutrient	Specified Amount (Unit) [2]		Std. MVM [2]	Lipo. MVM [2]
Vegan Beta Carotene	A	900	mcg	950	1098
Methylcobalamin	B12	1000	mcg	1258	1210
Ascorbic Acid	C	90	mg	108.4	112.7
Vegan Vitamin D-3	D3	20	mcg	22.2	23.9
(6S)-5-Methyltetrahydrofolate	Folic Acid [3]	235.3	mcg	320.6	326.5
Ferrous Glycinate	Iron *	9	mg	10.1	9.37
Magnesium Glycinate	Magnesium *	20	mg	23.3	22.0
Manganese Citrate	Manganese	2.3	mg	2.6	2.6
Zinc Citrate	Zinc	11	mg	12.8	15.0
Benfotiamine	B1	1.2	mg	1.0	0.9
Riboflavin-5-Phosphate	B2	1.3	mg	0.9	1.1
Niacinamide	B3	16	mg	18.1	17.1
Calcium d-Pantothenate	B5	5	mg	6.4	5.1
Pyridoxine-5-Phospate	B6	1.7	mg	1.0	0.6
Biotin	Biotin	30	mcg	56.8	33.0
Choline Bitartrate	Choline	25	mg	57.6	44.8
Chromium Glycinate	Chromium	35	mcg	54.7	58.0
Co-Q10	Co-Q10	5	mg	4.3	3.8
Sunflower Vitamin E (a-Tocopherol)	E	15	mg	16.4	16.3
Inositol	Inositol	25	mg	19.2	22.3
Potassium Iodide	Iodine	150	mcg	171.9	244.0
K1	K1	120	mcg	129	74.5
Vegan Lutein Beadlet	Lutein	1	mg	1.27	1.25
Molybdenum Glycinate	Molybdenum	45	mcg	54.1	53.6
P-Aminobenzoic Acid	PABA	5	mg	3.4	4.1
Selenium	Selenium	55	mg	54.7	51.8
Acerola Cherry Juice Powder	-	2	mg	NT	NT
Trace Minerals	-	5	mg	NT	NT
Rosehips	-	2	mg	NT	NT

Standard MVM and liposomal MVM nutrient content based on certificate of analysis for each product. Nutrient content based on one serving (2 capsules). NT: not tested; [1] other ingredients: Veg 00 capsule, cellulose, stearic acid, and silica (manufacturing aids). [2] Specified amount and quantities listed for standard and liposomal MVM indicate the quantity of the specified nutrient, not the quantity of raw material. [3] Folic Acid is not converted to DFE units; multiply by 1.7 for DFE units. * Study outcome.

2.4. Nutrient Analysis

After collection, all samples were shipped on dry ice to a partner laboratory (Heartland Assays, Ames, IA, USA) for analysis. Magnesium and iron were analyzed using colorimetric assay (HM929, Pointe Scientific, Canton, MI and MAK025, Sigma-Aldrich, St. Louis, MO, USA, respectively). All samples were blinded for analysis, and output was provided to the principal investigator for statistical analysis. Complete data were available for magnesium and iron (n = 25 participants; n = 200 samples).

2.5. Statistical Analysis

Initially, data were examined for extreme outliers (i.e., values above Q3 + 3xIQR or below Q3–3xIQR). No extreme outliers were observed at any time point for raw iron concentrations. One extreme outlier was observed for iron changes from baseline to 6 h, with no extreme outliers identified at other time intervals. No extreme outliers for magnesium concentrations at baseline, 4 h, and 6 h were observed. Two extreme outliers were present for magnesium concentrations at 2 h after MVM ingestion. For changes in magnesium concentrations, five extreme outliers were identified for the interval from baseline to 2 h, four extreme outliers were identified for the interval from baseline to 4 h, and two extreme outliers were identified for the interval from baseline to 6 h. Based on these findings, rank-based tests were performed to allow for preservation of the entire sample size without concerns regarding the distribution of the data. Additionally, non-rank-based tests were performed with and without inclusion of extreme outliers. For completeness, analysis of raw iron and magnesium concentrations are presented in Appendix A, although these analyses were viewed as secondary due to variation in baseline concentrations rendering their comparison less informative than relative changes.

Data were analyzed in R (version 4.1.2; R Foundation for Statistical Computing, Vienna, Austria). Percent changes from baseline concentrations and ranks of percent changes from baseline were visualized using the *ggplot2* [26] package (v. 3.3.5) with within-subject error bars [27,28]. These data were analyzed using linear mixed-effects models (*nlme* package [29], v. 3.1-153) with a random intercept for participant and a first-order autoregressive (AR1) variance-covariance matrix. These models were fit by maximizing the restricted log-likelihood (REML). In all models, the reference groups were the standard MVM for condition, female for sex, and the baseline time point for time. The fixed effects of condition, time, sex, and their interactions were examined, and significant effects were followed up with pairwise comparisons using the *emmeans* [30] package (v. 1.7.2). Multiple comparisons were accounted for using the Benjamini and Hochberg correction [31].

The incremental area under the concentration vs. time curve (iAUC) was calculated using the method of Brouns et al. [32]. As the 2, 4, and 6 h time points were specified relative to MVM ingestion, and a mean $\pm$ SD of 9.5 $\pm$ 1.8 min elapsed between the baseline blood draw and MVM ingestion, values of 0, 2.158, 4.158, and 6.158 h were used for iAUC and other pharmacokinetic calculations. The *PKNCA* [33] package (v. 0.9.5) was used to establish the maximum observed concentration (Cmax) and time of maximum observed concentration (Tmax). These values were calculated for the entire sample, females only, and males only. The iAUC and Cmax values were examined for extreme outliers, as well as for normality of differences between conditions. When no extreme outliers were present and normality of differences was observed (via visual inspection of QQ plots and Shapiro–Wilk tests), data were analyzed using paired samples *t*-tests. When extreme outliers were present and/or probable normality violations were observed, Wilcoxon signed-rank tests were performed. Accordingly, paired-samples *t*-tests were performed for iron iAUC values in the entire sample, males only, and females only; magnesium iAUC values in females only; and iron and magnesium Cmax values in the entire sample and females only. Wilcoxon signed-rank tests were performed for magnesium iAUC values in the entire sample and males only, as well as magnesium and iron Cmax values in males only. These analyses were performed using the *rstatix* [34] package (v. 0.7.0). Associated metrics of effect size (i.e., Cohen's *d* for paired *t*-tests (d) and Wilcoxon *r* for Wilcoxon signed-rank tests [*r*]) were also calculated. Cohen's *d* effect sizes can be interpreted as: <0.2 (negligible), 0.2 to <0.5 (small), 0.5 to <0.8 (medium), and >0.8 (large), and Wilcoxon *r* effect sizes can be interpreted as: 0.1 to <0.3 (small), 0.3 to <0.5 (moderate), and $\geq$0.5 (large) [34]. Due to the nature of the data, all Tmax values were analyzed using Wilcoxon signed-rank tests. Statistical significance was accepted at $p < 0.05$.

3. Results

3.1. Mixed Models

3.1.1. Iron

For iron, statistically significant condition $\times$ time interactions were observed for rank of percent change from baseline ($p = 0.01$; Figure 2A), percent change from baseline ($p = 0.002$; Figure 2B), and raw concentrations ($p = 0.02$; Appendix A).

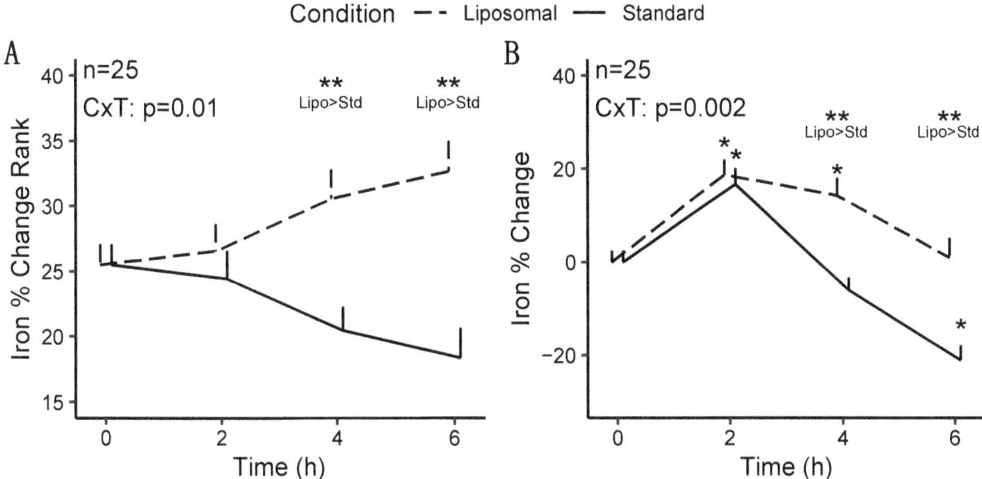

Figure 2. Changes in iron concentrations following multivitamin/mineral ingestion. The rank of percent changes from baseline (**A**) and percent changes from baseline (**B**) are presented. CxT indicates the *p*-value for the condition × time interaction. ** indicates a statistically significant difference between conditions at a particular time point. * indicates a statistically significant difference from baseline concentrations in a particular condition.

Follow-up testing for the significant condition × time interaction for rank of percent change from baseline indicated that the liposomal condition exhibited greater values at 4 ($p = 0.01$) and 6 h ($p = 0.0003$) following MVM ingestion, as compared to the standard MVM condition, without difference at 2 h ($p = 0.66$). A significant time × sex interaction was also present ($p = 0.006$); however, follow-up testing did not reveal any statistically significant pairwise comparisons after correction for multiple comparisons.

Follow-up testing for the significant condition × time interaction for percent change from baseline indicated that the liposomal condition exhibited greater values than the standard MVM condition at 4 ($p = 0.0001$; +14.3 ± 18.5% vs. −6.0 ± 13.1% [mean ± SD]) and 6 h ($p = 0.0002$; +1.0 ± 20.9% vs. −21.0 ± 15.3%) following MVM ingestion, without difference at 2 h ($p = 0.84$; 18.6 ± 16.2% vs. 16.7 ± 16.9%). A statistically significant condition × time interaction ($p = 0.002$) was also observed in the sensitivity analysis ($n = 24$; removal of one extreme outlier), with statistically significant differences between conditions still observed at the 4 and 6 h time points. Based on the lack of difference in condition effects and interactions between analyses, the results of the full sample are presented. A significant time × sex interaction was also present ($p = 0.006$). Follow-up testing indicated that females presented larger changes in iron at 2 h ($p = 0.004$), as compared to males, but not at other time points ($p = 0.08$ to 1.0).

3.1.2. Magnesium

For magnesium, no statistically significant effects of condition, time, or sex were observed in any model (Figure 3), except for a condition × sex × time interaction ($p = 0.04$) for magnesium changes from baseline in the sensitivity analysis (i.e., following removal of extreme outliers; Table 3). However, follow-up testing revealed no significant two-way interactions or pairwise comparisons after correction for multiple comparisons. Based on the difference in statistical significance of the condition × time × sex interaction, the results of the sensitivity analysis are presented.

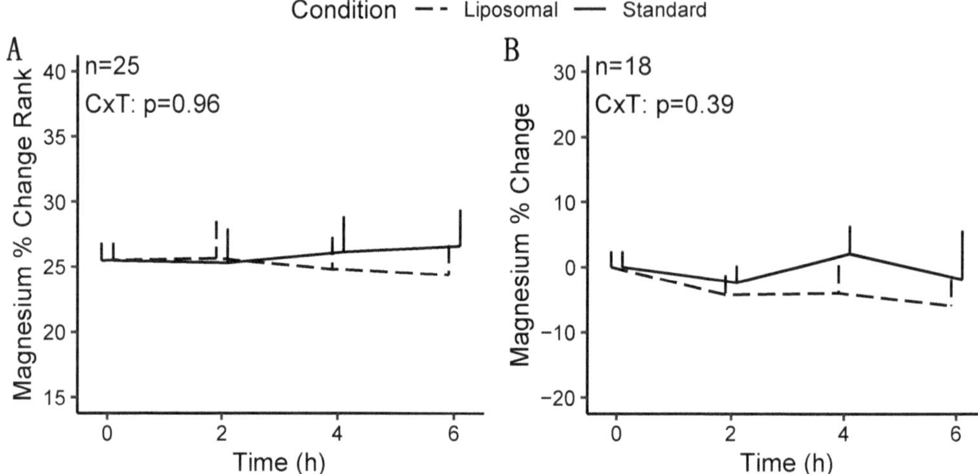

Figure 3. Changes in magnesium concentrations following multivitamin/mineral ingestion. The rank of percent changes from baseline (**A**) and percent changes from baseline after removal of extreme outliers (sensitivity analysis) (**B**) are presented. CxT indicates the *p*-value for the condition × time interaction.

Table 3. *p*-Values from mixed models.

	Iron			Magnesium		
	Δ Rank	Δ	Δ (S)	Δ Rank	Δ	Δ (S)
n	25	25	24	25	25	18
Intercept	**<0.001 ***	0.72	0.98	**<0.001 ***	0.13	0.14
Condition	**<0.001 ***	**0.005 ***	**0.02 ***	0.66	0.93	1.00
Time	1.00	**<0.001 ***	**<0.001 ***	1.00	0.14	0.12
Sex	0.07	**0.045 ***	0.08	0.25	0.80	0.77
Condition × Time	**0.01 ***	**0.002 ***	**0.004 ***	0.96	0.79	0.74
Condition × Sex	0.80	0.72	0.64	0.34	0.41	0.44
Time × Sex	**0.006 ***	**0.006 ***	**0.001 ***	0.58	0.36	0.41
Condition × Time × Sex	0.37	0.27	0.29	0.16	0.06	**0.04 ***

"Δ Rank" indicates rank of percent change from baseline; "Δ" indicates percent change from baseline; "S" indicates sensitivity analysis results (removal of extreme outliers). "*" and bold text indicate statistical significance (i.e., $p < 0.05$).

3.2. Pharmacokinetic Analysis

3.2.1. Iron

Greater iAUC values were observed for iron in the liposomal MVM condition in the entire sample ($p = 0.016$, $d = 0.52$; Figure 4), with a 50% difference in mean values. Greater iAUC was also observed in males only ($p = 0.03$, $d = 0.68$; Table 4). In females only, the difference in iAUC was not statistically significant ($p = 0.13$, $d = 0.47$), although a similar magnitude of effect size as in the entire sample was observed.

There was no difference in iron Cmax in the entire sample ($p = 0.51$, $d = 0.13$; Table 5) and females only ($p = 0.25$, $d = 0.35$). In males only, a greater Cmax was observed for iron in the standard MVM condition ($p = 0.01$, $r = 0.69$). However, examination of the data indicated this was due to higher baseline concentrations in the standard MVM condition rather than an increase after ingestion, as further evidenced by the greater iAUC in the liposomal MVM condition. A difference in Tmax for iron was observed in the entire sample ($p = 0.002$, $r = 0.68$; Table 6) and males only ($p = 0.01$, $r = 0.77$), indicating lower (earlier) Tmax values for the standard MVM condition. Tmax values did not significantly differ between conditions in females only ($p = 0.10$, $r = 0.57$).

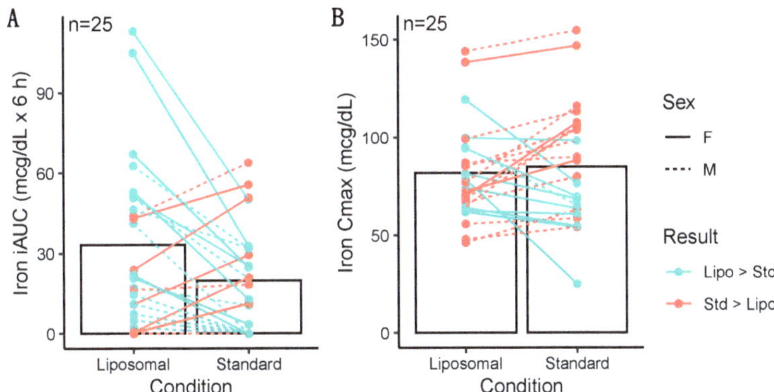

Figure 4. Individual differences in incremental area under the curve (iAUC; panel **A**) and maximal observed concentration (Cmax; panel **B**) values for iron. Iron iAUC values were significantly greater after ingestion of the liposomal product as compared to the standard product (p = 0.016 via paired samples t-test), with no difference observed for Cmax (p = 0.51 via paired samples t-test). Individual responses are represented by lines, and bars represent mean values in each condition.

Table 4. Incremental area under the concentration vs. time curve (iAUC) comparison.

	Liposomal MVM				Standard MVM				All		Females		Males	
	Mean	**sd**	**Med**	**IQR**	**Mean**	**sd**	**Med**	**IQR**	p	**ES**	p	**ES**	p	**ES**
Iron	33.22	30.90	22.00	39.84	19.84	19.75	18.40	32.01	**0.02 ***	0.52	0.13	0.47	**0.03 ***	0.68
Mag.	0.67	1.15	0.07	0.67	0.84	1.13	0.43	1.20	0.30	0.22	0.09	0.54	1.00	0.02

Descriptive data for each condition are presented for all participants. Values are in units of incremental area under the time (hours) vs. the concentration curve. Concentration units are mcg/dL for iron and mg/dL for magnesium. p-values were generated by a paired samples t-test or Wilcoxon signed-rank test, as appropriate (see text for details). Effect sizes (ES) correspond to Cohen's d when paired samples t-tests were performed and Wilcoxon r when Wilcoxon signed-rank tests were performed. "*" and bold text indicate statistical significance (i.e., $p < 0.05$). sd: standard deviation; med: median; IQR: interquartile range; ES: effect size.

Table 5. Maximal observed concentration (Cmax) comparison.

	Liposomal MVM				Standard MVM				All		Females		Males	
	Mean	**sd**	**Med**	**IQR**	**Mean**	**sd**	**Med**	**IQR**	p	**ES**	p	**ES**	p	**ES**
Iron	81.72	24.72	77.07	29.60	84.94	30.39	79.87	41.89	0.51	0.13	0.25	0.35	**0.01 ***	0.69
Mag.	2.49	0.84	2.10	1.20	2.59	0.88	2.20	1.40	0.41	0.17	0.27	0.34	0.46	0.21

Descriptive data for each condition are presented for all participants. Units are mcg/dL for iron and mg/dL for magnesium; p-values were generated by a paired samples t-test or Wilcoxon signed-rank test, as appropriate (see text for details). Effect sizes (ES) correspond to Cohen's d when paired samples t-tests were performed and Wilcoxon r when Wilcoxon signed-rank tests were performed. "*" and bold text indicate statistical significance (i.e., $p < 0.05$). sd: standard deviation; med: median; IQR: interquartile range; ES: effect size.

Table 6. Time of maximal observed concentration (Tmax) comparison.

	Liposomal MVM				Standard MVM				All		Females		Males	
	Mean	**sd**	**Med**	**IQR**	**Mean**	**sd**	**Med**	**IQR**	p	**ES**	p	**ES**	p	**ES**
Iron	3.0	1.5	2.2	2.0	1.5	1.0	2.2	2.2	**0.002 ***	0.68	0.10	0.57	**0.01 ***	0.77
Mag.	3.2	2.7	4.2	6.2	3.6	2.5	4.2	4.0	0.62	0.08	0.48	0.13	1.00	0.05

Descriptive data for each condition are presented for all participants in units of hours. p-values were generated by Wilcoxon signed-rank test, and effect sizes (ES) correspond to Wilcoxon r. "*" and bold text indicate statistical significance (i.e., $p < 0.05$). sd: standard deviation; med: median; IQR: interquartile range; ES: effect size.

3.2.2. Magnesium

No differences in magnesium iAUC values were observed in the entire sample ($p = 0.30$, $r = 0.22$; Table 4; Figure 5), males only ($p = 1.0$, $r = 0.02$), or females only ($p = 0.09$, $d = 0.54$). Additionally, no differences between conditions for Cmax values were observed in the entire sample ($p = 0.41$, $d = 0.17$; Table 5), males only ($p = 0.46$, $r = 0.21$), or females only ($p = 0.27$, $d = 0.34$). No differences in Tmax were observed for magnesium in the entire sample ($p = 0.62$, $r = 0.08$; Table 6), males only ($p = 1.0$, $r = 0.05$), or females only ($p = 0.48$, $r = 0.13$).

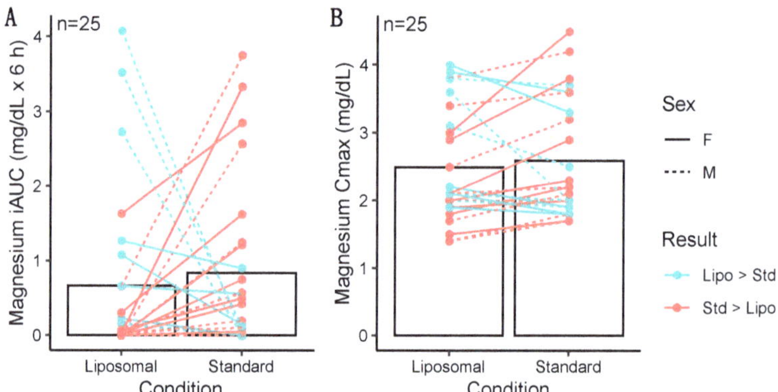

Figure 5. Individual differences in incremental area under the curve (iAUC; panel **A**) and maximal observed concentration (Cmax; panel **B**) values for magnesium. Magnesium iAUC values ($p = 0.30$ via Wilcoxon signed-rank test) and Cmax values ($p = 0.41$ via paired samples t-test) did not differ between conditions. Individual responses are represented by lines, and bars represent mean values in each condition.

3.3. Side Effects

No side effects related to MVM consumption were reported by participants in either condition.

4. Discussion

The present randomized crossover trial investigated mineral absorption from liposomal and non-liposomal MVM products. As hypothesized, improved iron absorption was observed following ingestion of the liposomal product. Specifically, larger changes in iron from baseline—using both percent changes and ranks of percent changes—were observed in the liposomal condition at 4 and 6 h after MVM ingestion. Additionally, the iAUC for iron was 50% greater following ingestion of the liposomal MVM product. No differences between conditions were observed for magnesium absorption. Importantly, the dose of iron contained in the MVM product represented a relevant dose, whereas the relative dose of magnesium was much lower. The quantity of elemental iron in the MVM products represented ~50% of the 18-mg Daily Value used by the U.S. Food and Drug Administration (FDA) for nutrition labeling purposes [18]. Additionally, the doses of elemental iron used in the present study (9.4 to 10.1 mg) meet the Recommended Dietary Allowance (RDA) for males of all ages, except 14–18 years (RDA: 11 mg/d), and meet or make a substantive contribution to the recommended intake for adult females (RDA: 8 to 18 mg/d, depending on age, in non-pregnant females; 27 mg/d in pregnant females). This indicates that the quantity of iron contained in the MVM product is meaningful relative to daily intake recommendations and supports the relevance of the improved absorption seen with the liposomal formulation. In contrast, the doses of elemental magnesium in the present study were low relative to daily intakes (22 to 23.3 mg vs. U.S. FDA Daily Value of 420 mg; ~5% of Daily Value) [19]. This may indicate that the influence of liposomal delivery on magnesium absorption should be further investigated with higher doses, perhaps through a standalone magnesium supplement due to practical limitations on absolute quantities of ingredients in MVM products. Preliminary research with a form of liposomal magnesium supports this contention, as one study indicated enhanced absorption as compared to non-liposomal forms following ingestion of 350 mg magnesium (i.e., ~15-fold higher than the dose in the present investigation) [35]. This research was conducted within the context of single-nutrient products, in contrast to the present investigation, which examined nutrient absorption from MVM products.

The observed improvement in iron absorption with liposomal packaging is notable for several reasons. First, iron is essential for a host of physiological functions, ranging from oxygen handling as part of hemoglobin and myoglobin, to hormone synthesis and support of normal cellular function [18]. Second, while previous research has indicated benefits of liposomal delivery for absorption of vitamins [14–17], there is little information regarding mineral absorption. As such, the present investigation demonstrates the promise of liposomal technology in this regard. Third, the benefit to iron absorption was observed in the context of a relatively bioavailable source of iron, ferrous glycinate. Previous research has demonstrated superior bioavailability of ferrous glycinate as compared to iron salts, such as ferrous sulfate [36–38]. As such, it is notable that liposomal packaging further improved absorption. While speculative, it is possible that the enhancement of absorption with liposomal packaging would be even more evident with less bioavailable forms of nutrients. Fourth, the global impact of inadequate iron intake and international recommendations for iron supplementation indicate the importance of using the most effective supplement form. The WHO estimated that one in three non-pregnant women, corresponding to ~500 million individuals, were anemic in 2011, and that iron deficiency likely contributed to at least half of these cases [20]. Correspondingly, a 2016 WHO report recommends daily iron supplementation in menstruating adult women and adolescent girls living in settings where anemia is prevalent [20]. It is well established that groups at risk for inadequate iron intake include adolescent, pregnant, and premenopausal women, as well as infants and children [18]. Racial disparities have also been reported, with higher rates of depleted iron stores in Mexican American and non-Hispanic Black pregnant women [39]. Additionally, those in food-insecure homes are more likely to experience inadequate iron intake [40]. Collectively, these and other lines of evidence indicate the importance of iron supplementation in several contexts and demonstrate the need for effective supplementation formulations.

While few investigations have examined the acute absorption properties of minerals encapsulated in liposomes, some have indicated promise for these products for health improvements following chronic supplementation in clinical conditions [41–44]. A randomized trial in chronic kidney disease patients demonstrated similar increases in hemoglobin after 3 months of treatment with oral liposomal iron supplements or intravenous iron administration, along with a lower incidence of adverse effects with oral supplementation [41]. A separate single-arm trial in patients with chronic kidney disease indicated that a liposomal iron preparation was well tolerated and increased hemoglobin, relative to baseline, after 12 months of supplementation [42]. In a single-arm trial conducted in anemic patients with inflammatory bowel disease, 62% of patients completing treatment with oral liposomal iron supplements increased hemoglobin above a prespecified threshold or presented with hemoglobin normalization after 8 weeks of treatment; improvements in quality of life and reductions in fatigue were also noted [43]. Finally, a randomized trial in pregnant, non-anemic women indicated that liposomal iron was effective for elevating hemoglobin and ferritin concentrations as compared with control [44]. While these investigations demonstrate the potential utility of liposomal iron formulations, several trials are limited by a lack of control or comparison groups. As such, additional research is needed to investigate the potential for unique health effects in various clinical populations, as well as in the general population.

While both multi-nutrient (e.g., MVM) and single-nutrient supplements may exhibit benefits in specific contexts, the investigation of individual nutrient absorption from an

MVM product is relevant due to the notable prevalence of MVM supplementation. As previously noted, NHANES data indicate that MVM products are consumed more frequently than any other type of dietary supplement [3]. Importantly, the bioavailability of individual nutrients from an MVM product may be dissimilar to absorption from single-nutrient products due to the specific matrix of nutrients and corresponding nutrient–nutrient interactions within an MVM [10]. As such, and due to the dearth of bioavailability data for MVM products, investigations that directly quantity single nutrient absorption from an MVM product—such as the present report—are warranted.

A major strength of the present investigation is the use of COA to objectively verify the nutrient content of the liposomal and standard MVM products. Interestingly, the benefits of liposomal iron delivery are further highlighted by the fact that the elemental iron content was 7.5% (0.73 mg) lower in the liposomal MVM than the standard MVM, based on the COA. Similarly, the elemental magnesium content was 5.7% lower in the liposomal MVM product. Additional strengths of this study include the rigorous procedural standardization during data collection and multifaceted statistical analysis. Limitations of the present work include the inability to examine additional outcomes; the low dose of magnesium, which limits the relevance of this nutrient as a study outcome; the relatively young, healthy, and homogenous sample; and the limited number of time points utilized. Additionally, the monitoring period of 6 h following MVM ingestion could be a limitation, as a statistically significant difference in iron changes between conditions was present at this time point; therefore, the full duration of differential iron changes could have persisted beyond this point. However, interestingly, iron had returned to near baseline levels in the liposomal condition by 6 h after MVM ingestion, whereas iron concentrations fell below baseline levels in the standard MVM condition. An additional consideration is that while the two MVM products were designed with identical specified amounts of each nutrient, minor differences between the standard and liposomal MVM products were observed through the laboratory COA results. While this may be unavoidable for products with many ingredients and small doses of each individual ingredient, it is worth noting. However, as mentioned, the dose of iron (as well as magnesium) was actually lower in the liposomal MVM, indicating the positive results observed were not due to this slightly different dose. Lastly, the potential influence of compounds within the MVM products on iron and magnesium absorption should be considered. Based on the potential for complex, and largely unknown, interactions between compounds within MVM products [10], the results of the present study could have differed if compounds were studied in isolation.

5. Conclusions

In conclusion, the present randomized crossover trial demonstrated improved iron absorption following ingestion of iron from a novel liposomal MVM as compared to a standard MVM. This finding helps to determine optimal iron supplementation strategies and demonstrates the potential for liposomal packaging to benefit mineral absorption. Future research should continue to examine the potential utility of liposomal delivery of micronutrients in a variety of populations, for both MVM formulations and single-nutrient products.

Author Contributions: Conceptualization, G.M.T., P.S.H., M.T.S., M.R.S. and C.R.; methodology, G.M.T., P.S.H., M.T.S., M.R.S. and C.R.; software, G.M.T.; formal analysis, G.M.T.; investigation, G.M.T., P.S.H., M.T.S., M.R.S. and C.R.; resources, G.M.T.; data curation, G.M.T., P.S.H., M.T.S., M.R.S. and C.R.; writing—original draft preparation, G.M.T.; writing—review and editing, G.M.T., P.S.H., M.T.S., M.R.S. and C.R.; visualization, G.M.T.; supervision, G.M.T.; project administration, G.M.T., P.S.H., M.T.S., M.R.S. and C.R.; funding acquisition, G.M.T. All authors have read and agreed to the published version of the manuscript.

Funding: This research was funded by Nutraceutical International Corporation, grant number A21-0282. The APC was also funded through this award.

Institutional Review Board Statement: The study was conducted according to the guidelines of the Declaration of Helsinki and approved by the Institutional Review Board of Texas Tech University (protocol code: 2021-527; date of approval: 22 July 2021).

Informed Consent Statement: Informed consent was obtained from all subjects involved in the study.

Data Availability Statement: Requests for data will be reviewed by the researchers and associated personnel.

Acknowledgments: The researchers would like to acknowledge the participants for their time and dedication during this study.

Conflicts of Interest: The authors declare no conflict of interest. The funders provided general input on study outcomes of interest, but did not play a role in the details of study design. Scientific advisory board members from the sponsor provided technical expertise related to laboratory analyses and interpretation, but did not play any role in laboratory assays, data handling, statistical analysis, or writing of the manuscript.

Appendix A

Table A1. p-values from mixed models using raw concentrations, uncorrected for baseline values.

	Iron	Magnesium	
	Raw	Raw	Raw (S)
n	25	25	23
Intercept	<0.001 *	<0.001 *	<0.001 *
Condition	0.89	0.58	0.34
Time	<0.001	0.29	0.14
Sex	0.32	0.88	0.66
Condition × Time	0.02 *	0.83	0.74
Condition × Sex	<0.001 *	0.38	0.79
Time × Sex	0.40	0.38	0.34
Condition × Time × Sex	0.97	0.09	0.04 *

"S" indicates sensitivity analysis results. "*" and bold text indicate statistical significance (i.e., $p < 0.05$).

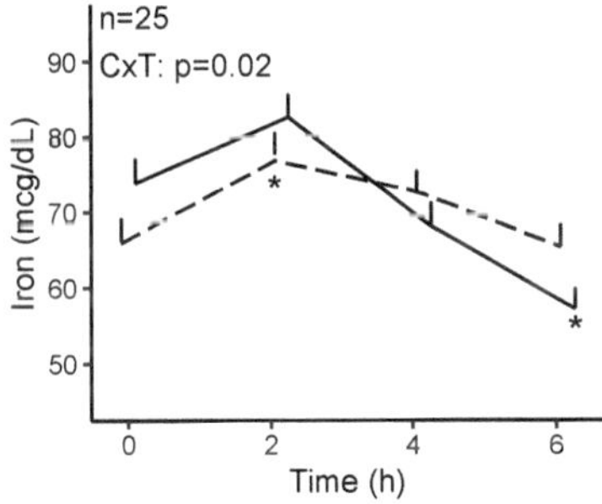

Figure A1. Raw Iron Concentrations. A statistically significant condition × time interaction was observed for raw iron concentrations ($p = 0.02$; Table A1), uncorrected for baseline concentrations. Follow-up testing did not reveal significant differences between conditions at any individual time point ($p = 0.08$ to 0.30). However, iron concentrations in the liposomal MVM condition increased from baseline to 2 h ($p = 0.03$), but were not elevated relative to baseline at any subsequent time point ($p = 0.14$ to $p = 0.93$). In contrast, within the standard MVM condition, the sole significant difference from baseline values was that the final observed value (6 h after MVM ingestion) fell below baseline values ($p = 0.001$), unlike in the liposomal condition ($p = 0.93$). Due to variation in raw baseline concentrations between conditions, the raw changes presented above are viewed as supplemental/secondary, while changes from the baseline and pharmacokinetic parameters (i.e., iAUC) were the primary outcomes. "*" indicate statistical significance (i.e., $p < 0.05$).

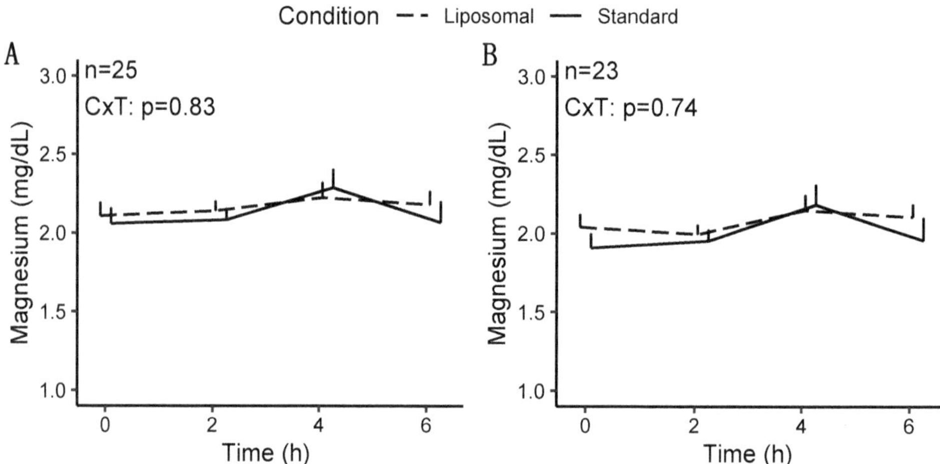

Figure A2. Raw Magnesium Concentrations. No significant condition * time interaction was observed for raw magnesium concentrations, uncorrected for baseline values, in the full sample (**A**) or in a sensitivity analysis with extreme outliers removed (**B**). A significant condition * time * sex interaction was observed in the sensitivity analysis (Table A1); however, follow-up testing did not reveal any statistically significant pairwise comparisons. Due to variation in raw baseline concentrations between conditions, the raw changes presented above are viewed as supplemental/secondary, while changes from the baseline and pharmacokinetic parameters (i.e., iAUC) were the primary outcomes.

References

1. National Institutes of Health Office of Dietary Supplements. Office of Dietary Supplements—Multivitamin/mineral Supplements. Available online: https://ods.od.nih.gov/factsheets/MVMS-Consumer (accessed on 26 January 2022).
2. Cowan, A.E.; Jun, S.; Gahche, J.J.; Tooze, J.A.; Dwyer, J.T.; Eicher-Miller, H.A.; Bhadra, A.; Guenther, P.M.; Potischman, N.; Dodd, K.W.; et al. Dietary Supplement Use Differs by Socioeconomic and Health-Related Characteristics among U.S. Adults, NHANES 2011–2014. *Nutrients* **2018**, *10*, 1114. [CrossRef] [PubMed]
3. Mishra, S.; Stierman, B.; Gahche, J.J.; Potischman, N. *Dietary Supplement Use among Adults: United States, 2017–2018*; National Center for Health Statistics: Hyattsville, MD, USA, 2021. [CrossRef]
4. Murphy, S.P.; White, K.K.; Park, S.Y.; Sharma, S. Multivitamin-multimineral supplements' effect on total nutrient intake. *Am. J. Clin. Nutr.* **2007**, *85*, 280s–284s. [CrossRef] [PubMed]
5. Bailey, R.L.; Pac, S.G.; Fulgoni, V.L., 3rd; Reidy, K.C.; Catalano, P.M. Estimation of Total Usual Dietary Intakes of Pregnant Women in the United States. *JAMA Netw. Open* **2019**, *2*, e195967. [CrossRef] [PubMed]
6. Harris, E.; Macpherson, H.; Vitetta, L.; Kirk, J.; Sali, A.; Pipingas, A. Effects of a multivitamin, mineral and herbal supplement on cognition and blood biomarkers in older men: A randomised, placebo-controlled trial. *Hum. Psychopharmacol.* **2012**, *27*, 370–377. [CrossRef] [PubMed]
7. Özkaya, M.O.; Nazıroğlu, M.; Barak, C.; Berkkanoglu, M. Effects of multivitamin/mineral supplementation on trace element levels in serum and follicular fluid of women undergoing in vitro fertilization (IVF). *Biol. Trace Elem. Res.* **2011**, *139*, 1–9. [CrossRef] [PubMed]
8. Maraini, G.; Williams, S.L.; Sperduto, R.D.; Ferris, F.L.; Milton, R.C.; Clemons, T.E.; Rosmini, F.; Ferrigno, L. Effects of multivitamin/mineral supplementation on plasma levels of nutrients. Report No. 4 of the Italian-American clinical trial of nutritional supplements and age-related cataract. *Ann. Ist. Super. Sanita* **2009**, *45*, 119–127.
9. Yetley, E.A. Multivitamin and multimineral dietary supplements: Definitions, characterization, bioavailability, and drug interactions. *Am. J. Clin. Nutr.* **2007**, *85*, 269s–276s. [CrossRef]
10. Comerford, K.B. Recent developments in multivitamin/mineral research. *Adv. Nutr.* **2013**, *4*, 644–656. [CrossRef] [PubMed]
11. Shade, C.W. Liposomes as Advanced Delivery Systems for Nutraceuticals. *Integr. Med.* **2016**, *15*, 33–36.
12. Akbarzadeh, A.; Rezaei-Sadabady, R.; Davaran, S.; Joo, S.W.; Zarghami, N.; Hanifehpour, Y.; Samiei, M.; Kouhi, M.; Nejati-Koshki, K. Liposome: Classification, preparation, and applications. *Nanoscale Res. Lett.* **2013**, *8*, 102. [CrossRef]
13. Daraee, H.; Etemadi, A.; Kouhi, M.; Alimirzalu, S.; Akbarzadeh, A. Application of liposomes in medicine and drug delivery. *Artif. Cells Nanomed. Biotechnol.* **2016**, *44*, 381–391. [CrossRef] [PubMed]

14. Davis, J.L.; Paris, H.L.; Beals, J.W.; Binns, S.E.; Giordano, G.R.; Scalzo, R.L.; Schweder, M.M.; Blair, E.; Bell, C. Liposomal-encapsulated Ascorbic Acid: Influence on Vitamin C Bioavailability and Capacity to Protect Against Ischemia-Reperfusion Injury. *Nutr. Metab. Insights* **2016**, *9*, 25–30. [CrossRef] [PubMed]
15. Łukawski, M.; Dałek, P.; Borowik, T.; Foryś, A.; Langner, M.; Witkiewicz, W.; Przybyło, M. New oral liposomal vitamin C formulation: Properties and bioavailability. *J. Liposome Res.* **2020**, *30*, 227–234. [CrossRef] [PubMed]
16. Gopi, S.; Balakrishnan, P. Evaluation and clinical comparison studies on liposomal and non-liposomal ascorbic acid (vitamin C) and their enhanced bioavailability. *J. Liposome Res.* **2021**, *31*, 356–364. [CrossRef]
17. Nowak, J.K.; Sobkowiak, P.; Drzymała-Czyż, S.; Krzyżanowska-Jankowska, P.; Sapiejka, E.; Skorupa, W.; Pogorzelski, A.; Nowicka, A.; Wojsyk-Banaszak, I.; Kurek, S.; et al. Fat-Soluble Vitamin Supplementation Using Liposomes, Cyclodextrins, or Medium-Chain Triglycerides in Cystic Fibrosis: A Randomized Controlled Trial. *Nutrients* **2021**, *13*, 4554. [CrossRef]
18. National Institutes of Health Office of Dietary Supplements. Office of Dietary Supplements—Iron. Available online: https://ods.od.nih.gov/factsheets/Iron-HealthProfessional (accessed on 10 June 2022).
19. National Institutes of Health Office of Dietary Supplements. Office of Dietary Supplements—Magnesium. Available online: https://ods.od.nih.gov/factsheets/Magnesium-HealthProfessional (accessed on 10 June 2022).
20. World Health Organization. Guideline: Daily Iron Supplementation in Adult Women and Adolescent Girls. Available online: https://apps.who.int/iris/bitstream/handle/10665/204761/9789241510196_eng.pdf?sequence=1&isAllowed=y (accessed on 9 June 2022).
21. Cabrera, C.C.; Ekström, M.; Linde, C.; Persson, H.; Hage, C.; Eriksson, M.J.; Wallén, H.; Persson, B.; Tornvall, P.; Lyngå, P. Increased iron absorption in patients with chronic heart failure and iron deficiency. *J. Card. Fail.* **2020**, *26*, 440–443. [CrossRef] [PubMed]
22. Jensen, N.M.; Brandsborg, M.; Boesen, A.M.; Yde, H.; Dahlerup, J.F. Low-dose oral iron absorption test: Establishment of a reference interval. *Scand J. Clin. Lab. Investig.* **1998**, *58*, 511–519. [CrossRef] [PubMed]
23. Weiss, D.; Brunk, D.K.; Goodman, D.A. Scottsdale Magnesium Study: Absorption, Cellular Uptake, and Clinical Effectiveness of a Timed-Release Magnesium Supplement in a Standard Adult Clinical Population. *J. Am. Coll. Nutr.* **2018**, *37*, 316–327. [CrossRef] [PubMed]
24. Karagülle, O.; Kleczka, T.; Vidal, C.; Candir, F.; Gundermann, G.; Külpmann, W.R.; Gehrke, A.; Gutenbrunner, C. Magnesium absorption from mineral waters of different magnesium content in healthy subjects. *Complement. Med. Res.* **2006**, *13*, 9–14. [CrossRef] [PubMed]
25. Uschner, D.; Schindler, D.; Hilgers, R.-D.; Heussen, N. randomizeR: An R Package for the Assessment and Implementation of Randomization in Clinical Trials. *J. Stat. Softw.* **2018**, *85*, 1–22. [CrossRef]
26. Wickham, H. *ggplot2: Elegant Graphics for Data Analysis*; Springer: New York, NY, USA, 2016.
27. Cousineau, D. Confidence intervals in within-subject designs: A simpler solution to Loftus and Masson's method. *Tutor. Quant. Methods Psychol.* **2005**, *1*, 42–45. [CrossRef]
28. Morey, R.D. Confidence intervals from normalized data: A correction to Cousineau (2005). *Tutor. Quant. Methods Psychol.* **2008**, *4*, 61–64. [CrossRef]
29. Pinheiro, J.; Bates, D.; DebRoy, S.; Sarkar, D.; R Core Team. *Software: Nlme—Linear and Nonlinear Mixed Effects Models*, 2021.
30. Lenth, R. *Software: Emmeans—Estimated Marginal Means, aka Least-Squares Means*, 2020.
31. Benjamini, Y.; Hochberg, Y. Controlling the False Discovery Rate: A Practical and Powerful Approach to Multiple Testing. *J. R. Stat. Soc. Ser. B* **1995**, *57*, 289–300. [CrossRef]
32. Brouns, F.; Bjorck, I.; Frayn, K.N.; Gibbs, A.L.; Lang, V.; Slama, G.; Wolever, T.M. Glycaemic index methodology. *Nutr. Res. Rev.* **2005**, *18*, 145–171. [CrossRef]
33. Denney, W.S.; Duvvuri, S.; Buckeridge, C. Simple, Automatic Noncompartmental Analysis: The PKNCA R Package. *J. Pharmacokinet. Pharmacodyn.* **2015**, *42*, S65. [CrossRef]
34. Kassambara, A. *rstatix: Pipe-Friendly Framework for Basic Statistical Tests*, 2020.
35. Brilli, E.; Khadge, S.; Fabiano, A.; Zambito, Y.; Williams, T.; Tarantino, G. Magnesium bioavailability after administration of sucrosomial® magnesium: Results of an ex-vivo study and a comparative, double-blinded, cross–over study in healthy subjects. *Eur. Rev. Med. Pharmacol. Sci.* **2018**, *22*, 1843–1851. [CrossRef]
36. Pineda, O.; Ashmead, H.D. Effectiveness of treatment of iron-deficiency anemia in infants and young children with ferrous bis–glycinate chelate. *Nutrition* **2001**, *17*, 381–384. [CrossRef]
37. Zhuo, Z.; Fang, S.; Yue, M.; Zhang, Y.; Feng, J. Kinetics Absorption Characteristics of Ferrous Glycinate in SD Rats and Its Impact on the Relevant Transport Protein. *Biol. Trace Elem. Res.* **2014**, *158*, 197–202. [CrossRef] [PubMed]
38. Bovell-Benjamin, A.C.; Viteri, F.E.; Allen, L.H. Iron absorption from ferrous bisglycinate and ferric trisglycinate in whole maize is regulated by iron status. *Am. J. Clin. Nutr.* **2000**, *71*, 1563–1569. [CrossRef] [PubMed]
39. Mei, Z.; Cogswell, M.E.; Looker, A.C.; Pfeiffer, C.M.; Cusick, S.E.; Lacher, D.A.; Grummer-Strawn, L.M. Assessment of iron status in US pregnant women from the National Health and Nutrition Examination Survey (NHANES), 1999–2006. *Am. J. Clin. Nutr.* **2011**, *93*, 1312–1320. [CrossRef] [PubMed]
40. Eicher-Miller, H.A.; Mason, A.C.; Weaver, C.M.; McCabe, G.P.; Boushey, C.J. Food insecurity is associated with iron deficiency anemia in US adolescents. *Am. J. Clin. Nutr.* **2009**, *90*, 1358–1371. [CrossRef] [PubMed]

41. Pisani, A.; Riccio, E.; Sabbatini, M.; Andreucci, M.; Del Rio, A.; Visciano, B. Effect of oral liposomal iron versus intravenous iron for treatment of iron deficiency anaemia in CKD patients: A randomized trial. *Nephrol. Dial. Transplant.* **2015**, *30*, 645–652. [CrossRef] [PubMed]
42. Montagud-Marrahi, E.; Arrizabalaga, P.; Abellana, R.; Poch, E. Liposomal iron in moderate chronic kidney disease. *Nefrologia* **2020**, *40*, 446–452. [CrossRef] [PubMed]
43. de Alvarenga Antunes, C.V.; de Alvarenga Nascimento, C.R.; Campanha da Rocha Ribeiro, T.; de Alvarenga Antunes, P.; de Andrade Chebli, L.; Martins Gonçalves Fava, L.; Malaguti, C.; Maria Fonseca Chebli, J. Treatment of iron deficiency anemia with liposomal iron in inflammatory bowel disease: Efficacy and impact on quality of life. *Int. J. Clin. Pharm.* **2020**, *42*, 895–902. [CrossRef] [PubMed]
44. Parisi, F.; Berti, C.; Mandò, C.; Martinelli, A.; Mazzali, C.; Cetin, I. Effects of different regimens of iron prophylaxis on maternal iron status and pregnancy outcome: A randomized control trial. *J. Matern. Fetal Neonatal Med.* **2017**, *30*, 1787–1792. [CrossRef] [PubMed]

 *nutrients*

Review

The Dark Side of Iron: The Relationship between Iron, Inflammation and Gut Microbiota in Selected Diseases Associated with Iron Deficiency Anaemia—A Narrative Review

Ida J. Malesza [1,†], Joanna Bartkowiak-Wieczorek [2,†], Jakub Winkler-Galicki [2], Aleksandra Nowicka [2], Dominika Dzięciołowska [2], Marta Błaszczyk [2], Paulina Gajniak [2], Karolina Słowińska [2], Leszek Niepolski [2], Jarosław Walkowiak [1] and Edyta Mądry [2,*]

[1] Department of Pediatric Gastroenterology and Metabolic Diseases, Poznan University of Medical Sciences, 61-701 Poznan, Poland
[2] Department of Physiology, Poznan University of Medical Sciences, 61-701 Poznan, Poland
* Correspondence: edytamadry@gmail.com
† These authors contributed equally to this work.

Abstract: Iron is an indispensable nutrient for life. A lack of it leads to iron deficiency anaemia (IDA), which currently affects about 1.2 billion people worldwide. The primary means of IDA treatment is oral or parenteral iron supplementation. This can be burdened with numerous side effects such as oxidative stress, systemic and local-intestinal inflammation, dysbiosis, carcinogenic processes and gastrointestinal adverse events. Therefore, this review aimed to provide insight into the physiological mechanisms of iron management and investigate the state of knowledge of the relationship between iron supplementation, inflammatory status and changes in gut microbiota milieu in diseases typically complicated with IDA and considered as having an inflammatory background such as in inflammatory bowel disease, colorectal cancer or obesity. Understanding the precise mechanisms critical to iron metabolism and the awareness of serious adverse effects associated with iron supplementation may lead to the provision of better IDA treatment. Well-planned research, specific to each patient category and disease, is needed to find measures and methods to optimise iron treatment and reduce adverse effects.

Keywords: inflammatory bowel disease; colorectal cancer; obesity; oxidative stress; dysbiosis; anaemia

Citation: Malesza, I.J.; Bartkowiak-Wieczorek, J.; Winkler-Galicki, J.; Nowicka, A.; Dzięciołowska, D.; Błaszczyk, M.; Gajniak, P.; Słowińska, K.; Niepolski, L.; Walkowiak, J.; et al. The Dark Side of Iron: The Relationship between Iron, Inflammation and Gut Microbiota in Selected Diseases Associated with Iron Deficiency Anaemia—A Narrative Review. *Nutrients* **2022**, *14*, 3478. https://doi.org/10.3390/nu14173478

Academic Editor: Dietmar Enko

Received: 31 July 2022
Accepted: 19 August 2022
Published: 24 August 2022

Publisher's Note: MDPI stays neutral with regard to jurisdictional claims in published maps and institutional affiliations.

1. Introduction

Iron deficiency (ID) and iron deficiency anaemia (IDA) are important causes of diseases and disabilities worldwide. In 2016 there were over 1.2 billion cases of IDA, of which 41.7% of children (younger than 5 years), 40.1% of pregnant women, and 32.5% of non-pregnant women were anaemic worldwide; thus, IDA is one of the world's primary public health issues (Figure 1), [1–3]. For the World Health Organization (WHO), controlling anaemia is a global health priority with the purpose of a 50% reduction in anaemia prevalence in women by 2025 [4]. IDA mainly affects premenopausal and pregnant women, growing children and the elderly, however, it is also increasingly recognised as a complication of multiple diseases associated with lifestyle and dietary patterns, such as obesity, inflammatory bowel disease, colorectal cancer or chronic kidney disease [5,6].

There are two goals of the IDA treatment, to replenish iron stores and normalise haemoglobin concentration. Individual assessment of IDA causes and amount of dietary iron intake is recommended. The holistic nutritional strategy to treat and avoid iron deficiency in the future includes the two paths regarding iron intake and absorption. First is to increase the iron intake by promoting the consumption of iron-rich food such as meat, poultry, fish, and seafood. To optimise iron absorption it is recommended to increase vitamin C and fermented product intake and avoid the consumption of iron absorption inhibitors (phytates in cereals and nuts, tannins from coffee or tea and calcium-rich products) together with iron-rich food [7]. Although a described holistic approach to clinical management of ID is available, oral or parenteral iron products with varying doses and formulations are the primary means of supplementation [8]. However, there is a growing body of evidence for side effects of iron supplementation concerning the exacerbation of inflammation, alterations in gut microbiota, and gastrointestinal adverse events [9]. Therefore, this review aimed to investigate the knowledge regarding the relationship between iron supplementation, inflammatory status and changes in the gut microbiota milieu in diseases typically complicated with IDA and considered as having an inflammatory background, including inflammatory bowel disease, colorectal cancer, and obesity.

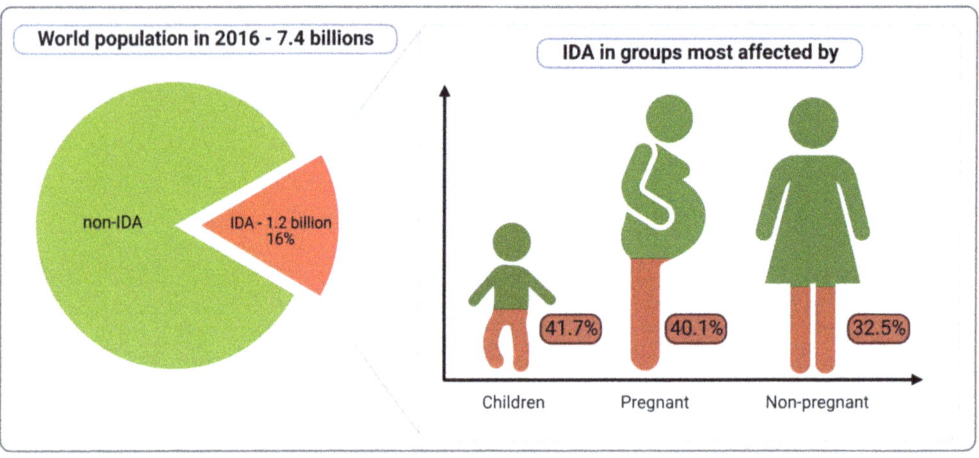

Figure 1. Occurrence of iron deficiency anemia. IDA—iron deficiency anemia; Non-IDA—anemia without iron-deficiency.

2. Search Strategy and Selection Criteria

PubMed/Medline and Cochrane databases were searched for articles about iron deficiency anaemia, iron and its relationship to inflammatory bowel disease, colorectal cancer and obesity. The following English terms and their combinations were used: iron, iron metabolism, iron deficiency, iron deficiency anaemia, microbiota, inflammatory bowel disease, colorectal cancer, colon cancer, and obesity. The titles and abstracts were screened for the 282 papers, and 97 were chosen for thorough study. In addition, a review of their reference lists identified further relevant articles, so in total, 141 articles were selected for review.

2.1. Iron in the Human Body

Iron is a micronutrient fundamental for many biochemical reactions. It acts primarily as a component of enzymes or other proteins involved in oxygen transport, particularly in haemoglobin and myoglobin. Moreover, it is involved in energy metabolism in mitochondria, cell proliferation and growth, and DNA processing [10,11]. In contrast to other nutrients, iron has no active excretory mechanisms; therefore, its resource control must be provided in the small intestine [12]. Iron is lost mainly with shedded enterocytes, epidermic cells, and menstrual blood in women, and less with urine or faeces [10]. Most body iron stores are regained from senescent and damaged erythrocytes as an endogenous source [13]. Due to the modest iron loss, absorbing 1 to 2 milligrams daily is enough to compensate for iron decline and maintain a concentration of 3–5 g [10,14].

Dietary iron heme is provided in the form of myoglobin and haemoglobin in animal food sources or as non-heme iron in plant and animal food sources [13]. Due to the difference in the biochemical structure, heme and non-heme iron vary in bioavailability. Non-heme iron tends to sequestrate and bind to other food components and this causes lower absorption in comparison to heme iron [13,15,16]. Although heme iron does not make up the majority of iron diet intake in carnivores, this form is most easily absorbed [13]. Iron deficiency and excess lead to disordered homeostasis, with excess iron increasing the pool of labile iron in plasma. Unbound iron causes an overload in tissues and further damage by generating reactive oxygen species (ROS) via the Fenton or Haber-Weiss reactions [17]. Subsequently, lipid and amino acid peroxidation and DNA damage occur leading to cell death [10,18].

2.2. Iron Absorption and Metabolism

Iron absorption occurs mainly in the duodenum and proximal part of the jejunum, which is mostly adjusted to this role due to the high expression of proteins involved with non-heme iron absorption [15,16,19]. Regarding heme iron, there are two prevailing hypotheses of its absorption: the receptor-mediated endocytosis of heme and direct transport into the intestinal enterocyte by heme transporters [20]. Non-heme iron uses divalent metal-ion transporter-1 (DMT1) to enter the apical membrane of the enterocyte. DMT1 function is conjugated with the electrochemical gradient of protons that generate the driving force for iron transfer from the extracellular space to the cytoplasm of the cells of the intestinal lining [21,22]. The importance of the DMT1 symporter was confirmed in animal models with an intestinal DMT1 knockout that caused a severe iron absorption disorder [23]. Furthermore, a rare DMT1 mutation was also detected in humans with microcytic hypochromic anaemia [24]. The DMT1 function must be assisted by an enzyme called duodenal cytochrome B (DcytB) located in the brush border [15].

Iron exists in an oxidised, insoluble ferric (Fe^{3+}) form at physiological pH and to be absorbed, it must be reduced into a soluble ferrous (Fe^{2+}) state or be bound by a protein, such as heme [25]. Hence, non-heme iron is transported via DMT1 in the form of Fe^{2+} ions. As DcytB executes the non-heme iron reduction from Fe^{3+} to Fe^{2+}, it forms a substrate for DMT [14,25]. Higher DcytB expression in the enterocyte lining leads to increased iron uptake, which indicates DcytB's major role in iron absorption [26].

If iron is not currently required for any ongoing processes, it is temporarily housed in ferritin. Ferritin is a cytosolic protein capable of sheltering up to 4500 $Fe3^{+}$ atoms in one molecule [15,27]. Increased iron demand activates ferroportin, the transmembrane protein located in the basolateral side of duodenal enterocytes. It facilitates the transfer of enterocyte iron stores into the bloodstream, where iron is bound by transferrin. This step needs iron to be oxidised back to Fe^{3+}, which is catalysed by the enzyme hephaestin. The iron–transferrin complex is then recognised by transferrin receptors leading to iron endocytosis in the body cells and subsequent iron use in cellular processes [15,28].

Iron metabolism is precisely regulated both on systemic and cellular levels. The first one involves mainly a liver-derived peptide, hepcidin, the expression of which depends on the body's iron demand [15]. When the organism is saturated with iron, hepcidin expression increases and it interacts with ferroportin to cause ferroportin internalisation and subsequent degradation, which decreases iron transfer into the bloodstream. In contrast, the amount of hepcidin decreases during ID and facilitates the alignment of iron stores by unimpeded ferroportin function [29].

The cellular iron regulation is based mainly on two regulatory proteins—iron regulatory protein 1 (IRP1) and iron regulatory protein 2 (IRP2). They are responsible for post-transcriptional regulation of the iron homeostatic gene expression. The IRP action mechanism involves binding to iron-responsive elements (IREs) in the untranslated target mRNA regions. So, IRPs control the translation and stability of mRNA for proteins related to iron, oxygen and energy metabolism. Interestingly IRP2 can also act as cytosolic aconitase. IRP1 controls the expression of hypoxia-inducible factor 2α (HIF2α). Thus, it is considered a key regulator of erythropoiesis. IRP2 regulates the expression of transferrin receptor 1 (TfR1) and 5-aminolevulinic acid synthase 2. It is responsible for iron uptake and heme biosynthesis in erythroid progenitor cells (for further information, see [15,30]). The main aspects of iron metabolism in the human body are presented in Figure 2.

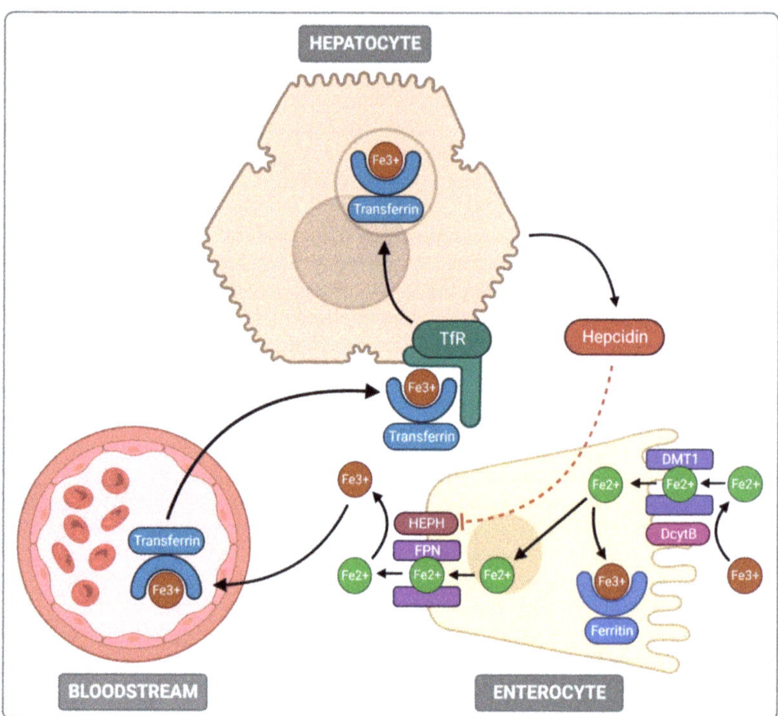

Figure 2. Iron metabolism. DcytB—duodenal cytochrome B; DMT1—divalent metal-ion transporter-1; FPN—ferroportin; HEPH—hephaestin; TfR—transferrin receptor; dashed line–inhibition.

2.3. Iron Toxicity

Iron is both essential and potentially harmful. Therefore, its homeostasis requires precise regulatory mechanisms to provide cells with the necessary element supply and concurrently prevent the unfavourable effects of excess iron [17,31,32]. Transferrin and ferritin bind and store iron, respectively, to protect against the adverse effects of labile iron [31]. Iron toxicity occurs because the iron-binding protein capacity is restricted. Its toxicity stems from two distinct qualities of the metal, the ability to generate free radicals and serving as an essential growth factor for nearly all pathogenic bacteria, fungi, and protozoa, as well as for all neoplastic cells [33]. It is also worth mentioning that excess dietary iron may affect the ingestion of other divalent metals. For example, in murine and rat models, the high nutritional iron intake contributed to copper deficiency [34–37]. However, the clinical significance of this finding in humans requires further investigation.

2.4. Iron and Oxidative Stress

A parameter that reflects excessive iron accumulation is a pule pool of labile iron, which physiologically accounts for about 5% of total cellular iron. Iron overload increases the non-transferrin-bound iron. Unbound iron is particularly hazardous because it readily accepts and donates electrons while switching between the two forms, Fe^{2+} and Fe^{3+}, leading to free radical generation [38]. Hence, "free iron" is highly reactive and can participate in Fenton's or Haber-Weiss's reactions, generating ROS, such as highly dangerous hydroxyl radicals [39]. ROS initiate lipid and amino acid peroxidation, enzyme denaturation, polysaccharide depolymerisation and DNA damage [39,40]. An insufficiency of antioxidative mechanisms due to iron excess and oxidative stress triggers cell damage [39,41–43]; therefore, iron must be sequestered by ferritin or transferrin to prevent ROS production [28,39].

Since the small intestine is the leading site for iron absorption and regulation, ROS-generating responses primarily affect this part of the gastrointestinal tract. Overproduction of free radicals causes cellular changes and increases intestinal barrier permeability. These changes are exacerbated by the onset of inflammation as expressed by hepcidin and/or calprotectin levels, which increase with oral iron supplementation [44–46]. ROS generate stress in the mitochondria and endoplasmic reticulum, causing swelling. This is the first step in the initiation of the cell death pathway [47]. Moreover, this process is facilitated by the increased peroxidation of lipids leading to the destruction of cell membranes [48]. Enterocyte apoptosis finally violates gut barrier function and contributes to the increased permeation of substances such as bacterial toxins into the bloodstream [47].

These mechanisms were demonstrated in Chinese Yellow broilers treated with a high dose of dietary iron, leading to increased malondialdehyde (MDA), the main product of lipid peroxidation. Alterations within intestinal villi were also reported [49]. The research on rats also demonstrated adverse intestinal effects of the iron supply in the context of oxidative stress [17]. The connection between iron-dependent oxidative stress and the aggravation of inflammation from the viewpoint of disorders and diseases is described in detail below.

2.5. Iron and Microbiota

Although commonly used, iron supplementation may have adverse effects on the gastrointestinal tract, mainly on the intestinal microbiota [50]. This article focuses mainly on the mutual interaction between the oral iron supply and gut microbiota. The excess unabsorbed iron passes through the colon and is involved in Fenton and Haber-Weiss reactions, with adverse effects on the intestinal structure. Iron supplementation also affects the microbiota [51], with the composition of intestinal flora affecting iron absorption [52]; therefore, iron and microbiota are in a complex and bilateral relationship [53].

The amount of available "free iron" in the intestines is significantly lower than the optimal level required for the proper functioning and replication of bacterial cells [52,54–56]. Some pathogenic bacteria like *Salmonella* or the pathogenic strain of *Escherichia coli* are equipped with siderophores, extracellular ferric chelators that enable bacteria to capture trivalent iron (Fe^{+3}) from ferritin and transferrin [56,57]. Siderophores are considered virulence factors, as they help bacteria to survive in an iron-deficient environment [58]. Interestingly, some commensal bacteria such as the genera *Lactobacillus* and *Bifidobacterium* do not require a high portion of iron to grow and expand [59].

The intestinal microbiota controls systemic iron homeostasis in two ways, first, by inhibiting the intestinal iron absorption pathways via hypoxia-induced factor (HIF-2α). In a state of increased iron demand (ID, hypoxia, augmented erythropoiesis), HIF-2α increases the expression of DMT1, DcytB at the enterocyte apical brush border membrane, and ferroportin at the basal membrane [60–62]. Second, iron homeostasis is controlled by increasing cellular iron storage induced by higher ferritin expression [63]. Moreover, it was demonstrated that *Lactobacillus plantarum* increases iron absorption in the intestine of iron-deficient females [64], so could be used as a component of iron supplements to alleviate the negative side effects of oral iron administration [65]. Some studies have shown the reduction of the abundance of beneficial microbes simultaneously with an increased abundance of deleterious microbes after oral iron supplementation (Table 1), [66]. For example, African children with anaemia, fed with iron-fortified biscuits for six months, presented an unfavourable ratio of pathogenic to commensal bacteria and increased calprotectin concentration in the stools, which is indicative of intestinal inflammation. [67]. The other authors suggested that iron supplementation at the physiological level does not lead to mucositis unless other factors such as pathogen invasion or systemic inflammation are present [59]. Similar results concerning changes in microbiota were also obtained by studies conducted in children from Kenya, which is also suggestive of the negative impact of excess dietary iron on the intestinal microbiome [68,69].

Table 1. Impact of oral iron supplementation on gut microbiota composition (↓—decrease, ↑—increase).

Bacteria	Iron Supplementation
Phylum: Firmicutes	↓
Genus: Enterococcus	↑
Genus: Lactobacillus	↓
Genus: Roseburia	↑
Genus: Clostridium	↑
Phylum: Proteobacteria	↑
Family: Enterobacteriaceae	↑
Species: E. coli	↑
Genus: Salmonella	↑
Genus: Shigella	↑
Genus: Citrobacter	↑
Order: Bacteroidales	↑
Genus: Bacteroides	↑
Genus: Campylobacter	↑
Genus: Bifidobacterium	↓
Genus: Prevotella	↓
Genus: Rothia	↓

However, some researchers found an entirely diverse outcome in rat experiments, with oral iron supplementation promoting commensal and dominant bacteria, and increasing the intestinal microbiota metabolic activity as evidenced by the increased concentration of beneficial short-chain fatty acids in the colon [70]. Furthermore, in a randomised controlled trial in South African school-aged children, no significant changes in the intestinal microbiome or inflammatory markers were observed due to oral iron supplementation [71].

In summary, studies concerning the relationship between microbiota and iron have shown inconsistent results, reporting negative and positive impacts, highlighting the need for further investigation on this topic. In particular, there is a lack of high-quality data examining the potentially harmful effect of untargeted iron supplementation, for example, in women of childbearing age [50,72].

3. Inflammatory Bowel Diseases (IBDs)

IBDs, including ulcerative colitis (UC) and Crohn's disease (CD), are chronic, relapsing inflammatory conditions of the gastrointestinal (GI) tract. Interactions between the environmental factors and commensal intestinal microflora in genetically predisposed individuals are considered the leading cause of an inappropriate immune response and as a result, the development of inflammatory disease [73]. In recent years, the incidence of IBD has increased in highly industrialised western countries, mainly affecting people between 16 and 30 years old. It is associated with lifestyle changes, including a Western-style high-fat diet, cigarette smoking, distress, as well as taking oral contraceptives, hormone replacement therapy, and non-steroidal anti-inflammatory drugs [74].

UC primarily affects the colon and the rectum, whereas CD can involve any part of the GI tract, from the mouth to the anus, but most commonly involves the distal part of the small intestine and colon. IBDs may present with symptoms such as abdominal cramps and pain, persistent diarrhoea, fatigue, weight loss or bleeding.

3.1. Anaemia as a Complication of IBD

Anaemia is the most prevalent extraintestinal complication of IBD, with an estimated 70% of inpatients and 20% of outpatients with IBD developing anaemia, which is believed to affect one-third of the IBD patients at any one time [75]. Traditionally, anaemia was classified as IDA and anaemia of chronic disease (ACD). In IBD patients, both the aforementioned types of anaemia may occur because the local inflammation in the intestines contributes to the onset of systemic inflammation [76].

IDA is a consequence of chronic haemorrhages from the ulcerated mucosa, impaired dietary iron absorption, as well as self-imposed dietary restrictions relating to gastrointestinal symptoms. Moreover, appetite loss during an exacerbation of the disease and a range of other factors such as medicines used for IBD treatment (e.g., proton pump inhibitors, methotrexate, thiopurines and sulfasalazine) also negatively impact iron absorption and erythropoiesis [75]. In turn, ACD is connected with the systemic immune response that accompanies inflammatory diseases such as IBD [77]. Immune cells release pro-inflammatory cytokines, mainly interleukin-6 (IL-6), which upregulates the expression of liver hepcidin, the main regulator of iron homeostasis, thereby decreasing iron uptake from the enterocytes and a reduced ability to take advantage of sufficient iron for effective erythropoiesis [78]. This issue is described in more detail in the section about iron metabolism.

3.2. Iron Replacement Therapy in IBD

As described in the previous paragraph, the high incidence of anaemia in patients with IBD requires therapeutic intervention, either intravenous or oral iron administration. The European Crohn's and Colitis Organization [79] guidelines recommend intravenous iron supply as a mainstay treatment for IBD patients and front-line therapy for haemoglobin (Hb) levels < 10 g/dL (e.g., iron sucrose, ferric gluconate, ferric carboxymaltose, iron isomaltoside). Another indication for intravenous iron is the active phase of the disease because inflammation impairs iron absorption in the intestines [80]. An intravenous iron supply is also recommended in patients with a poor iron tolerance and a previous unsuccessful attempt at oral iron treatment [81,82]. If oral iron supplementation must be used, it should be limited to IBD patients with mild anaemia (Hb $\geq$ 11.0 g/dL), with an inactive disease and no prior intolerance to oral iron [83].

3.3. Negative Consequences of Oral Iron

Oral iron administration can cause side effects due to a large amount of non-absorbed iron (about 90%) remaining in the intestines, including gastroduodenitis, nausea, bloating, vomiting, dyspepsia, constipation, diarrhoea, abdominal pain or darkening of the stools [47]. Thus, oral iron therapy is not optimal with as many as 50% of patients discontinuing the therapy [8]. The aforementioned gastrointestinal symptoms most likely result from a combination of a high concentration of free iron radicals induced by redox cycling in the gut lumen and at the mucosal surface which can promote inflammation and alterations in the gut microbiota composition [9,45].

Moreover, non-absorbed iron can be toxic and exacerbates disease activity in IBD. A study conducted in rats with dextran sulphate sodium (DSS)-induced colitis indicated that dietary iron administration aggravates colitis and is associated with oxidative stress, neutrophil infiltration and NF-kappaB pathway activation which increases the expression of pro-inflammatory cytokines such as interferon-γ (IFN-γ), tumour necrosis factor α (TNF-α), and inducible nitric oxide synthase (iNOS). These negative effects can be ameliorated by vitamin E [84]. Another animal study showed that a diet without iron sulphate combined with intravenous iron administration prevents the development of chronic ileitis in a mouse model of CD, suggesting that oral iron sulphate replacement therapy may trigger the inflammatory processes linked to the progression of CD-like ileitis [85]. There is an imbalance between pro-oxidative and antioxidant mechanisms in CD, with patients having increased levels of reactive oxygen intermediates (ROI) and DNA oxidation products, as well as markedly raised iron levels in combination with reduced copper and activity of zinc superoxide dismutase (Cu/Zn SOD) [86]. Oral iron due to its pro-oxidative capacity may promote a pro-inflammatory effect.

3.4. Impact of Iron on the Intestine

Iron has a significant impact on intestinal functioning, potentially via various mechanisms. Experimental evidence suggests that excessive iron in the lumen may be also harmful to the intestinal mucous membrane. Despite its key role in cellular processes, free iron in the colon can generate toxic free radicals that may directly impair the integrity of the intestinal epithelium via oxidative stress. The relationship between iron and redox stress is discussed in more detail in the section about iron and redox stress.

A key element of the physical intestinal barrier is a single layer of epithelial cells, mainly consisting of enterocytes, which in addition to absorbing nutrients, play an important role in immune activity, mediating the release of cytokines and the expression of receptors engaged in the immune response [87]. The epithelial layer also includes the goblet cells secreting mucus, Paneth cells synthesising defensins, enterochromatophilic cells releasing hormones and neuropeptides, and M cells that capture antigens from the intestinal lumen. The epithelial integrity and as well as selective permeability are dependent on appropriate connections between cells, mainly tight junctions, including claudins, occludins, protein junctional adhesion molecules, and tricellulins [88].

Oral iron supplementation contributes to lipid peroxidation of cellular membranes and disrupts energy processes in the cell due to mitochondrial damage, as well as promoting endoplasmic reticulum dysfunction [89]. Consequently, these cellular disorders trigger death pathways, thereby destroying the mechanical intestinal barrier and increasing permeability, known as a leaky gut [47]. There are also changes in the enterocyte shedding-proliferation axis [90]. This impairment of the intestinal barrier increases the risk of exposure to bacterial endotoxins (e.g., LPS), leading to metabolic endotoxemia and microinflammation (gut-derived inflammation). LPS binding to its receptor complex on macrophages results in markedly increased production of pro-inflammatory cytokines such as IL-1β, IL-6, IL-12, IFN-β or IFN-γ, which may worsen intestinal inflammation in IBD patients. This impaired integrity has been demonstrated in an in vitro study with Caco-2 cells exposed to iron [91].

The proper functioning of the intestinal barrier also involves short-chain fatty acids (SCFAs) including butyrate, acetate and propionate, produced by the microbiota in the colon through the anaerobic fermentation of indigestible polysaccharides such as dietary fibre and resistant starch. There are two main pathways for the conversion of butyrate–CoA into butyrate. The first pathway includes a two-step reaction using butyrate kinase and phosphate butyryltransferase. The second pathway is a single-step reaction conducted by butyryl–CoA: acetate Co-A transferase [92]. SCFAs are the main source of energy to colonocytes and modulate the immune response through inhibition of the LPS-induced NF-κB pathway, and reduced production of pro-inflammatory cytokines and chemokine by the epithelial intestinal cells (EIC). Besides, SCFAs increase the secretion of antimicrobial peptides (LL-37 and CAP-18) and IL-18, a cytokine that maintains homeostasis in EIC, thereby protecting against colitis [93,94].

The influence of oral iron supply on SCFAs production remains still unclear. Some research indicates that iron can contribute to increases in gut SCFAs production, thus positively affecting gut health [95]. Some in vitro studies compared the impact on SCFAs production between the normal iron condition and iron deficiency, which does not fully reflect increased iron content in the gut lumen on SCFAs production [96]. Another study investigated the influence of oral iron supplementation on colitis exacerbation and the composition of the gut microbiome. A decreased abundance of SCFA-producing genera was shown without assessing the levels of faecal SCFAs [97]. However, the literature also provides evidence for the adverse impact of oral heme iron on murine colitis model, reducing the level of butyrate production and expression of butyrate kinase, phosphate butyryltransferase and the α subunit of butyryl–CoA: acetate Co-A transferase [92].

In conclusion, the loss of intestinal barrier integrity is an early event which contributes to chronic inflammation (Figure 3).

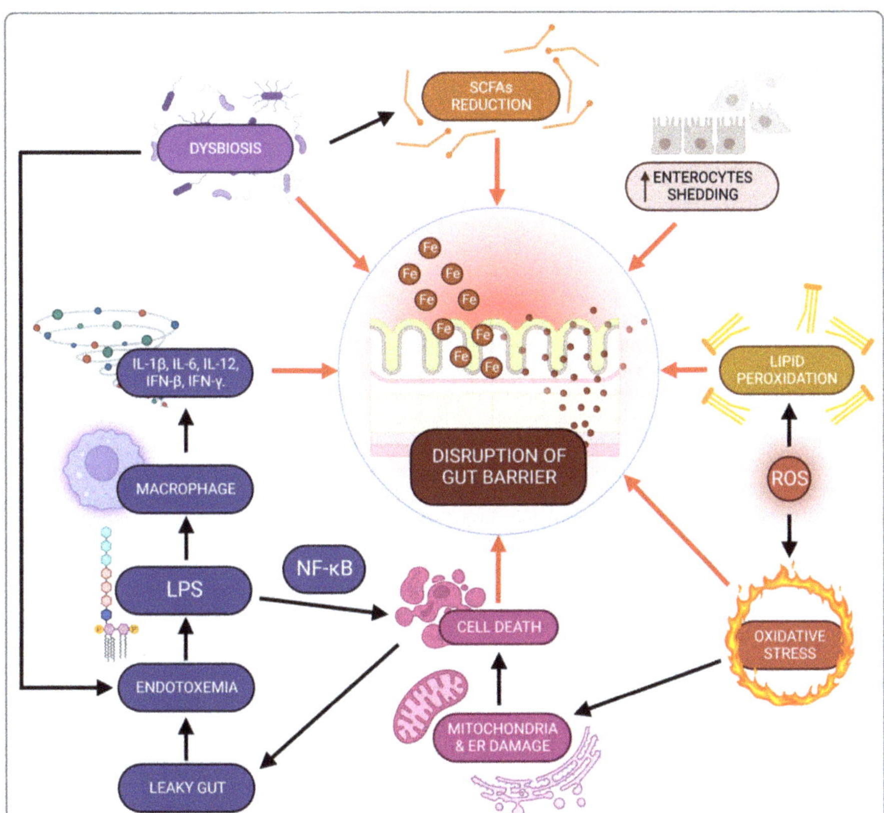

Figure 3. Iron-related gut barrier disruption mechanisms. SCFAs—short-chain fatty acids; IFN-β—interferon β; IFN-γ—interferon-γ; IL-1β—interleukin 1β; IL-6—interleukin 6; IL-12—interleukin 12; LPS—lipopolysaccharide; NF-kβ—nuclear factor kappa-light-chain-enhancer of activated B cells; ROS—reactive oxygen species.

3.5. Impact of Iron on the Microbiota in IBD Patients

Besides the unfavourable impact of iron on the intestinal epithelial barrier via oxidative stress, iron may also affect the gut microbiota [98]. As mentioned above, the altered gut microbiota is a crucial factor in driving inflammation in IBD but it is unclear if dysbiosis is the cause or the outcome of mucosal inflammation. Numerous studies have indicated differences in the composition and diversity of the intestinal microbiota among IBD patients in comparison to healthy individuals. Characteristic changes observed in patients with IBD include increased bacteria such as Proteobacteria, Fusobacterium species, and *Ruminococcus gnavus* and in turn, decreased protective groups such as Lachnospiraceae, Bifidobacterium species, *Roseburia*, and *Sutterella* [99,100]. Iron leads to a shift in the microbiota composition and exacerbation of dysbiosis in IBD. Low iron bioavailability results in a high concentration in the intestinal lumen and accessibility to the gut microflora. Lee et al. [101] designed an open-labelled clinical trial with iron-deficient participants with or without IBD and performed pre-iron therapy and post-iron therapy measurements whereby the individuals served as the controls. This study compared the effects of *per oral* (p.o) versus *intravenous* (i.v) iron replacement therapy (IRT), indicating that oral iron supplementation decreased the diversity of intestinal microflora in patients with IBD and ID, especially *Faecalibacterium prausnitzii*, *Ruminococcus bromii*, *Dorea* sp. and *Collinsella aerofaciens*. It is worth noting that dysbiosis aggravated by iron can also be a causative factor disrupting the integrity of the EIC and promoting an oxidative pro-inflammatory microenvironment.

In summary, the need for iron supplementation in IBD patients who additionally suffer from anaemia exacerbates the dysbiosis caused by the disease, thus worsening their clinical condition.

4. Colorectal Cancer

Colorectal cancer (CRC) is the third most commonly diagnosed cancer globally and the second major cause of mortality (935,000 deaths in 2020) [76]. It affects mainly older adults from highly developed countries, as it is a multifactorial disease closely associated with lifestyle and risk factors like obesity, lack of physical activity, alcohol consumption, cigarette smoking or low fibre, fruit and vegetable supply [102]. The American Cancer Society estimates the lifetime risk of developing colorectal cancer is about 4.3% for men and 4% for women. Increasing research indicates that iron supplementation and the resulting changes in the composition of the intestinal microbiota are essential factors causing CRC development.

4.1. Impact of Heme Iron on Cancerogenesis

This section discusses the influence of heme iron on CRC development [103]. The relationship between CRC and non-heme iron has not been established and requires further investigation. The International Agency for Research on Cancer classified the consumption of processed meat as carcinogenic and red meat as probably carcinogenic in humans in 2015. There are several reasons for such classification, but most importantly, heme iron plays a crucial role in promoting colorectal carcinogenesis. Heme iron contributes to DNA damage through the increased formation of lipid peroxyl radicals (MDA and 4-hydroxynonenal) and catalysing ROS production, as well as increasing the endogenous formation of N-nitroso compounds (NOCs), all of which are cytotoxic and genotoxic. There is a growing body of evidence that heme iron indirectly contributes to a weakened gut mucosal barrier and proliferation of the colon epithelium. This process is stimulated by a large amount of heme iron found in red meat, which is an essential factor in the development of pathogenic microbiota, reducing the number of probiotic bacteria such as lactobacilli that do not require iron. These pathobionts are sulphate-reducing and mucin-degrading bacteria, especially *Akkermansia mucinphila*, which reduces disulphide bonds between mucin proteins, thus reducing the mucus layer. This process increases the intestinal wall permeability to luminal cytotoxic compounds and products of bacterial degradation, which in turn, leads to compensatory hyperproliferation, hyperplasia and finally, neoplasm [104,105].

Bastide et al. showed that heme iron plays a leading role in mucin-depleted foci (MDF) formation [106]. Moreover, some gut strains, such as *Bacteroides fragilis*, can use heme iron directly [107]. Various studies have shown the effect of iron intake on intestinal microbiota composition; for example, the six months consumption of iron-supplemented biscuits by anaemic African children caused a potentially more pathogenic gut microbiota profile and increased inflammation. There was also a significant growth of enterobacteria and a reduced amount of lactobacilli when compared to the control group receiving non-supplemented biscuits [46]. In Kenya, infants were given iron-fortified micronutrient powder, which also caused significant changes in the intestinal microbiota, increasing enterobacteria and decreasing lactobacilli, resulting in inflammation as evidenced by an increase in faecal calprotectin [69]. The described mechanisms are summarised in Figure 4.

4.2. Dysbiosis and Cancerogenesis

The bacterial strains that dominate in the colon during dysbiosis contribute to DNA damage, suppression of apoptosis by affecting multiple signalling pathways, and the production of pro-inflammatory cytokines or toxic metabolites. The microbes activate the NF-kB (nuclear factor-κB) pathway via the host Toll-like receptors (TLRs), initiating carcinogenesis due to the inhibition of apoptosis in enterocytes. Moreover, the activated NF-kB pathway causes an increase in the production of pro-inflammatory cytokines, TNF-α, IL-1, IL-6, IL-8 and anti-apoptotic genes [108]. NF-kB activation is also possible via the increased production of secondary bile acids, which cause an increase in reactive oxygen, the direct activator of the NF-kB signalling pathway [109].

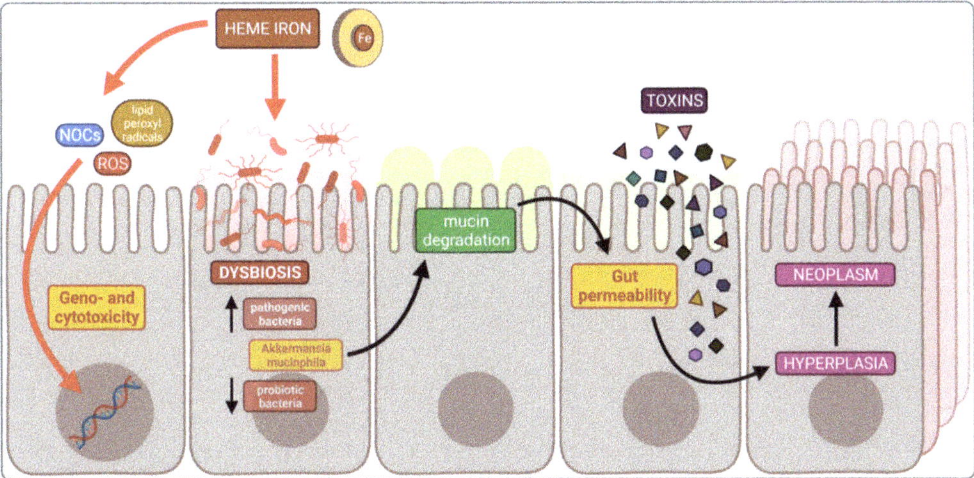

Figure 4. Mechanisms of heme iron induction of colorectal carcinogenesis. NOCs—*N*-nitroso compounds; ROS—reactive oxygen species.

The adenomatous polyposis coli (Apc) gene is mutated in both types of CRC, familial sporadic and colitis-associated CRC. Apc is the primary regulator of the Wnt/β-catenin signalling pathway responsible for the physiological maturation of enterocytes. The activation of Wnt signalling leads to the nuclear accumulation of β-catenin and begins continuous stimulation of the transcription T-cell factor/lymphoid enhancer factor (TCF/LEF), in turn, activating the target genes, c-myc and cyclin D1, initiating uncontrolled cell proliferation. Iron and iron-related-dysbiosis can increase Wnt signalling following the loss of Apc function [99,100].

The activation of the Wnt/β-catenin pathway is also possible through substances produced directly by the bacteria that colonise the gut. For example, studies have shown that *Bacteroides fragilis* toxin enhances cell signalling via the Wnt/β-catenin pathway [97]. Furthermore, in a mouse model, *Fusobacterium nucleatum*, a gram negative oral anaerobe, activates the E-cadherin/β-catenin pathway via the adhesin factor FadA leading to upregulation of chk2, increasing DNA damage and enhancing cancer tumor growth [101]. The described mechanisms are summarised in Figure 5.

4.3. Microbiota Composition and Bacterial Metabolites

Bacterial drivers of colorectal cancer are defined as intestinal bacteria with carcinogenic characteristics that may initiate tumour development. The accumulation of specific genera of the bacteria has been observed in colon cancer and currently there are three strains of bacteria with proven cancerogenic potential: enterotoxigenic *Bacteroides fragilis* (ETBF), Fusobacterium and certain *Escherichia coli* strains [110]. ETBF produces *Bacteroides fragilis* toxin (BFT) also known as fragilysin, which upregulates spermine oxidase (SMO) in colonic epithelial cells, causing an increase in ROS. ETDF also stimulates the production of IL-17 by Th-17 cells to support the growth and survival of cancer cells and suppresses T-cell proliferation [111]. Fusobacterium also increases IL-17, TNF-α, IL-6, IL-8, and IL-12 production and is associated with high cyclooxygenase 2 (COX-2) activity and a 3.5-fold increased risk of colonic adenomas [112].

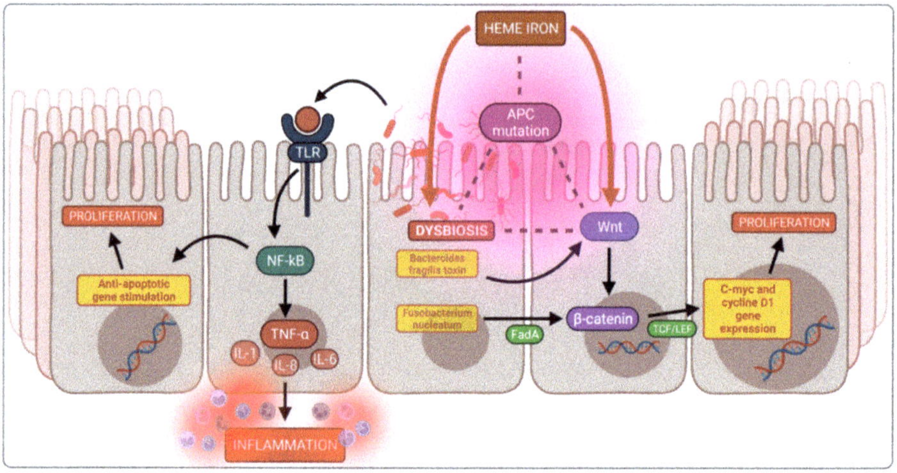

Figure 5. Mechanisms of heme iron induction of colorectal carcinogenesis. APC—adenomatous polyposis coli gene; C-myc, c-Myc–oncogenes; FadA—Fusobacterium nucleatum adhesin; IL-1—interleukin 1; IL-6—interleukin 6; IL-8—interleukin 8; NF-kβ—nuclear factor kappa-light-chain-enhancer of activated B cells; TCF/LEF—T-cell factor/lymphoid enhancer factor family; TLR—Toll-like receptor; TNF-α—tumor necrosis factor α; Wnt—Wnt signaling pathways.

The most genotoxic substance involved in CRC development, colibactin, is produced by the pathogenic *E. coli* strains possessing the polyketide synthase (pks) island. Colibactin contributes to chromosome aberrations or double-strand DNA breaks (Dubinsky b.d.). It also suppresses the mutL homologue 1 (MLH1), the mismatch repair protein. There are suggestions that the use of inhibitors of colibactin synthesis could stop the proliferation of tumour cells and be used as a new form of treatment [113].

The *E. coli* strains in CRC are more invasive than strains in other diseases. They support SUMO (small ubiquitin-like modifier) conjugation to the p53 tumour suppressor protein and cause cell senescence [109]. *Enterococcus faecalis* can also produce secondary bile salts and hydrogen sulphide, promoting inflammation and cancerogenesis. Hydrogen sulphide is toxic to the intestinal epithelium and causes damage to it, as well as contributes to the formation of mutations [114].

4.4. Comparison of the Intestinal Microflora of Healthy Subjects and CRC Patients

Studies in mice have proved the differences between the healthy versus CRC gut microflora. When compared to a healthy gut, a CRC gut is more abundant in *E. coli* and Proteobacteria with fewer Bacteroidetes [111]. Another study has revealed a profusion of Fusobacterium, Campylobacter and Leptotrichia in CRC patients. It is worth noting that these are oral bacteria; therefore, the critical question is whether these bacteria are from the oral cavity, belong to the colon microbiota, or are a cancer strain [115].

4.5. Driver–Passenger Model

Tlajsma et al. developed a model in which pathogenic bacteria (termed bacterial drivers) cause permanent changes in the intestinal epithelium leading to CRC development. The alteration of the environment enables colonisation by the colon passenger bacteria, which take advantage of the tumour microenvironment. By competing with driver bacteria, they reduce their abundance, suggesting that the microenvironment changes as the tumour progresses. CRC development increases the number of passenger bacteria while the number of driver bacteria decreases [110]. By proving that passenger bacterial growth depends on CRC metabolites, Gorza et al. confirmed the driver–passenger hypothesis [116]. Research by Wang et al. made it possible to distinguish species referred to as drivers and passengers, with potential driver bacteria including Bacillus, Bradyrhizobium, Methylobacterium, Streptomyces, Intrasporangiaceae and Sinobacterace located on off-tumour sites. The on-tumour sites harboured fourteen species of potential passengers, Fusobacterium, Campylobacter, Streptococcus, Schwartzia, Parvimonas, Dethiosulfatibacter, Selenomonas, Peptostreptococus, Leptotrichia, Granulicatella, Shewanella, Mogibacterium, Eikenella, and Anaerococus. The authors suggest that the driver or passenger species can be used as a biomarker for estimating the risk of initiation of CRC or patients with CRC [117].

4.6. Probiotic Bacteria and Butyrate

Probiotic bacteria maintain favourable intestinal microbiota by preserving the integrity of the intestinal barrier and reducing bacterial translocation. Attempts have been made to use probiotics in the treatment of colorectal cancer and to reduce the risk of its development. The analysis of randomised trials shows that using probiotics, especially *Lactobacilli* and *Bifidobacterium* species, reduces postoperative complications and accelerates patient recovery [118]. Faghfoori et al. proved that some *Bifidobacteria* species produce substances that affect the activation of apoptotic pathways in human colorectal cancer cells (HT-29 and Caco-2 cell lines) but highlighted the need for further research in this area [119].

Low fibre intake is one of the risk factors for CRC development. During fibre fermentation, the gut microbiota produces SCFAs, such as butyrate, the crucial energy source for intestinal epithelial cells. Besides, butyrate over-activates the Wnt signalling pathway and promotes cancer cell apoptosis. It has been shown that CRC patients have fewer butyrate-producing bacteria than healthy controls [120]. Research in mice has demonstrated that butyrate inhibits the proliferation of cancer cells without affecting healthy colon cells. Moreover, butyrate reduces the number of free radicals [117]. The knowledge in this area can contribute to proposing the appropriate composition of new probiotic therapy that can be used in treating CRC [111].

4.7. Iron Deficiency (ID) and Supplementation

Initially, in the course of colorectal cancer, excess iron contributes to its progression. ID occurs as the disease progresses due to frequent bleeding from the gastrointestinal tract, reduces hematopoiesis and decreases the immune response, allowing tumour cells to survive. Furthermore, ID contributes to the modification of macrophage polarisation and Treg populations, favouring a carcinogenic tumour immune microenvironment. These processes often contribute to a poor treatment response and decreased survival after surgery [103]. Studies by Kam et al. proved that intravenous iron administration before surgery increases the level of Hb, lowering the incidence of blood transfusion and reducing the risk of complications associated with the surgery [121]. Another study has shown that postoperative intravenous iron administration to anaemic patients increases the concentration of Hb without postoperative complications [122]. Oral iron supplementation is not recommended due to contributing to the progression of CRC, as it increases the iron concentration in the colon and causes significant gastrointestinal side effects such as diarrhoea, constipation, nausea or abdominal pain [9].

5. Obesity

Obesity, called an epidemic of the 21st century, is a condition of pathological adipose tissue accumulation in the body and is associated with many disorders such as cardiovascular disease, Type 2 diabetes, chronic kidney disease, retinopathy and several cancers [123]. Due to the chronic inflammation accompanying obesity, it can be considered a disease by itself. The devastation of physiological processes caused by obesity also affects iron management, with numerous studies confirming that obese individuals are characterised by a lower iron concentration when compared to people of a normal body weight [123–125].

5.1. Adipose Tissue as an Endocrine Organ

Apart from providing thermal isolation and storing energy, adipose tissue is the largest endocrine organ in terms of mass. Its cell composition, tissue weaving, and ability to produce and secrete adipokines differ depending on its location in the body [126]. Visceral adipose tissue is more metabolically active, better vascularised and has more macrophages than subcutaneous fat [126,127]. Adipocytes secret pro-inflammatory adipokines and cytokines, such as adiponectin, leptin, resistin, PAI-1, TNF-α, IL-6, CRP, and more cytokines involved in insulin resistance [127].

5.2. Chronic Low-Grade Inflammation: An Inherent Consequence of Obesity

Currently, obesity is considered a low-grade systemic, chronic inflammation. The fact that adipose tissue is significantly involved in the production of pro-inflammatory cytokines makes it an indispensable marker of obesity [126].

A high-calorie diet, rich in fat and sugar, causes dysbiotic changes in the intestinal microbiota, becoming much less biodiverse with an increased ratio of Gram-positive Firmicutes to Gram-negative Bacteroidetes, thereby increasing energy production from indigestible carbohydrates [128,129]. Moreover, dysbiosis contributes to an increased "porosity" of the intestinal wall, which leads to expanded contact of microorganisms with the intestinal mucosa, activation of TLRs and inflammation. Increased intestinal permeability also promotes an increase in blood lipopolysaccharides (LPS), which enhance the secretion of IL-6 and CRP [129]. Dysbiosis in obese people also enhances further weight gain. The relationship between microbiota and obesity is two-sided. In studies in mice, the transfer of microbiota from obese to lean animals resulted in an increase in energy absorption from food and weight gain. Conversely, the use of prebiotics in mice to improve microflora decreased gut permeability and overall inflammation. Moreover, as a result of prolonged antibiotic therapy in the treatment of endocarditis, patients gained weight. Bell et al. emphasised the role of dysbiosis in the development of obesity, suggesting that only the combination of genes related to obesity with dysbiosis leads to fat deposition and inflammation [128]. The high ratio of saturated to unsaturated fatty acids in the diet promotes TLR activation and thus the expression of IL-6, TNF-α, and chemokines. Another issue is adipocyte hyperplasia as the need for fat storage increases. Since the possibilities of their growth are limited, the cells may break, promoting inflammation. Visceral adipocytes are even more unstable because they are exposed to sudden changes in pressure when coughing or exercising [127]. Another mechanism observed in obese patients is the infiltration of macrophages into adipose tissue, especially visceral fat. They are formed because of the transformation of monocytes from the bloodstream and their concentration in adipose tissue positively correlates with the amount of fat. Macrophages are believed to be the main source of inflammatory cytokines such as TNF-α, produced by adipose tissue [123,126].

5.3. Iron Deficiency (ID) Is Common in Obese Individuals

ID often accompanies obesity, with many studies showing a negative correlation between body mass index (BMI) and serum iron levels [124,130,131]. One of the mechanisms leading to ID in obesity is the increase in the expression of hepcidin, a protein involved in regulating iron homeostasis [123]. In a Swiss study, overweight children consumed similar amounts of bioavailable iron in their diet as children with a normal BMI but had lower iron levels and higher hepcidin, Il-6, CRP, and leptin concentrations [132]. Although the liver is the main site of hepcidin production, it is also expressed in adipose tissue, the weight of which in obese people can be twenty times greater than that of the liver [133]. Many factors stimulate its expression but apart from iron overload, other mechanisms are also of particular importance in obesity. These include cytokines secreted by adipose tissue, such as IL-6, IL-1, TNF-α, CRP, and leptin [123]. Additionally, a high leptin level stimulates the transcription of hepcidin genes via the JAK/STAT signalling pathway in hepatocytes. In addition, a study in mice showed that a diet that provides the body with excess nutrients could induce endoplasmic reticulum stress, which also increases the secretion of hepcidin [134]. Moreover, it often leads to damage of hepatocytes and further overproduction of hepcidin by the Stat3 and C/EBP pathways [124]. Subsequently, hepcidin is degraded by binding to ferroportin, thereby preventing the transport of iron into the bloodstream, thus reducing its concentration in the body. As a result, iron provided in the diet, despite its bioavailable form, is not absorbed [123–125].

Ferroportin degradation also disrupts iron transport from the macrophages of the spleen, liver and bone marrow into the bloodstream, resulting in an accumulation of iron in the organs mentioned above [135]. Furthermore, hemojuvelin expression occurs in adipose tissue, which stimulates the local expression of hepcidin via the BMP–HJV pathway. Its concentration increases significantly in obese patients affecting general iron homeostasis [134]. Numerous studies show that disturbances in iron balance in obesity also occur independently of hepcidin. The impairment of iron uptake by enterocytes is caused by the disturbed expression of oxidoreductases, despite the deficiency of iron in enterocytes [136,137]. A study in mice showed that a diet rich in fat decreased mRNA levels of duodenal cytochrome B (DcytB) oxidoreductase and hepcidin, leading to the impairment of redox processes necessary for the transport of iron through enterocyte membranes [136]. The above-described mechanism involved in the development of iron deficiency in people with obesity is pictured in Figure 6.

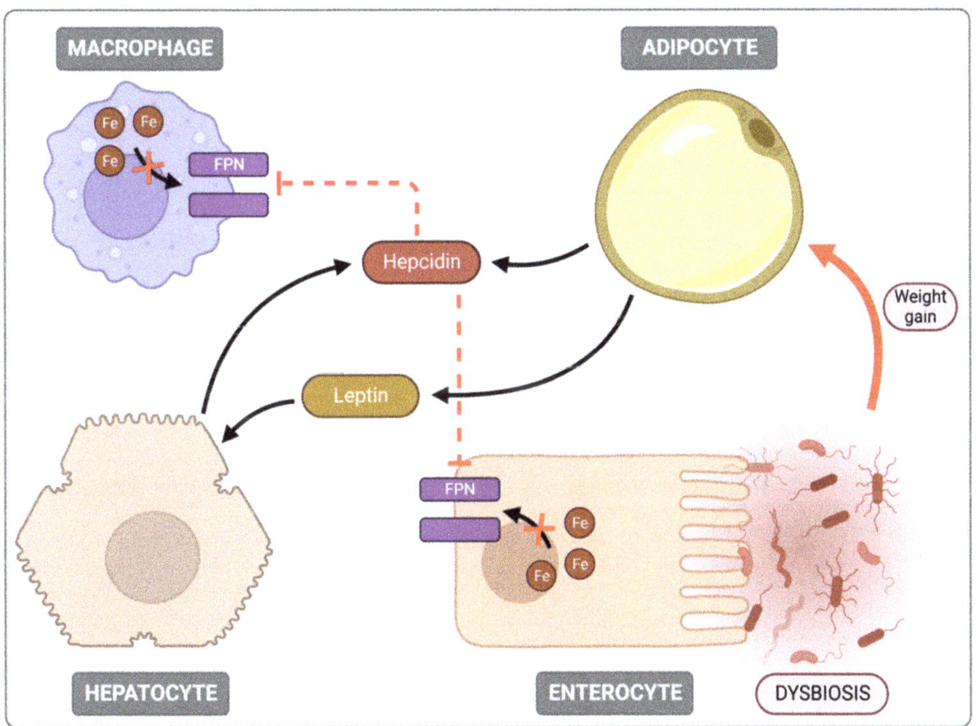

Figure 6. Iron deficiency in people with obesity. FPN—ferroportin, dashed line—inhibition.

5.4. Dysmetabolic Iron Overload Syndrome (DIOS)

The disturbances in iron metabolism in one-third of obese patients take the form of hyperferritinemia. Typical symptoms are increased ferritin levels and normal or slightly increased transferrin saturation. The condition is usually associated with metabolic syndrome (MS) components: high blood pressure, high blood triglycerides, low levels of HDL cholesterol and insulin resistance. The increase in ferritin positively correlates with the number of MS characteristics. Both ID and DIOS are characterised by an increased concentration of hepcidin and a lower expression of ferroportin, suggesting that both disease entities are different symptoms of the same underlying pathology. Factors such as age, gender, and the degree of BMI increase the influence of the form of iron disturbance in obesity. ID occurs in adolescents and adults with morbid obesity, in which iron is lost as a result of menstruation or inflammation of adipose tissue, whereas iron accumulation becomes a problem in postmenopausal women and insulin-resistant men. Contrary to many diseases in the case of obesity, the literature provides little data on iron supplementation, its effectiveness and its impact on the microbiota. This issue seems to be significant considering the scale of the problem [123,138].

5.5. Treatment of Iron Deficiency (ID) in Obesity

There is little well-designed research regarding oral iron supplementation in obese subjects. Nevertheless, Hurrel et al. [135] speculate that it is problematic due to impaired iron absorption caused by general inflammation and rising hepcidin levels. However, Rabindrakumar et al. [139] reported an increase in iron concentration in the intestines accompanying daily supplementation which may stimulate hepcidin expression. Some studies suggest that weight loss reduces inflammation and hepcidin levels and improves iron absorption from the diet. There was also an increase in transferrin saturation and the restoration of normal iron homeostasis [123,135,140]. Due to the proven relationship between dysbiosis, obesity, and iron status, one of the methods of treating ID in obese patients has become probiotic therapy. Several studies have been conducted with promising results. Iron management has been proven to improve with single strain therapy through various mechanisms. A 2020 study described the function of probiotics as iron carriers, demonstrating that probiotics convert inaccessible forms of iron into absorbable forms, producing metabolites that stimulate iron absorption and reduce the Fe accumulation in the liver [141]. A multi-strain supplementation study showed reductions in hair iron and blood FAM levels leading to the conclusion that Fe shifted from hair to bone marrow. However, too few volunteers took part in the study to confirm the effectiveness of the therapy in this case [125]. The described results show a relationship between the microbiota status in obese people and iron metabolism.

6. Summary

The side effects of iron supplementation are indisputable and often irreversible, whether iron is delivered to the gastrointestinal tract or the blood (Figure 7).

In the blood, only protein-bound iron is safe. If the transferrin and ferritin binding capacity is exceeded, the serum's free iron concentration, the labile Fe pool increases, generating reactive oxygen species (ROS) [31,33]. Subsequently, lipid and amino acid peroxidation, enzyme denaturation, polysaccharide depolymerisation and DNA damage lead to endothelial cell dysfunction [10,18,39–43]. Reactive oxygen species are also critical for activating the physiological signalling pathway NF-κB and producing pro-inflammatory cytokines such as IFN-γ, TNF-α, and iNOS. All the processes mentioned lead to systemic inflammation.

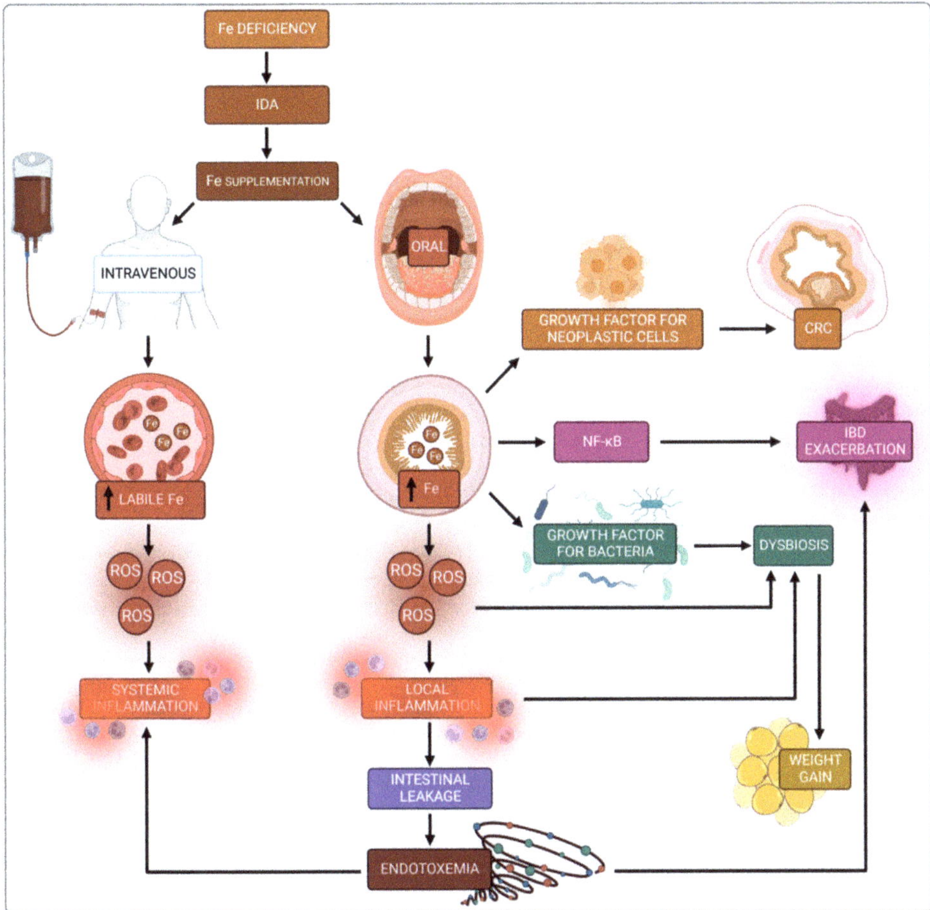

Figure 7. Impact of oral and intravenous iron supplementation on development and course in chosen clinical conditions. IDA—iron deficiency anemia; CRC—colorectal cancer; IBD—inflammatory bowel disease; NF-kβ—nuclear factor kappa-light-chain-enhancer of activated B cells; ROS reactive oxygen species.

Oral iron supplementation activates ROS production simultaneously in the gut lumen and the enterocytes. It is the beginning of the local inflammatory process leading to damage to the intestinal wall integrity and resulting in a leaky gut syndrome. The increased permeability of the intestinal wall leads to the leakage of metabolites and bacterial toxins into the blood resulting in endotoxaemia and contributing to systemic inflammation. In the case of IBD, it worsens clinical symptoms [47,88].

Iron in the intestinal lumen directly activates the NF-κB pathway and stimulates the expression of pro-inflammatory cytokines and iNOS. This contributes to the deterioration of symptoms and is the reason why oral iron administration in inflammatory bowel disease (IBD) is not recommended [82,83].

The excess unabsorbed iron supplemented orally passes through the colon and becomes an essential growth factor for nearly all pathogenic bacteria, fungi and protozoa, as well as for all neoplastic cells [32]. It reduces the abundance of beneficial microbes simultaneously with an increased abundance of deleterious microbes, leading to dysbiosis [66,68].

The numerical advantage of unfavourable bacterial strains leads to increased energy absorption from the food and increased body weight, which is particularly detrimental for obese individuals [128].

7. Conclusions

Despite severe side effects, supplementation of iron deficiencies in the case of anaemia is necessary. Well-planned research, specific to each patient category and disease, is needed to find measures and methods to optimise iron treatment and reduce adverse effects.

Author Contributions: Conceptualization, I.J.M.; investigation, J.W.-G, A.N., D.D., M.B., P.G. and K.S.; writing—original draft preparation, I.J.M. and E.M.; writing—review and editing, J.B.-W., J.W.-G., L.N., J.W. and E.M.; visualization, I.J.M.; supervision, E.M. and J.W.; project administration, I.J.M. and E.M. All authors have read and agreed to the published version of the manuscript.

Funding: This research was founded by Physiology Department Poznan University of Medical Sciences.

References

1. GBD 2016 Disease and Injury Incidence and Prevalence Collaborators. Global, Regional, and National Incidence, Prevalence, and Years Lived with Disability for 328 Diseases and Injuries for 195 Countries, 1990–2016: A Systematic Analysis for the Global Burden of Disease Study 2016. *Lancet* **2017**, *390*, 1211–1259. [CrossRef]
2. GHO | By Category | Anaemia in Children <5 Years—Estimates by WHO Region. Available online: https://apps.who.int/gho/data/view.main.ANEMIACHILDRENREGv?lang=en (accessed on 20 July 2022).
3. GHO | By Category | Prevalence of Anaemia in Women. Available online: https://apps.who.int/gho/data/node.main.ANAEMIAWOMEN?lang=en (accessed on 20 July 2022).
4. Global Nutrition Targets 2025: Policy Brief Series. Available online: https://www.who.int/publications-detail-redirect/WHO-NMH-NHD-14.2 (accessed on 20 July 2022).
5. Cappellini, M.D.; Musallam, K.M.; Taher, A.T. Iron Deficiency Anaemia Revisited. *J. Intern. Med.* **2020**, *287*, 153–170. [CrossRef] [PubMed]
6. Lopez, A.; Cacoub, P.; Macdougall, I.C.; Peyrin-Biroulet, L. Iron Deficiency Anaemia. *Lancet* **2016**, *387*, 907–916. [CrossRef]
7. Aspuru, K.; Villa, C.; Bermejo, F.; Herrero, P.; López, S.G. Optimal Management of Iron Deficiency Anemia Due to Poor Dietary Intake. *Int. J. Gen. Med.* **2011**, *4*, 741–750. [CrossRef] [PubMed]
8. Pasricha, S.-R.; Tye-Din, J.; Muckenthaler, M.U.; Swinkels, D.W. Iron Deficiency. *Lancet* **2021**, *397*, 233–248. [CrossRef]
9. Tolkien, Z.; Stecher, L.; Mander, A.P.; Pereira, D.I.A.; Powell, J.J. Ferrous Sulfate Supplementation Causes Significant Gastrointestinal Side-Effects in Adults: A Systematic Review and Meta-Analysis. *PLoS ONE* **2015**, *10*, e0117383. [CrossRef]
10. Duck, K.A.; Connor, J.R. Iron Uptake and Transport across Physiological Barriers. *Biometals* **2016**, *29*, 573–591. [CrossRef]
11. Abbaspour, N.; Hurrell, R.; Kelishadi, R. Review on Iron and Its Importance for Human Health. *J. Res. Med. Sci. Off. J. Isfahan Univ. Med. Sci.* **2014**, *19*, 164–174.
12. Gulec, S.; Anderson, G.J.; Collins, J.F. Mechanistic and Regulatory Aspects of Intestinal Iron Absorption. *Am. J. Physiol.-Gastrointest. Liver Physiol.* **2014**, *307*, G397–G409. [CrossRef]
13. Hurrell, R.; Egli, I. Iron Bioavailability and Dietary Reference Values. *Am. J. Clin. Nutr.* **2010**, *91*, 1461S–1467S. [CrossRef]
14. Lane, D.J.R.; Bae, D.-H.; Merlot, A.M.; Sahni, S.; Richardson, D.R. Duodenal Cytochrome b (DCYTB) in Iron Metabolism: An Update on Function and Regulation. *Nutrients* **2015**, *7*, 2274–2296. [CrossRef]
15. Anderson, G.J.; Frazer, D.M. Current Understanding of Iron Homeostasis. *Am. J. Clin. Nutr.* **2017**, *106*, 1559S–1566S. [CrossRef]
16. Fuqua, B.K.; Vulpe, C.D.; Anderson, G.J. Intestinal Iron Absorption. *J. Trace Elem. Med. Biol.* **2012**, *26*, 115–119. [CrossRef]
17. Fang, S.; Zhuo, Z.; Yu, X.; Wang, H.; Feng, J. Oral Administration of Liquid Iron Preparation Containing Excess Iron Induces Intestine and Liver Injury, Impairs Intestinal Barrier Function and Alters the Gut Microbiota in Rats. *J. Trace Elem. Med. Biol.* **2018**, *47*, 12–20. [CrossRef]
18. Galaris, D.; Pantopoulos, K. Oxidative Stress and Iron Homeostasis: Mechanistic and Health Aspects. *Crit. Rev. Clin. Lab. Sci.* **2008**, *45*, 1–23. [CrossRef]
19. Mackenzie, B.; Garrick, M.D. Iron Imports. II. Iron Uptake at the Apical Membrane in the Intestine. *Am. J. Physiol.-Gastrointest. Liver Physiol.* **2005**, *289*, G981–G986. [CrossRef]
20. West, A.R.; Oates, P.S. Mechanisms of Heme Iron Absorption: Current Questions and Controversies. *World J. Gastroenterol. WJG* **2008**, *14*, 4101–4110. [CrossRef]
21. Shawki, A.; Knight, P.B.; Maliken, B.D.; Niespodzany, E.J.; Mackenzie, B. H+-Coupled Divalent Metal-Ion Transporter-1: Functional Properties, Physiological Roles and Therapeutics. *Curr. Top. Membr.* **2012**, *70*, 169–214. [CrossRef]
22. Collins, J.F.; Prohaska, J.R.; Knutson, M.D. Metabolic Crossroads of Iron and Copper. *Nutr. Rev.* **2010**, *68*, 133–147. [CrossRef]

23. Shawki, A.; Anthony, S.R.; Nose, Y.; Engevik, M.A.; Niespodzany, E.J.; Barrientos, T.; Öhrvik, H.; Worrell, R.T.; Thiele, D.J.; Mackenzie, B. Intestinal DMT1 Is Critical for Iron Absorption in the Mouse but Is Not Required for the Absorption of Copper or Manganese. *Am. J. Physiol.-Gastrointest. Liver Physiol.* **2015**, *309*, G635–G647. [CrossRef]
24. Mims, M.P.; Guan, Y.; Pospisilova, D.; Priwitzerova, M.; Indrak, K.; Ponka, P.; Divoky, V.; Prchal, J.T. Identification of a Human Mutation of DMT1 in a Patient with Microcytic Anemia and Iron Overload. *Blood* **2005**, *105*, 1337–1342. [CrossRef] [PubMed]
25. Ems, T.; St Lucia, K.; Huecker, M.R. Biochemistry, Iron Absorption. In *StatPearls*; StatPearls Publishing: Treasure Island, FL, USA, 2022.
26. McKie, A.T. The Role of Dcytb in Iron Metabolism: An Update. *Biochem. Soc. Trans.* **2008**, *36*, 1239–1241. [CrossRef] [PubMed]
27. Moreira, A.C.; Mesquita, G.; Gomes, M.S. Ferritin: An Inflammatory Player Keeping Iron at the Core of Pathogen-Host Interactions. *Microorganisms* **2020**, *8*, 589. [CrossRef] [PubMed]
28. Farina, M.; Avila, D.S.; Da Rocha, J.B.T.; Aschner, M. Metals, Oxidative Stress and Neurodegeneration: A Focus on Iron, Manganese and Mercury. *Neurochem. Int.* **2013**, *62*, 575–594. [CrossRef]
29. Ganz, T. Hepcidin and Iron Regulation, 10 Years Later. *Blood* **2011**, *117*, 4425–4433. [CrossRef]
30. Wilkinson, N.; Pantopoulos, K. The IRP/IRE System in Vivo: Insights from Mouse Models. *Front. Pharmacol.* **2014**, *5*, 176. [CrossRef]
31. Galaris, D.; Barbouti, A.; Pantopoulos, K. Iron Homeostasis and Oxidative Stress: An Intimate Relationship. *Biochim. Biophys. Acta Mol. Cell Res.* **2019**, *1866*, 118535. [CrossRef]
32. Fleming, R.E.; Ponka, P. Iron Overload in Human Disease. *N. Engl. J. Med.* **2012**, *366*, 348–359. [CrossRef]
33. Jomova, K.; Valko, M. Advances in Metal-Induced Oxidative Stress and Human Disease. *Toxicology* **2011**, *283*, 65–87. [CrossRef]
34. Lee, J.K.; Ha, J.-H.; Collins, J.F. Dietary Iron Intake in Excess of Requirements Impairs Intestinal Copper Absorption in Sprague Dawley Rat Dams, Causing Copper Deficiency in Suckling Pups. *Biomedicines* **2021**, *9*, 338. [CrossRef]
35. Ha, J.-H.; Doguer, C.; Flores, S.R.L.; Wang, T.; Collins, J.F. Progressive Increases in Dietary Iron Are Associated with the Emergence of Pathologic Disturbances of Copper Homeostasis in Growing Rats. *J. Nutr.* **2018**, *148*, 373–378. [CrossRef]
36. Ha, J.-H.; Doguer, C.; Collins, J.F. Consumption of a High-Iron Diet Disrupts Homeostatic Regulation of Intestinal Copper Absorption in Adolescent Mice. *Am. J. Physiol.-Gastrointest. Liver Physiol.* **2017**, *313*, G353–G360. [CrossRef]
37. Ha, J.-H.; Doguer, C.; Wang, X.; Flores, S.R.; Collins, J.F. High-Iron Consumption Impairs Growth and Causes Copper-Deficiency Anemia in Weanling Sprague-Dawley Rats. *PLoS ONE* **2016**, *11*, e0161033. [CrossRef]
38. Pouillevet, H.; Soetart, N.; Boucher, D.; Wedlarski, R.; Jaillardon, L. Inflammatory and Oxidative Status in European Captive Black Rhinoceroses: A Link with Iron Overload Disorder? *PLoS ONE* **2020**, *15*, e0231514. [CrossRef]
39. Nakamura, T.; Naguro, I.; Ichijo, H. Iron Homeostasis and Iron-Regulated ROS in Cell Death, Senescence and Human Diseases. *Biochim. Biophys. Acta Gen. Subj.* **2019**, *1863*, 1398–1409. [CrossRef]
40. Patel, M.; Ramavataram, D.V.S.S. Non Transferrin Bound Iron: Nature, Manifestations and Analytical Approaches for Estimation. *Indian J. Clin. Biochem.* **2012**, *27*, 322–332. [CrossRef]
41. Cabantchik, Z.I.; Sohn, Y.S.; Breuer, W.; Espósito, B.P. The Molecular and Cellular Basis of Iron Toxicity in Iron Overload (IO) Disorders. Diagnostic and Therapeutic Approaches. *Thalass. Rep.* **2013**, *3*, e3. [CrossRef]
42. Pietrangelo, A. Iron and the Liver. *Liver Int. Off. J. Int. Assoc. Study Liver* **2016**, *36* (Suppl 1), 116–123. [CrossRef]
43. D'Mello, S.R.; Kindy, M.C. Overdosing on Iron: Elevated Iron and Degenerative Brain Disorders. *Exp. Biol. Med. Maywood NJ* **2020**, *245*, 1444–1473. [CrossRef]
44. Moretti, D.; Goede, J.S.; Zeder, C.; Jiskra, M.; Chatzinakou, V.; Tjalsma, H.; Melse-Boonstra, A.; Brittenham, G.; Swinkels, D.W.; Zimmermann, M.B. Oral Iron Supplements Increase Hepcidin and Decrease Iron Absorption from Daily or Twice-Daily Doses in Iron-Depleted Young Women. *Blood* **2015**, *126*, 1981–1989. [CrossRef]
45. Stoffel, N.U.; Cercamondi, C.I.; Brittenham, G.; Zeder, C.; Geurts-Moespot, A.J.; Swinkels, D.W.; Moretti, D.; Zimmermann, M.B. Iron Absorption from Oral Iron Supplements given on Consecutive versus Alternate Days and as Single Morning Doses versus Twice-Daily Split Dosing in Iron-Depleted Women: Two Open-Label, Randomised Controlled Trials. *Lancet Haematol.* **2017**, *4*, e524–e533. [CrossRef]
46. Zimmermann, M.B.; Chassard, C.; Rohner, F.; N'Goran, E.K.; Nindjin, C.; Dostal, A.; Utzinger, J.; Ghattas, H.; Lacroix, C.; Hurrell, R.F. The Effects of Iron Fortification on the Gut Microbiota in African Children: A Randomized Controlled Trial in Côte d'Ivoire. *Am. J. Clin. Nutr.* **2010**, *92*, 1406–1415. [CrossRef]
47. Qi, X.; Zhang, Y.; Guo, H.; Hai, Y.; Luo, Y.; Yue, T. Mechanism and Intervention Measures of Iron Side Effects on the Intestine. *Crit. Rev. Food Sci. Nutr.* **2020**, *60*, 2113–2125. [CrossRef]
48. Gaschler, M.M.; Stockwell, B.R. Lipid Peroxidation in Cell Death. *Biochem. Biophys. Res. Commun.* **2017**, *482*, 419–425. [CrossRef]
49. Gou, Z.Y.; Li, L.; Fan, Q.L.; Lin, X.J.; Jiang, Z.Y.; Zheng, C.T.; Ding, F.Y.; Jiang, S.Q. Effects of Oxidative Stress Induced by High Dosage of Dietary Iron Ingested on Intestinal Damage and Caecal Microbiota in Chinese Yellow Broilers. *J. Anim. Physiol. Anim. Nutr.* **2018**, *102*, 924–932. [CrossRef]
50. Yang, Q.; Liang, Q.; Balakrishnan, B.; Belobrajdic, D.P.; Feng, Q.-J.; Zhang, W. Role of Dietary Nutrients in the Modulation of Gut Microbiota: A Narrative Review. *Nutrients* **2020**, *12*, 381. [CrossRef]
51. Seyoum, Y.; Baye, K.; Humblot, C. Iron Homeostasis in Host and Gut Bacteria—A Complex Interrelationship. *Gut Microbes* **2021**, *13*, 1874855. [CrossRef]

52. Sousa Gerós, A.; Simmons, A.; Drakesmith, H.; Aulicino, A.; Frost, J.N. The Battle for Iron in Enteric Infections. *Immunology* **2020**, *161*, 186–199. [CrossRef]
53. Malesza, I.J.; Malesza, M.; Walkowiak, J.; Mussin, N.; Walkowiak, D.; Aringazina, R.; Bartkowiak-Wieczorek, J.; Mądry, E. High-Fat, Western-Style Diet, Systemic Inflammation, and Gut Microbiota: A Narrative Review. *Cells* **2021**, *10*, 3164. [CrossRef] [PubMed]
54. Richard, K.L.; Kelley, B.R.; Johnson, J.G. Heme Uptake and Utilization by Gram-Negative Bacterial Pathogens. *Front. Cell. Infect. Microbiol.* **2019**, *9*, 81. [CrossRef] [PubMed]
55. Nairz, M.; Ferring-Appel, D.; Casarrubea, D.; Sonnweber, T.; Viatte, L.; Schroll, A.; Haschka, D.; Fang, F.C.; Hentze, M.W.; Weiss, G.; et al. Iron Regulatory Proteins Mediate Host Resistance to Salmonella Infection. *Cell Host Microbe* **2015**, *18*, 254–261. [CrossRef] [PubMed]
56. Drakesmith, H.; Prentice, A.M. Hepcidin and the Iron-Infection Axis. *Science* **2012**, *338*, 768–772. [CrossRef] [PubMed]
57. Benjamin, W.H.; Hall, P.; Briles, D.E. A HemA Mutation Renders Salmonella Typhimurium Avirulent in Mice, yet Capable of Eliciting Protection against Intravenous Infection with S. Typhimurium. *Microb. Pathog.* **1991**, *11*, 289–295. [CrossRef]
58. Sarowska, J.; Futoma-Koloch, B.; Jama-Kmiecik, A.; Frej-Madrzak, M.; Ksiazczyk, M.; Bugla-Ploskonska, G.; Choroszy-Krol, I. Virulence Factors, Prevalence and Potential Transmission of Extraintestinal Pathogenic Escherichia Coli Isolated from Different Sources: Recent Reports—PMC. *Gut Pathog.* **2019**, *11*, 10. Available online: https://www.ncbi.nlm.nih.gov/pmc/articles/PMC6383261/ (accessed on 21 July 2022). [CrossRef]
59. Paganini, D.; Zimmermann, M.B. The Effects of Iron Fortification and Supplementation on the Gut Microbiome and Diarrhea in Infants and Children: A Review. *Am. J. Clin. Nutr.* **2017**, *106*, 1688S–1693S. [CrossRef]
60. Schwartz, A.J.; Das, N.K.; Ramakrishnan, S.K.; Jain, C.; Jurkovic, M.T.; Wu, J.; Nemeth, E.; Lakhal-Littleton, S.; Colacino, J.A.; Shah, Y.M. Hepatic Hepcidin/Intestinal HIF-2α Axis Maintains Iron Absorption during Iron Deficiency and Overload. *J. Clin. Investig.* **2018**, *129*, 336–348. [CrossRef]
61. Xu, M.-M.; Wang, J.; Xie, J.-X. Regulation of Iron Metabolism by Hypoxia-Inducible Factors. *Sheng Li Xue Bao* **2017**, *69*, 598–610.
62. Mastrogiannaki, M.; Matak, P.; Peyssonnaux, C. The Gut in Iron Homeostasis: Role of HIF-2 under Normal and Pathological Conditions. *Blood* **2013**, *122*, 885–892. [CrossRef]
63. Das, N.K.; Schwartz, A.J.; Barthel, G.; Inohara, N.; Liu, Q.; Sankar, A.; Hill, D.; Ma, X.; Lamberg, O.; Schnizlein, M.K.; et al. Microbial Metabolite Signaling Is Required for Systemic Iron Homeostasis. *Cell Metab.* **2020**, *31*, 115–130.e6. [CrossRef]
64. Axling, U.; Önning, G.; Combs, M.A.; Bogale, A.; Högström, M.; Svensson, M. The Effect of Lactobacillus Plantarum 299v on Iron Status and Physical Performance in Female Iron-Deficient Athletes: A Randomized Controlled Trial. *Nutrients* **2020**, *12*, 1279. [CrossRef]
65. Sandberg, A.-S.; Önning, G.; Engström, N.; Scheers, N. Iron Supplements Containing Lactobacillus Plantarum 299v Increase Ferric Iron and Up-Regulate the Ferric Reductase DCYTB in Human Caco-2/HT29 MTX Co-Cultures. *Nutrients* **2018**, *10*, 1949. [CrossRef]
66. Yilmaz, B.; Li, H. Gut Microbiota and Iron: The Crucial Actors in Health and Disease. *Pharmaceuticals* **2018**, *11*, 98. [CrossRef]
67. Kim, S.Y.; An, S.; Park, D.K.; Kwon, K.A.; Kim, K.O.; Chung, J.-W.; Kim, J.H.; Kim, Y.J. Efficacy of Iron Supplementation in Patients with Inflammatory Bowel Disease Treated with Anti-Tumor Necrosis Factor-Alpha Agents. *Ther. Adv. Gastroenterol.* **2020**, *13*, 1756284820961302. [CrossRef]
68. Tang, M.; Frank, D.N.; Hendricks, A.E.; Ir, D.; Esamai, F.; Liechty, E.; Hambidge, K.M.; Krebs, N.F. Iron in Micronutrient Powder Promotes an Unfavorable Gut Microbiota in Kenyan Infants. *Nutrients* **2017**, *9*, 776. [CrossRef]
69. Jaeggi, T.; Kortman, G.A.M.; Moretti, D.; Chassard, C.; Holding, P.; Dostal, A.; Boekhorst, J.; Timmerman, H.M.; Swinkels, D.W.; Tjalsma, H.; et al. Iron Fortification Adversely Affects the Gut Microbiome, Increases Pathogen Abundance and Induces Intestinal Inflammation in Kenyan Infants. *Gut* **2015**, *64*, 731–742. [CrossRef]
70. Dostal, A.; Lacroix, C.; Pham, V.T.; Zimmermann, M.B.; Del'homme, C.; Bernalier-Donadille, A.; Chassard, C. Iron Supplementation Promotes Gut Microbiota Metabolic Activity but Not Colitis Markers in Human Gut Microbiota-Associated Rats. *Br. J. Nutr.* **2014**, *111*, 2135–2145. [CrossRef]
71. Dostal, A.; Baumgartner, J.; Riesen, N.; Chassard, C.; Smuts, C.M.; Zimmermann, M.B.; Lacroix, C. Effects of Iron Supplementation on Dominant Bacterial Groups in the Gut, Faecal SCFA and Gut Inflammation: A Randomised, Placebo-Controlled Intervention Trial in South African Children. *Br. J. Nutr.* **2014**, *112*, 547–556. [CrossRef]
72. Fischer, J.A.; Pei, L.X.; Goldfarb, D.M.; Albert, A.; Elango, R.; Kroeun, H.; Karakochuk, C.D. Is Untargeted Iron Supplementation Harmful When Iron Deficiency Is Not the Major Cause of Anaemia? Study Protocol for a Double-Blind, Randomised Controlled Trial among Non-Pregnant Cambodian Women. *BMJ Open* **2020**, *10*, e037232. [CrossRef]
73. Younis, N.; Zarif, R.; Mahfouz, R. Inflammatory Bowel Disease: Between Genetics and Microbiota. *Mol. Biol. Rep.* **2020**, *47*, 3053–3063. [CrossRef]
74. Frolkis, A.D.; Dykeman, J.; Negrón, M.E.; Debruyn, J.; Jette, N.; Fiest, K.M.; Frolkis, T.; Barkema, H.W.; Rioux, K.P.; Panaccione, R.; et al. Risk of Surgery for Inflammatory Bowel Diseases Has Decreased over Time: A Systematic Review and Meta-Analysis of Population-Based Studies. *Gastroenterology* **2013**, *145*, 996–1006. [CrossRef]
75. Nielsen, O.H.; Soendergaard, C.; Vikner, M.E.; Weiss, G. Rational Management of Iron-Deficiency Anaemia in Inflammatory Bowel Disease. *Nutrients* **2018**, *10*, 82. [CrossRef] [PubMed]

76. Mahadea, D.; Adamczewska, E.; Ratajczak, A.E.; Rychter, A.M.; Zawada, A.; Eder, P.; Dobrowolska, A.; Krela-Kaźmierczak, I. Iron Deficiency Anemia in Inflammatory Bowel Diseases-A Narrative Review. *Nutrients* **2021**, *13*, 4008. [CrossRef] [PubMed]
77. Weiss, G.; Goodnough, L.T. Anemia of Chronic Disease. *N. Engl. J. Med.* **2005**, *352*, 1011–1023. [CrossRef] [PubMed]
78. Iqbal, T.; Stein, J.; Sharma, N.; Kulnigg-Dabsch, S.; Vel, S.; Gasche, C. Clinical Significance of C-Reactive Protein Levels in Predicting Responsiveness to Iron Therapy in Patients with Inflammatory Bowel Disease and Iron Deficiency Anemia. *Dig. Dis. Sci.* **2015**, *60*, 1375–1381. [CrossRef]
79. European Crohn's and Colitis Organisatio—ECCO—Industry Exhibition. Available online: https://www.ecco-ibd.eu/exhibit-sponsor-2015/industry-exhibition-2015.html (accessed on 21 July 2022).
80. D'Amico, F.; Peyrin-Biroulet, L.; Danese, S. Oral Iron for IBD Patients: Lessons Learned at Time of COVID-19 Pandemic. *J. Clin. Med.* **2020**, *9*, E1536. [CrossRef]
81. Niepel, D.; Klag, T.; Malek, N.P.; Wehkamp, J. Practical Guidance for the Management of Iron Deficiency in Patients with Inflammatory Bowel Disease. *Ther. Adv. Gastroenterol.* **2018**, *11*, 1756284818769074. [CrossRef]
82. Stein, J.; Walper, A.; Klemm, W.; Farrag, K.; Aksan, A.; Dignass, A. Safety and Efficacy of Intravenous Iron Isomaltoside for Correction of Anaemia in Patients with Inflammatory Bowel Disease in Everyday Clinical Practice. *Scand. J. Gastroenterol.* **2018**, *53*, 1059–1065. [CrossRef]
83. Stein, J.; Dignass, A.U. Management of Iron Deficiency Anemia in Inflammatory Bowel Disease—A Practical Approach. *Ann. Gastroenterol.* **2013**, *26*, 104–113.
84. Carrier, J.C.; Aghdassi, E.; Jeejeebhoy, K.; Allard, J.P. Exacerbation of Dextran Sulfate Sodium-Induced Colitis by Dietary Iron Supplementation: Role of NF-KappaB. *Int. J. Colorectal Dis.* **2006**, *21*, 381–387. [CrossRef]
85. Werner, T.; Wagner, S.J.; Martínez, I.; Walter, J.; Chang, J.-S.; Clavel, T.; Kisling, S.; Schuemann, K.; Haller, D. Depletion of Luminal Iron Alters the Gut Microbiota and Prevents Crohn's Disease-like Ileitis. *Gut* **2011**, *60*, 325–333. [CrossRef]
86. Piechota-Polanczyk, A.; Fichna, J. Review Article: The Role of Oxidative Stress in Pathogenesis and Treatment of Inflammatory Bowel Diseases. *Naunyn. Schmiedebergs Arch. Pharmacol.* **2014**, *387*, 605–620. [CrossRef]
87. Yu, L.C.-H.; Wei, S.-C.; Ni, Y.-H. Impact of Microbiota in Colorectal Carcinogenesis: Lessons from Experimental Models. *Intest. Res.* **2018**, *16*, 346–357. [CrossRef]
88. Suzuki, T. Regulation of the Intestinal Barrier by Nutrients: The Role of Tight Junctions. *Anim. Sci. J. Nihon Chikusan Gakkaiho* **2020**, *91*, e13357. [CrossRef]
89. Li, Y.; Hansen, S.L.; Borst, L.B.; Spears, J.W.; Moeser, A.J. Dietary Iron Deficiency and Oversupplementation Increase Intestinal Permeability, Ion Transport, and Inflammation in Pigs. *J. Nutr.* **2016**, *146*, 1499–1505. [CrossRef]
90. Rohr, M.W.; Narasimhulu, C.A.; Rudeski-Rohr, T.A.; Parthasarathy, S. Negative Effects of a High-Fat Diet on Intestinal Permeability: A Review. *Adv. Nutr.* **2020**, *11*, 77–91. [CrossRef]
91. Ferruzza, S.; Scarino, M.L.; Gambling, L.; Natella, F.; Sambuy, Y. Biphasic Effect of Iron on Human Intestinal Caco-2 Cells: Early Effect on Tight Junction Permeability with Delayed Onset of Oxidative Cytotoxic Damage. *Cell. Mol. Biol. Noisy-Gd. Fr.* **2003**, *49*, 89–99.
92. Constante, M.; Fragoso, G.; Calvé, A.; Samba-Mondonga, M.; Santos, M.M. Dietary Heme Induces Gut Dysbiosis, Aggravates Colitis, and Potentiates the Development of Adenomas in Mice. *Front. Microbiol.* **2017**, *8*, 1809. [CrossRef]
93. Lewis, G.; Wang, B.; Shafiei Jahani, P.; Hurrell, B.P.; Banie, H.; Aleman Muench, G.R.; Maazi, H.; Helou, D.G.; Howard, E.; Galle-Treger, L.; et al. Dietary Fiber-Induced Microbial Short Chain Fatty Acids Suppress ILC2 Dependent Airway Inflammation. *Front. Immunol.* **2019**, *10*, 2051. [CrossRef]
94. Dostal, A.; Chassard, C.; Hilty, F.M.; Zimmermann, M.B.; Jaeggi, T.; Rossi, S.; Lacroix, C. Iron Depletion and Repletion with Ferrous Sulfate or Electrolytic Iron Modifies the Composition and Metabolic Activity of the Gut Microbiota in Rats. *J. Nutr.* **2012**, *142*, 271–277. [CrossRef]
95. Botta, A.; Barra, N.G.; Lam, N.H.; Chow, S.; Pantopoulos, K.; Schertzer, J.D.; Sweeney, G. Iron Reshapes the Gut Microbiome and Host Metabolism. *J. Lipid Atheroscler.* **2021**, *10*, 160–183. [CrossRef]
96. Dostal, A.; Fehlbaum, S.; Chassard, C.; Zimmermann, M.B.; Lacroix, C. Low Iron Availability in Continuous in Vitro Colonic Fermentations Induces Strong Dysbiosis of the Child Gut Microbial Consortium and a Decrease in Main Metabolites. *FEMS Microbiol. Ecol.* **2013**, *83*, 161–175. [CrossRef] [PubMed]
97. Mahalhal, A.; Williams, J.M.; Johnson, S.; Ellaby, N.; Duckworth, C.A.; Burkitt, M.D.; Liu, X.; Hold, G.L.; Campbell, B.J.; Pritchard, D.M.; et al. Oral Iron Exacerbates Colitis and Influences the Intestinal Microbiome. *PLoS ONE* **2018**, *13*, e0202460. [CrossRef] [PubMed]
98. Buret, A.G.; Motta, J.-P.; Allain, T.; Ferraz, J.; Wallace, J.L. Pathobiont Release from Dysbiotic Gut Microbiota Biofilms in Intestinal Inflammatory Diseases: A Role for Iron? *J. Biomed. Sci.* **2019**, *26*, 1. [CrossRef] [PubMed]
99. Rehman, A.; Rausch, P.; Wang, J.; Skieceviciene, J.; Kiudelis, G.; Bhagalia, K.; Amarapurkar, D.; Kupcinskas, L.; Schreiber, S.; Rosenstiel, P.; et al. Geographical Patterns of the Standing and Active Human Gut Microbiome in Health and IBD. *Gut* **2016**, *65*, 238–248. [CrossRef]
100. Halfvarson, J.; Brislawn, C.J.; Lamendella, R.; Vázquez-Baeza, Y.; Walters, W.A.; Bramer, L.M.; D'Amato, M.; Bonfiglio, F.; McDonald, D.; Gonzalez, A.; et al. Dynamics of the Human Gut Microbiome in Inflammatory Bowel Disease. *Nat. Microbiol.* **2017**, *2*, 17004. [CrossRef]

101. Lee, J.-Y.; Cevallos, S.A.; Byndloss, M.X.; Tiffany, C.R.; Olsan, E.E.; Butler, B.P.; Young, B.M.; Rogers, A.W.L.; Nguyen, H.; Kim, K.; et al. High-Fat Diet and Antibiotics Cooperatively Impair Mitochondrial Bioenergetics to Trigger Dysbiosis That Exacerbates Pre-Inflammatory Bowel Disease. *Cell Host Microbe* **2020**, *28*, 273–284.e6. [CrossRef]
102. Global Cancer Statistics 2020: GLOBOCAN Estimates of Incidence and Mortality Worldwide for 36 Cancers in 185 Countries—Sung—2021—CA: A Cancer Journal for Clinicians—Wiley Online Library. Available online: https://acsjournals.onlinelibrary.wiley.com/doi/10.3322/caac.21660 (accessed on 21 July 2022).
103. Phipps, O.; Brookes, M.J.; Al-Hassi, H.O. Iron Deficiency, Immunology, and Colorectal Cancer. *Nutr. Rev.* **2021**, *79*, 88–97. [CrossRef]
104. Ijssennagger, N.; Belzer, C.; Hooiveld, G.J.; Dekker, J.; Van Mil, S.W.C.; Müller, M.; Kleerebezem, M.; Van der Meer, R. Gut Microbiota Facilitates Dietary Heme-Induced Epithelial Hyperproliferation by Opening the Mucus Barrier in Colon. *Proc. Natl. Acad. Sci. USA* **2015**, *112*, 10038–10043. [CrossRef]
105. Seiwert, N.; Heylmann, D.; Hasselwander, S.; Fahrer, J. Mechanism of Colorectal Carcinogenesis Triggered by Heme Iron from Red Meat. *Biochim. Biophys. Acta BBA-Rev. Cancer* **2020**, *1873*, 188334. [CrossRef]
106. Bastide, N.M.; Chenni, F.; Audebert, M.; Santarelli, R.L.; Taché, S.; Naud, N.; Baradat, M.; Jouanin, I.; Surya, R.; Hobbs, D.A.; et al. A Central Role for Heme Iron in Colon Carcinogenesis Associated with Red Meat Intake. *Cancer Res.* **2015**, *75*, 870–879. [CrossRef]
107. Wandersman, C.; Stojiljkovic, I. Bacterial Heme Sources: The Role of Heme, Hemoprotein Receptors and Hemophores. *Curr. Opin. Microbiol.* **2000**, *3*, 215–220. [CrossRef]
108. Candela, M.; Turroni, S.; Biagi, E.; Carbonero, F.; Rampelli, S.; Fiorentini, C.; Brigidi, P. Inflammation and Colorectal Cancer, When Microbiota-Host Mutualism Breaks. *World J. Gastroenterol.* **2014**, *20*, 908–922. [CrossRef]
109. Park, C.H.; Eun, C.S.; Han, D.S. Intestinal Microbiota, Chronic Inflammation, and Colorectal Cancer. *Intest. Res.* **2018**, *16*, 338–345. [CrossRef]
110. Tjalsma, H.; Boleij, A.; Marchesi, J.R.; Dutilh, B.E. A Bacterial Driver-Passenger Model for Colorectal Cancer: Beyond the Usual Suspects. *Nat. Rev. Microbiol.* **2012**, *10*, 575–582. [CrossRef]
111. Chen, J.; Pitmon, E.; Wang, K. Microbiome, Inflammation and Colorectal Cancer. *Semin. Immunol.* **2017**, *32*, 43–53. [CrossRef]
112. Reis, S.A.D.; Da Conceição, L.L.; Peluzio, M.D.C.G. Intestinal Microbiota and Colorectal Cancer: Changes in the Intestinal Microenvironment and Their Relation to the Disease. *J. Med. Microbiol.* **2019**, *68*, 1391–1407. [CrossRef]
113. Faïs, T.; Delmas, J.; Barnich, N.; Bonnet, R.; Dalmasso, G. Colibactin: More Than a New Bacterial Toxin. *Toxins* **2018**, *10*, 151. [CrossRef]
114. Attene-Ramos, M.S.; Wagner, E.D.; Plewa, M.J.; Gaskins, H.R. Evidence That Hydrogen Sulfide Is a Genotoxic Agent. *Mol. Cancer Res. MCR* **2006**, *4*, 9–14. [CrossRef]
115. Warren, R.L.; Freeman, D.J.; Pleasance, S.; Watson, P.; Moore, R.A.; Cochrane, K.; Allen-Vercoe, E.; Holt, R.A. Co-Occurrence of Anaerobic Bacteria in Colorectal Carcinomas. *Microbiome* **2013**, *1*, 16. [CrossRef]
116. Garza, D.R.; Taddese, R.; Wirbel, J.; Zeller, G.; Boleij, A.; Huynen, M.A.; Dutilh, B.E. Metabolic Models Predict Bacterial Passengers in Colorectal Cancer. *Cancer Metab.* **2020**, *8*, 3. [CrossRef]
117. Wang, Y.; Zhang, C.; Hou, S.; Wu, X.; Liu, J.; Wan, X. Analyses of Potential Driver and Passenger Bacteria in Human Colorectal Cancer. *Cancer Manag. Res.* **2020**, *12*, 11553–11561. [CrossRef]
118. Pitsillides, L.; Pellino, G.; Tekkis, P.; Kontovounisios, C. The Effect of Perioperative Administration of Probiotics on Colorectal Cancer Surgery Outcomes. *Nutrients* **2021**, *13*, 1451. [CrossRef] [PubMed]
119. Faghfoori, Z.; Faghfoori, M.H.; Saber, A.; Izadi, A.; Yari Khosroushahi, A. Anticancer Effects of Bifidobacteria on Colon Cancer Cell Lines. *Cancer Cell Int.* **2021**, *21*, 258. [CrossRef] [PubMed]
120. Wu, X.; Wu, Y.; He, L.; Wu, L.; Wang, X.; Liu, Z. Effects of the Intestinal Microbial Metabolite Butyrate on the Development of Colorectal Cancer. *J. Cancer* **2018**, *9*, 2510–2517. [CrossRef] [PubMed]
121. Kam, P.M.-H.; Chu, C.W.-H.; Chan, E.M.-Y.; Liu, O.-L.; Kwok, K.-H. Use of Intravenous Iron Therapy in Colorectal Cancer Patient with Iron Deficiency Anemia: A Propensity-Score Matched Study. *Int. J. Colorectal Dis.* **2020**, *35*, 521–527. [CrossRef] [PubMed]
122. Laso-Morales, M.J.; Vives, R.; Gómez-Ramírez, S.; Pallisera-Lloveras, A.; Pontes, C. Intravenous Iron Administration for Post-Operative Anaemia Management after Colorectal Cancer Surgery in Clinical Practice: A Single-Centre, Retrospective Study. *Blood Transfus. Trasfus. Sangue* **2018**, *16*, 338–342. [CrossRef]
123. Aigner, E.; Feldman, A.; Datz, C. Obesity as an Emerging Risk Factor for Iron Deficiency. *Nutrients* **2014**, *6*, 3587–3600. [CrossRef]
124. Mujica-Coopman, M.F.; Brito, A.; López de Romaña, D.; Pizarro, F.; Olivares, M. Body Mass Index, Iron Absorption and Iron Status in Childbearing Age Women. *J. Trace Elem. Med. Biol.* **2015**, *30*, 215–219. [CrossRef]
125. Skrypnik, K.; Bogdański, P.; Sobieska, M.; Suliburska, J. The Effect of Multistrain Probiotic Supplementation in Two Doses on Iron Metabolism in Obese Postmenopausal Women: A Randomized Trial. *Food Funct.* **2019**, *10*, 5228–5238. [CrossRef]
126. Coelho, R.; Viola, T.W.; Walss-Bass, C.; Brietzke, E.; Grassi-Oliveira, R. Childhood Maltreatment and Inflammatory Markers: A Systematic Review. *Acta Psychiatr. Scand.* **2014**, *129*, 180–192. [CrossRef]
127. Monteiro, C.A.; Levy, R.B.; Claro, R.M.; Inês de Castro, R.R.; Cannon, G. A New Classification of Foods Based on the Extent and Purpose of Their Processing. *Cad. Saude Publica* **2010**, *26*, 2039–2049. [CrossRef] [PubMed]
128. Bell, D.S.H. Changes Seen in Gut Bacteria Content and Distribution with Obesity: Causation or Association? *Postgrad. Med.* **2015**, *127*, 863–868. [CrossRef]

129. Singer-Englar, T.; Barlow, G.; Mathur, R. Obesity, Diabetes, and the Gut Microbiome: An Updated Review. *Expert Rev. Gastroenterol. Hepatol.* **2019**, *13*, 3–15. [CrossRef]
130. Alshwaiyat, N.M.; Ahmad, A.; Wan Hassan, W.M.R.; Al-Jamal, H.A.N. Association between Obesity and Iron Deficiency (Review). *Exp. Ther. Med.* **2021**, *22*, 1268. [CrossRef]
131. Ortíz Pérez, M.; Vázquez López, M.A.; Ibáñez Alcalde, M.; Galera Martínez, R.; Martín González, M.; Lendínez Molinos, F.; Bonillo Perales, A. Relationship between Obesity and Iron Deficiency in Healthy Adolescents. *Child. Obes. Print* **2020**, *16*, 440–447. [CrossRef]
132. Aeberli, I.; Hurrell, R.F.; Zimmermann, M.B. Overweight Children Have Higher Circulating Hepcidin Concentrations and Lower Iron Status but Have Dietary Iron Intakes and Bioavailability Comparable with Normal Weight Children. *Int. J. Obes.* **2009**, *33*, 1111–1117. [CrossRef]
133. Gottardo, E.T.; Prokoski, K.; Horn, D.; Viott, A.D.; Santos, T.C.; Fernandes, J.I.M. Regeneration of the Intestinal Mucosa in Eimeria and E. Coli Challenged Broilers Supplemented with Amino Acids. *Poult. Sci.* **2016**, *95*, 1056–1065. [CrossRef]
134. Citelli, M.; Fonte-Faria, T.; Nascimento-Silva, V.; Renovato-Martins, M.; Silva, R.; Luna, A.S.; Vargas da Silva, S.; Barja-Fidalgo, C. Obesity Promotes Alterations in Iron Recycling. *Nutrients* **2015**, *7*, 335–348. [CrossRef]
135. Hurrell, R.F. Influence of Inflammatory Disorders and Infection on Iron Absorption and Efficacy of Iron-Fortified Foods. *Nestle Nutr. Inst. Workshop Ser.* **2012**, *70*, 107–116. [CrossRef]
136. Gotardo, É.M.F.; Caria, C.R.E.P.; De Oliveira, C.C.; Rocha, T.; Ribeiro, M.L.; Gambero, A. Effects of Iron Supplementation in Mice with Hypoferremia Induced by Obesity. *Exp. Biol. Med.* **2016**, *241*, 2049–2055. [CrossRef]
137. Sonnweber, T.; Ress, C.; Nairz, M.; Theurl, I.; Schroll, A.; Murphy, A.T.; Wroblewski, V.; Witcher, D.R.; Moser, P.; Ebenbichler, C.F.; et al. High-Fat Diet Causes Iron Deficiency via Hepcidin-Independent Reduction of Duodenal Iron Absorption. *J. Nutr. Biochem.* **2012**, *23*, 1600–1608. [CrossRef] [PubMed]
138. Deugnier, Y.; Bardou-Jacquet, É.; Lainé, F. Dysmetabolic Iron Overload Syndrome (DIOS). *Presse Medicale* **2017**, *46*, e306–e311. [CrossRef] [PubMed]
139. Rabindrakumar, M.S.K.; Wickramasinghe, V.P.; Arambepola, C.; Senanayake, H.; Karunaratne, V.; Thoradeniya, T. Baseline Iron and Low-Grade Inflammation Modulate the Effectiveness of Iron Supplementation: Evidence from Follow-up of Pregnant Sri Lankan Women. *Eur. J. Nutr.* **2021**, *60*, 1101–1109. [CrossRef] [PubMed]
140. Luciani, N.; Brasse-Lagnel, C.; Poli, M.; Anty, R.; Lesueur, C.; Cormont, M.; Laquerriere, A.; Folope, V.; LeMarchand-Brustel, Y.; Gugenheim, J.; et al. Hemojuvelin: A New Link between Obesity and Iron Homeostasis. *Obes. Silver* **2011**, *19*, 1545–1551. [CrossRef]
141. Rusu, I.G.; Suharoschi, R.; Vodnar, D.C.; Pop, C.R.; Socaci, S.A.; Vulturar, R.; Istrati, M.; Moroșan, I.; Fărcaș, A.C.; Kerezsi, A.D.; et al. Iron Supplementation Influence on the Gut Microbiota and Probiotic Intake Effect in Iron Deficiency—A Literature-Based Review. *Nutrients* **2020**, *12*, 1993. [CrossRef]

Iron Depletion in Systemic and Muscle Compartments Defines a Specific Phenotype of Severe COPD in Female and Male Patients: Implications in Exercise Tolerance

Maria Pérez-Peiró [1,2], Mariela Alvarado [1,2,3], Clara Martín-Ontiyuelo [2,4], Xavier Duran [5], Diego A. Rodríguez-Chiaradía [1,2] and Esther Barreiro [1,2,*]

[1] Muscle Wasting and Cachexia in Chronic Respiratory Diseases and Lung Cancer Research Group, Pulmonology Department, Department of Medicine and Life Sciences (MELIS), Hospital del Mar, Medical Research Institute (IMIM), Parc de Salut Mar, Universitat Pompeu Fabra (UPF), Barcelona Biomedical Research Park (PRBB), 08003 Barcelona, Spain

[2] Centro de Investigación en Red de Enfermedades Respiratorias (CIBERES), Instituto de Salud Carlos III (ISCIII), 08003 Barcelona, Spain

[3] Pulmonology Department, Hospital Universitario Sant Joan de Reus, 43204 Reus, Spain

[4] Department of Pulmonary Medicine, Hospital Clínic, Institut d'Investigacions Biomèdiques August Pi i Sunyer (IDIBAPS), University of Barcelona, 08036 Barcelona, Spain

[5] Scientific, Statistics, and Technical Department, Hospital del Mar-IMIM, Parc de Salut Mar, 08003 Barcelona, Spain

* Correspondence: ebarreiro@imim.es; Tel.: +34-93-316-0385; Fax: +34-93-316-0410

Citation: Pérez-Peiró, M.; Alvarado, M.; Martín-Ontiyuelo, C.; Duran, X.; Rodríguez-Chiaradía, D.A.; Barreiro, E. Iron Depletion in Systemic and Muscle Compartments Defines a Specific Phenotype of Severe COPD in Female and Male Patients: Implications in Exercise Tolerance. *Nutrients* **2022**, *14*, 3929. https://doi.org/10.3390/nu14193929

Academic Editor: Dietmar Enko

Received: 29 July 2022
Accepted: 17 September 2022
Published: 22 September 2022

Publisher's Note: MDPI stays neutral with regard to jurisdictional claims in published maps and institutional affiliations.

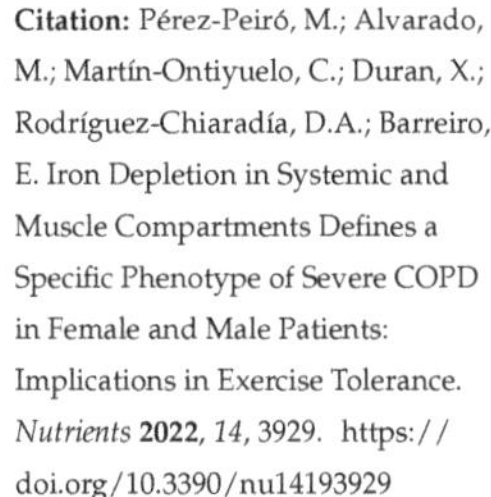

Abstract: We hypothesized that iron content and regulatory factors, which may be involved in exercise tolerance, are differentially expressed in systemic and muscle compartments in iron deficient severe chronic obstructive pulmonary disease (COPD) patients. In the vastus lateralis and blood of severe COPD patients with/without iron depletion, iron content and regulators, exercise capacity, and muscle function were evaluated in 40 severe COPD patients: non-iron deficiency (NID) and iron deficiency (ID) (20 patients/group). In ID compared to NID patients, exercise capacity, muscle iron and ferritin content, serum transferrin saturation, hepcidin-25, and hemojuvelin decreased, while serum transferrin and soluble transferrin receptor and muscle IRP-1 and IRP-2 increased. Among all COPD, a significant positive correlation was detected between FEV_1 and serum transferrin saturation. In ID patients, significant positive correlations were detected between serum ferritin, hepcidin, and muscle iron content and exercise tolerance and between muscle IRP-2 and serum ferritin and hepcidin levels. In ID severe COPD patients, iron content and its regulators are differentially expressed. A potential crosstalk between systemic and muscle compartments was observed in the ID patients. Lung function and exercise capacity were associated with several markers of iron metabolism regulation. Iron status should be included in the overall assessment of COPD patients given its implications in their exercise performance.

Keywords: severe chronic obstructive pulmonary disease; iron-deficient patients; iron metabolism regulation; pathophysiology of iron regulation; implications in exercise tolerance; systemic manifestations of chronic diseases

1. Introduction

Chronic obstructive pulmonary disease (COPD) is characterized by pulmonary and extrapulmonary manifestations. The extrapulmonary features of COPD include skeletal muscle dysfunction, nutritional abnormalities, cardiovascular alterations, and pulmonary hypertension, among others [1,2]. The systemic component of COPD worsens the patients' prognosis and their quality of life regardless of the severity of the lung disease [1,2]. Iron deficiency is recognized as an important comorbidity, even in the absence of anemia, in different chronic diseases including chronic heart failure and COPD. It has been estimated that iron deficiency may take place in approximately 40–50% of the COPD patients [3–5]. Alterations in iron homeostasis may negatively impact disease progression and functional status in COPD [5,6].

Iron metabolism, which is a tightly regulated process in humans, is an essential component that participates in many physiological processes such as energy metabolism, oxygen transport, and cell growth and differentiation [7,8]. Several molecules and hormones are involved in the regulation of iron release and maintenance of its reservoirs. Processes related to iron transport, uptake, and export to the bloodstream and those involving inflammation are regulated by several molecules such as the iron-sensitive element/iron-sensitive protein (IRE/IRP) system, hepcidin, and hemojuvelin [9–16]. Furthermore, iron content can also be measured in specimens from patients as another marker of iron metabolism [17,18].

In COPD cachectic patients, replicated genes that regulate heme metabolism were downregulated in blood samples [19], while mitochondrial breakdown signaling increased in the vastus lateralis muscle [20]. These findings were directly associated with both disease severity and the loss of mitochondrial content. Serum hepcidin was also shown to be a surrogate biomarker of iron status and metabolism in patients with chronic respiratory diseases including COPD [21]. Iron deficiency in COPD was also associated with reduced physical activity [22]. On the other hand, iron replacement improved exercise capacity and QoL in stable COPD patients with decreased iron content [23]. Studies conducted so far in COPD have focused on the quantification of iron content and regulation in either the blood or the muscle compartment. Assessment of iron metabolism in both muscle and systemic compartments within the same COPD patients is needed. Additionally, comparisons between patients with and without iron deficiency are also to be thoroughly analyzed in order to define a specific phenotypic profile of COPD patients. Moreover, potential associations between iron regulatory factors and exercise tolerance in patients with COPD, with a special emphasis on those with iron deficiency, also warrants a thorough analysis.

Thus, we hypothesized that iron content and regulatory factors are differentially expressed in both systemic and muscle compartments in patients with severe COPD and muscle dysfunction and iron deficiency compared to a cohort of COPD patients with identical disease severity and iron content levels within the normal range. This approach enabled us to define a specific phenotype of COPD patients characterized by significant alterations in iron regulation in both systemic and muscle compartments along with potential implications in their exercise tolerance and lung function status. Therefore, we sought to investigate in blood and muscle specimens from patients with severe COPD and muscle dysfunction with and without iron deficiency the following parameters and biomarkers: (1) lung function and exercise capacity; (2) hemogram, iron content, and regulators in the blood compartment; (3) regulators of iron homeostasis and iron content in the muscle specimens; and (4) potential associations between iron metabolism biomarkers and both exercise capacity and lung function parameters among all the patients, with a particular purpose in those with iron deficiency.

2. Materials and Methods

2.1. Study Population

This was a cross-sectional study, in which forty stable COPD patients (16 female patients) were prospectively and consecutively recruited over the years 2018–2021 from the Department of Respiratory Medicine at Hospital del Mar (Barcelona, Spain). COPD patients were further divided into two different groups: non-anemic iron deficiency (N = 20, 8 female patients) and normal iron content (N = 20, 8 female patients). All the participants were diagnosed according to the Global Strategy of Management of COPD patients (GOLD) criteria, in which the following parameters were included: spirometry values, number of exacerbations, and dyspnea score (modified medical research council, mMRC) [24]. Iron-deficient COPD patients (12 males) presented levels of hemoglobin > 12 g/dL in women and >13 g/dL in men, ferritin < 100 ng/mL, or ferritin 100–299 ng/mL with a transferrin saturation < 20% [21,23,25,26]. The study was approved on 17 January 2018 by the local Ethics Committee at Hospital del Mar (CEIm Parc de Salut Mar, registration # 2017/7691/I).

2.2. Exclusion Criteria

The following exclusion criteria were defined for this study: (1) acute exacerbations in the last three months; (2) other chronic respiratory disease or cardiovascular disorders; (3) neurological, metabolic, kidney, chronic liver disease, or uncontrolled psychiatric disorders; (4) known metabolic or neuromuscular myopathies; (5) treatment with drugs known to alter muscle structure and/or function (e.g., oral corticosteroids); (6) obesity (body mass index > 30 kg/m^2); (7) history of potentially bleeding conditions; and (8) active oncologic disease.

2.3. Anthropometric and Lung Function Assessment

Bioelectrical impedance was used to determined body mass index (BMI) and fat-free mass index (FFMI) of all the patients [27–29]. Lung function was assessed through spirometry, measuring the first second of forced expiratory volume (FEV$_1$) and forced vital capacity (FVC). Reference values were used to evaluate the resulting values [30–32].

2.4. Exercise Capacity and Muscle Function Assessment

The six-minute walk test (6-MWT) was performed to evaluate exercise capacity in all COPD patients following previous methodologies [33]. The 6-MWT was executed indoors along a flat, straight, 30 m walking course. During the 6-MWT, patients were encouraged every minute. Patients were allowed to stop to rest if needed. The test was resumed as soon as the patients were able to keep walking. The test lasted for six minutes for all the patients independently of whether they had to stop to rest.

Upper limb strength was evaluated analyzing hand grip strength using a specific hand dynamometer (Jamar 030J1, Chicago, IL, USA). Three consecutive measurements with the dynamometer were obtained from each patient, with a maximum of <5% variability between them. The higher measurement was used as the maximum voluntary contraction of the flexor muscles of the hands. In the analysis, reference values from Luna-Heredia et al. were used [34].

Lower limb strength was evaluated assessing quadriceps muscle strength through the determination of isometric maximum voluntary contraction (QMVC) of lower limbs as previously described [35]. For these measurements, a fixed handheld dynamometer (MicroFet 2^{TM}, Hoggan Scientific, Salt Lake City, UT, USA) was situated on the anterior tibia of the patient and the QMVC was recorded through the exerted compression force. Three different measurements were obtained for each subject (<5% variability among them), accepting the highest value as the QMVC. In the analysis, reference values from Seymour et al. [36] were used.

2.5. Blood Samples and Muscle Biopsies

Blood samples were obtained from the arm vein after an overnight fasting period. To evaluate the iron content in the patients, the following blood parameters were assessed: hemoglobin, hematocrit, mean corpuscular (erythrocyte) volume (MCV), mean corpuscular hemoglobin (MCH), mean corpuscular hemoglobin concentration (MCHC), serum iron, transferrin, transferrin saturation, serum ferritin, and soluble transferrin receptor. In order to determine levels of the serum hepcidin and hemojuvelin, blood was collected into serum tubes with clot activator (Vacuette®, Kremsmünster, Austria). Serum tubes with the blood samples were centrifuged at 1600× g for 15 min to obtain the serum. Immediately, serum samples were stored at −80 °C for further use.

As previously described, specimens from the vastus lateralis muscle from all study participants were acquired using the open biopsy technique [27–29]. Muscle samples were immediately frozen in liquid nitrogen and then stored at $-80\ ^\circ$C (temperature controlled with alarm control) for further molecular experiments. Additionally, a remaining specimen of muscle were immersed in an alcohol-formol bath to be thereafter embedded in paraffin. Paraffin-embedded samples were used for the assessment of structural modifications.

2.6. Biological Analyses

Muscle iron content analysis. Iron concentration in vastus lateralis specimens were measured using the Iron colorimetric assay kit (Elabscience, Houston, TX, USA) following the manufacturer's instructions. Vials 1 and 2 were provided by the manufactured kit and were used as recommended by the company. To analyze the iron content in the muscle specimens, 20 mg of frozen tissue was cleaned with PBS to remove traces of blood. Samples were immediately homogenized in PBS (0.01 M. pH 7.4) using a tissue homogenizer. The Bradford assay was applied in order to quantify protein concentration in the samples as previously described [28,29,37]. Briefly, protein quantification was quantified in triplicates for each sample and bovine serum albumin (BSA) (NZYtech, Lisbon, Portugal) was used as the standard. Briefly, the dye reagent concentrate (Bio-Rad Inc., Hercules, CA, USA) reacted with proteins, producing a change in absorbance at 595 nm that could be detected using a spectrophotometer. In order to quantify the iron content, 2 mg/L iron standard working solution was prepared by mixing reagent 1 in deionized water in a volume ratio of 1:49. An iron chromogenic agent was prepared following the kit instructions. Samples were subsequently 2:3 diluted in deionized water. The iron chromogenic agent was added to the tubes containing the samples, the standard, and the blank (500 μL deionized water). Tubes were vortexed for several seconds up until full homogenization was attained. Then, the tubes were incubated at 100 $^\circ$C in a water bath for 5 min. After cooling down the samples with running water, tubes were centrifuged at $2300\times g$ for 10 min. Then, 200 μL of the supernatant of each tube was transferred to the 96-well plates in a plate-reader (Infinite M Nano, Tecan Group Ltd., Zürich, Switzerland), and optical density (OD) values were measured at 520 nm wavelength. Intra-assay coefficients of variation for all the samples ranged from 0.07% to 7.60%. To calculate the iron concentration within the muscle specimens, the following mathematical formula was applied:

$$\text{Vastus lateralis iron content} \left(\frac{\text{mg}}{\text{gprot}}\right) = \frac{\text{OD}_{\text{sample}} - \text{OD}_{\text{blank}}}{\text{OD}_{\text{standard}} - \text{OD}_{\text{blank}}} \times 2\ \text{mg/L} \times \text{protein concentration}$$

Hemojuvelin. Determination of serum hemojuvelin concentration was assessed using Human HJV (Hemojuvelin) ELISA Kit (Elabscience, Houston, TX, USA) following the manufacturer's instructions. Briefly, 100 μL 50-fold diluted serum samples and standards were added in the corresponding wells of the hemojuvelin antibody pre-coated 96-well plate. Then, samples were incubated at 37 $^\circ$C for 90 min. After decanting the liquid, 100 μL of the biotinylated antibody was added to each well and samples were incubated 1 h at 37 $^\circ$C. Afterwards, samples were washed three times and consecutively 100 μL of HRP conjugate working solution was poured to each well. After an incubation of 30 min at 37 $^\circ$C, samples were washed five times. Then, 90 μL of substrate reagent was added to each well and samples were incubated 15 min at 37 $^\circ$C. Following this incubation, stop solution was added to each well to stop the enzyme substrate reaction. Absorbance of each sample was read at 450 nm wavelength. A standard curve was always generated with each assay run. Intra-assay coefficients of variation for all the samples ranged from 0.11% to 6.93%.

Hepcidin-25. Determination of serum Hepcidin-25 concentration was assessed using Human hepcidin (Hepc) ELISA kit (Biorbyt, Cambridgeshire, United Kingdom) following the manufacturer's instructions and previously described methodologies [21]. Briefly, in each well of the hepcidin antibody pre-coated microplate, 50 µL 5-fold diluted serum samples or standard and 50 µL HRP-conjugate were poured. The microplates containing the samples were then incubated at 37 °C for 1 h and, subsequently were washed three times. Successively, 50 µL substrate A and 50 µL substrate B were added and incubated at 37 °C for 15 min. Finally, the enzymatic reaction was stopped by adding 50 µL stop solution. Immediately afterwards, absorbance of each sample was read at 450 nm wavelength. A standard curve was always generated with each assay run. Intra-assay coefficients of variation for all the samples ranged from 0.11% to 10.98%.

Immunoblotting. Frozen muscle samples from vastus lateralis of all study patients were homogenized using a specific lysis buffer containing 50 mM 4-(2-hydroxyethyl)-1-piperazineethanesulfonic acid (HEPES), 150 mM NaCl, 100 nM NaF, 10 mM Na pyrophosphate, 5 mM EDTA, 0.5% Triton-X, 2 µg/mL leupeptin, 100 µg/mL phenylmethylsulfonyl fluoride (PMSF), 2 µg/mL aprotinin, and 10 µg/mL pepstatin A. Protein concentration of each muscle homogenate was determined using the Bradford method as previously described [27–29,37]. Between 5–20 micrograms of protein samples (according to antigen and antibody) were diluted 1:1 with 2X laemmli buffer (Bio-Rad Laboratories, Inc., Hercules, CA, USA) with 10% of 2-mercaptoethanol (Bio-Rad Laboratories, Inc., Hercules, CA, USA). Then, samples were boiled at 95 °C for five minutes and proteins were after separated by electrophoresis. Following the electrophoresis, proteins were transferred onto polyvinylidene difluoride (PVDF) membranes (Merck KGaA, Darmstadt, Germany). Prior to primary antibody incubation, membranes were blocked with bovine serum albumin (BSA) (NZYTech) or with 5% nonfat milk. The following primary antibodies were incubated overnight at 4 °C to analyze protein content of desired molecular markers, which include: ferritin (anti-ferritin antibody, Abcam Cambridge, UK), myoglobin (anti-myoglobin antibody, Santa Cruz Biotechnology, Dallas, TX, USA), ferroportin-1 (anti-ferroportin-1 antibody, Santa Cruz Biotechnology), transferrin receptor (anti-transferrin receptor antibody, Santa Cruz Biotechnology), IRP-1 (anti-IRP-1 antibody, Santa Cruz Biotechnology), IRP-2 (anti-IRP-2 antibody, Abcam), and glyceraldehyde-3-phosphate dehydrogenase (GAPDH, anti-GAPDH antibody, Santa Cruz Biotechnology). The following day, PVDF membranes were incubated for 1 h at room temperature with HRP-conjugated secondary antibodies (Jackson ImmunoResearch Inc., West Grove, PA, USA). In all samples, the desired antigens were detected using a chemiluminescence kit (Thermo Scientific, Rockford, IL, USA) and the Alliance Q9 Advanced (Uvitec Cambridge, UK) imager. Membranes from the different groups were jointly revealed under the same exposure conditions. OD from the resulting bands were quantified using the ImageJ software (National Institute of Health, available at http://rsb.info.nih.gov/ij/, accessed on 1 June 2022). Finally, optical density values corresponding on each protein of interest were normalized with optical density values of glycolytic enzyme GAPDH in all the immunoblots. Negative control experiments were conducted in this set of experiments. For this purpose, primary antibodies were omitted for each given marker and membranes were revealed with the corresponding secondary antibody only.

2.7. Statistical Analysis

Results are expressed in tables and graphs as mean (standard deviation). In the graphs, the green color dots or triangles indicate male patients in either study group, while the blue color dots or triangles indicate female patients in the same groups. In order to explore the potential influence of gender in the comparisons of variables between the two study groups, a linear regression analysis was calculated for all the variables (clinical and biological). Potential associations between clinical and biological variables were assessed using the Pearson's correlation coefficients. Such correlations were explored in all the COPD as a group and within each group of patients individually. Correlations were targeted specifically for variables whose mean values showed a statistically significant difference between the two study groups. Additionally, graphical correlation matrixes were depicted using the R package corrplot (https://cran.r-project.org/web/packages/corrplot/index.html, accessed on 15 June 2022). Blue dots indicated the existence of a positive correlation between two variables, while the red dots represented negative correlations. Additional multivariate linear regression analyses were used to test the associations between ferritin, hepcidin, and muscle iron content with the six-minute walk distance (meters and % predicted), in which age, sex, and FEV_1 were the adjusted variables. Sample size was calculated using the hepcidin as the target variable. In a two-sided test, accepting an alpha risk of 0.05 and a beta risk of 0.2 (80% power), a minimum of 16 patients in each group were required to detect a difference of at least 200 ng/mL of hepcidin. Statistical significance was established at $p \leq 0.05$. Actual p values are reported in both tables and figures. All the statistical analyses were carried out using the statistical software SPSS 23.0 (SPSS Inc., Chicago, IL, USA).

3. Results

3.1. General Clinical Features of the Study Patients

As can be seen in Table 1, clinical characteristics (anthropometry, smoking history, lung function, and GOLD classification) of the patients were similar in both study groups. Exercise capacity as measured by the six-minute walk test was significantly reduced in iron deficiency patients when compared to non-iron deficiency patients (Table 2). Nonetheless, no significant differences were detected in either upper or lower limb muscle function between the two groups of patients (Table 2). Iron metabolism parameters are shown in Table 3. Serum levels of ferritin, transferrin saturation, and muscle iron content were lower in the iron deficiency than in non-iron deficiency patients, while those of the soluble transferrin receptor and transferrin were higher in the former patients (Table 3).

Table 1. General clinical characteristics of the study patients.

	COPD Patients		
	Non-Iron Deficiency	**Iron Deficiency**	**p Value**
	N = 20	**N = 20**	
Anthropometry			
Age (years)	68 (8)	66 (8)	0.637
Males/Females	12/8	12/8	1.000
Body weight (kg)	60.3 (11.4)	64.0 (12.7)	0.291
BMI (kg/m^2)	22.8 (3.5)	24.1 (4.2)	0.287
FFMI (kg/m^2)	15.7 (2.2)	14.9 (2.2)	0.238
Smoking history			
Active, N (%)	9 (45)	12 (60)	0.525
Ex-smoker, N (%)	11 (55)	8 (40)	
Packs-year	52.5 (32.6)	42.94 (24.8)	0.327
Lung Function			
FEV_1 (L)	1.19 (0.4)	1.24 (0.4)	0.676
FEV_1 (% predicted)	46.7 (16.4)	44.0 (11.1)	0.485
FVC (L)	2.8 (0.5)	2.7 (0.8)	0.536
FVC (% predicted)	84.4 (14.5)	76.5 (12.0)	0.071
FEV_1/FVC	44.8 (11.52)	46.5 (11.1)	0.626

Table 1. *Cont.*

	COPD Patients		
	Non-Iron Deficiency	Iron Deficiency	*p* Value
	N = 20	N = 20	
GOLD classification			
1, N (%)	0 (0)	0 (0)	
2, N (%)	9 (45)	5 (25)	0.224
3, N (%)	9 (45)	12 (60)	
4, N (%)	2 (10)	3 (15)	
A, N (%)	11 (55)	9 (45)	
B, N (%)	7 (35)	9 (45)	0.699
C, N (%)	1 (5)	1 (5)	
D, N (%)	1 (5)	1 (5)	

Data are presented as mean (SD). Abbreviations: COPD, chronic obstructive pulmonary disease; BMI, body mass index; FFMI, fat-free mass index; N, number of patients; FEV_1, forced expiratory volume in one second; FVC, forced vital capacity; GOLD, Global Initiative for Chronic Obstructive Lung Disease. Adjusted *p* values for gender differences are shown in the table.

Table 2. Exercise and muscle function assessment of the study patients.

	COPD Patients		
	Non-Iron Deficiency	Iron Deficiency	*p* Value
	N = 20	N = 20	
Six-minute walk test			
Distance (m)	475.74 (63.29)	430.50 (64.77)	0.035
Distance (% predicted)	97.37 (17.58)	84.22 (17.14)	0.019
Upper limb muscle strength			
D-HGS (kg)	25.83 (7.38)	28.42 (8.42)	0.269
D-HGS (% predicted)	87.52 (18.30)	93.19 (20.42)	0.385
ND-HGS (kg)	23.64 (7.51)	25.16 (9.03)	0.648
ND-HGS (% predicted)	89.26 (23.68)	89.79 (21.12)	0.955
Lower limb muscle strength			
D-QMVC (kg)	22.38 (6.60)	22.54 (4.19)	0.332
D-QMVC (% predicted)	62.70 (21.37)	62.29 (12.38)	0.907
ND-QMVC (kg)	21.29 (5.69)	21.69 (4.80)	0.480
ND-QMVC (% predicted)	60.13 (21.54)	60.05 (13.97)	0.953

Data are presented as mean (SD). Abbreviations: COPD, chronic obstructive pulmonary disease; D, dominant; ND, non-dominant; HGS, hand grip strength; QMVC, quadriceps maximum voluntary contraction. Adjusted *p* values for gender differences are shown in the table.

Table 3. Iron metabolism parameters in the study patients.

	COPD Patients		
	Non-Iron Deficiency	Iron Deficiency	*p* Value
	N = 20	N = 20	
Iron status			
Hemoglobin (g/dL)	15.2 (1.4)	15.0 (1.6)	0.580
Hematocrit (%)	45.5 (4.7)	44.8 (4.4)	0.613

Table 3. *Cont.*

	COPD Patients		
	Non-Iron Deficiency	Iron Deficiency	*p* Value
	N = 20	N = 20	
MCV (fL)	93.7 (3.5)	91.7 (6.7)	0.250
MCH (pg)	31.4 (1.4)	30.7 (2.8)	0.311
MCHC (g/dL)	33.5 (1.3)	33.4 (1.2)	0.820
Ferritin (ng/mL)	214.8 (60.3)	66.5 (33.0)	0.000
Transferrin saturation (%)	31.0 (7.5)	25.4 (7.2)	0.024
Transferrin (g/dL)	236.6 (31.0)	260.7 (28.9)	0.012
Soluble transferrin receptor (mg/L)	2.2 (0.5)	3.0 (0.8)	0.002
Serum iron (µg/dL)	105.8 (26.3)	94.0 (28.8)	0.181
Muscle iron content (mg/g)	0.5 (0.2)	0.3 (0.1)	0.002

Data are presented as mean (SD). Abbreviations: COPD, chronic obstructive pulmonary disease; MCV, mean corpuscular (erythrocyte) volume; MCH, mean corpuscular hemoglobin; MCHC, mean corpuscular hemoglobin concentration. Adjusted *p* values for gender differences are shown in the table.

3.2. Associations between Clinical Parameters and Serum Iron Metabolism Markers

Among all patients, a significant positive correlation was observed between FEV_1 predicted and transferrin saturation (Figure 1A). Figure 1B shows the distribution of all the patients including both females and males in both groups (blue symbols represent the female while green symbols represented the male patients). Moreover, significant inverse correlations were detected between serum levels of ferritin, transferrin saturation, and serum iron and those of soluble transferrin receptor (Figure 1A and Figure S1A,B, respectively), whereas a positive association was seen between the latter parameter and transferrin levels (Figure 1A and Figure S1C–E, respectively). Furthermore, serum levels of transferrin positively correlated with soluble transferrin receptor, and negatively correlated with serum ferritin and transferrin saturation (Figure 1A and Figure S1F–H, respectively).

Serum levels of hepdicin-25 and hemojuvelin were significantly reduced in iron deficiency patients when compared to non-iron deficiency patients (Figure 2A,B, respectively). Furthermore, significant positive correlations were observed in these two serum parameters among all COPD patients, but not when analyzed independently (Figure 2C, respectively).

A significant correlation was found between serum ferritin levels and the walked distance in meters when all the patients were analyzed together (Figures 3 and 4A). Such a correlation was lost in non-iron deficiency patients (Figures 3 and 4A), while positive correlations were again detected between serum ferritin levels and the walked distance (meters and predicted variables) among iron deficiency patients. These correlations were maintained after adjusting for age, sex, and FEV_1 in these patients (Figures 3 and 4A,B, respectively). When all the patients were considered, positive correlations were detected between either serum hepcidin levels or muscle iron content and the walked distance (Figure 3, Figure 4A,B and Figure 5A,B, respectively). In non-iron deficiency patients, however, negative correlations were observed between serum hepcidin levels and the walked distance (predicted variables, Figures 3 and 4A,B, respectively) and between serum hemojuvelin levels and the latter parameter (meters and predicted variables, Figure 3). Additionally, among iron deficiency patients, positive associations were also seen between either serum hepcidin or iron content levels and the walked distance (meters and predicted variables, Figure 3, Figure 5A,B and Figure 6A,B, respectively). These correlations were maintained after adjusting for age, sex, and FEV_1 in these patients. Adjusted *p* values resulting from the multivariate linear regression showed statistically significant associations between distance 6 min walk test (m) and ferritin ($p = 0.001$), hepcidin ($p = 0.001$), and muscle iron content ($p = 0.028$), and between distance 6 min walk test (% predicted) and ferritin ($p = 0.023$), hepcidin ($p = 0.025$), and muscle iron content ($p = 0.032$) in the iron deficiency group. Among all the COPD patients, serum levels of hemojuvelin were also positively associated with those of ferritin, while they correlated negatively with those of soluble transferrin receptor (Figures 3 and 7A,B, respectively).

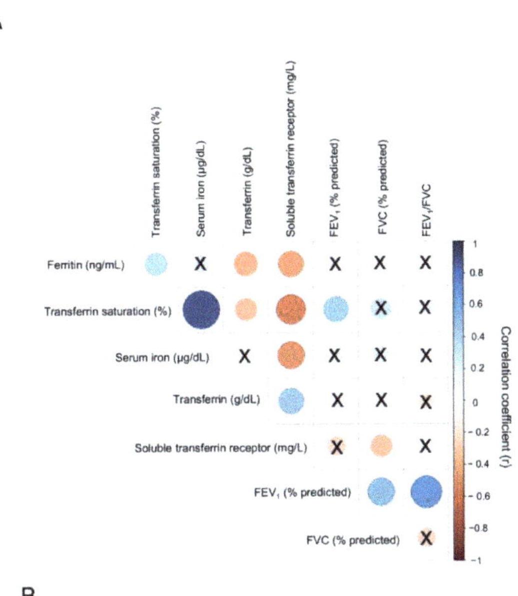

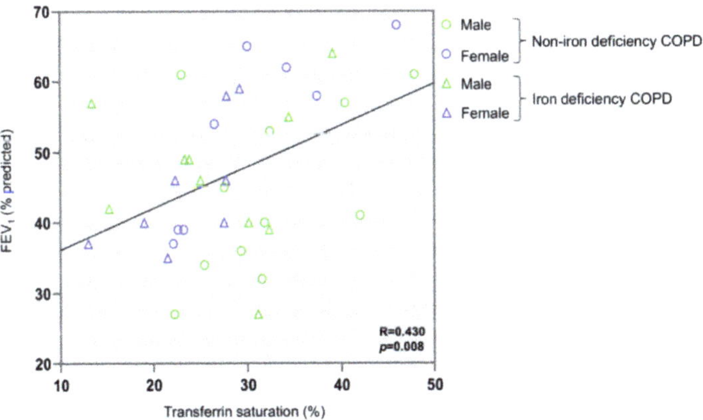

Figure 1. Correlation matrix of blood iron metabolism parameters and lung function parameters in all COPD patients (**A**) and scatter plot representation of correlation between FEV$_1$ and transferrin saturation in all COPD patients (**B**). In matrix, blue color represents positive correlations, while negative correlations are indicated in red. Crosses inside the circle mean that the correlation is not statistically significant (p value > 0.05). The correlation coefficients (Y axis of the graph) are proportional to the circle color intensity and size. In the scatter plot, the green color indicates male patients, while the blue color indicates female patients. The dots represent the patients with non-iron deficiency while the triangles represent the patients with iron deficiency. Abbreviations: FEV$_1$, first second of forced expiration; FVC, forced vital capacity; COPD, chronic obstructive pulmonary disease.

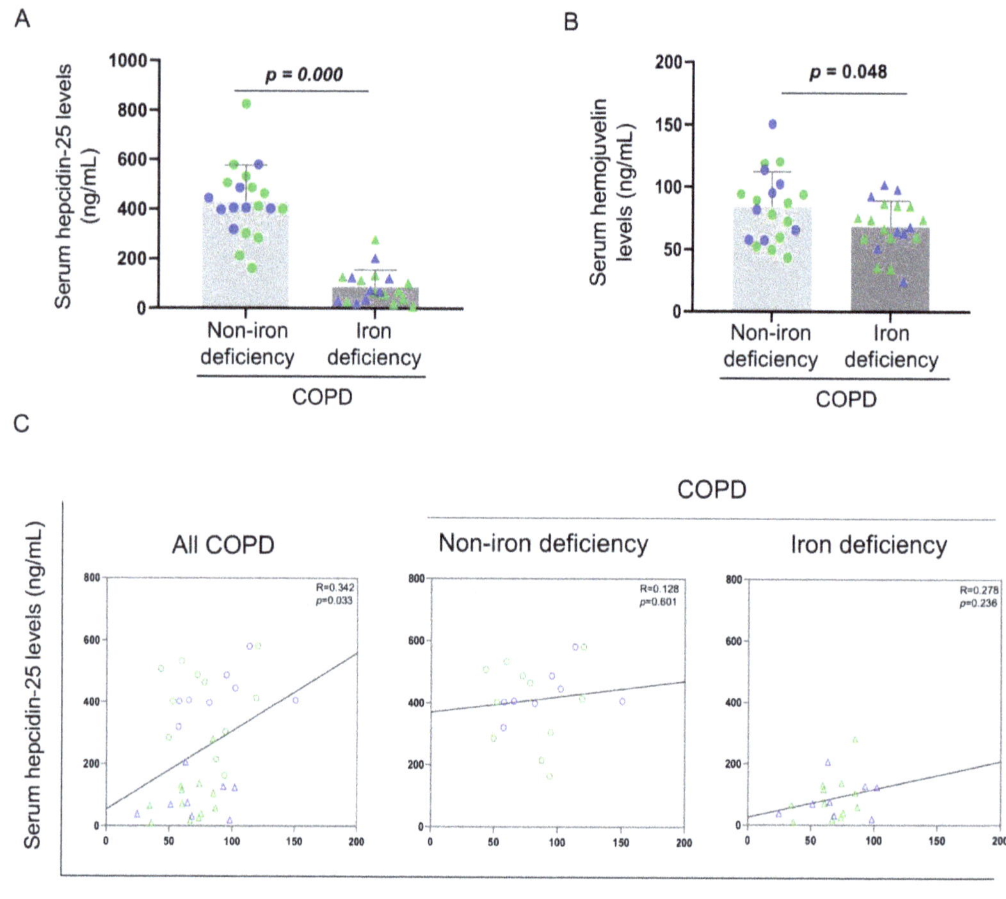

Figure 2. Mean values and standard deviation of (**A**) serum hepcidin-25 (ng/mL) and (**B**) hemojuvelin (ng/mL) variables in COPD patients. (**C**) Scatter plots representation of correlations between serum hepcidin and serum hemojuvelin in all COPD patients (**left** panel), non-iron deficiency (**middle** panel) and iron deficiency (**right** panel) COPD patients. In the graphs, the green color dots or triangles indicate male patients in either study group, while the blue color dots or triangles indicate female patients in the same groups. Definition of abbreviations: COPD, chronic obstructive pulmonary disease. Adjusted *p* values for gender differences are shown in the graphs.

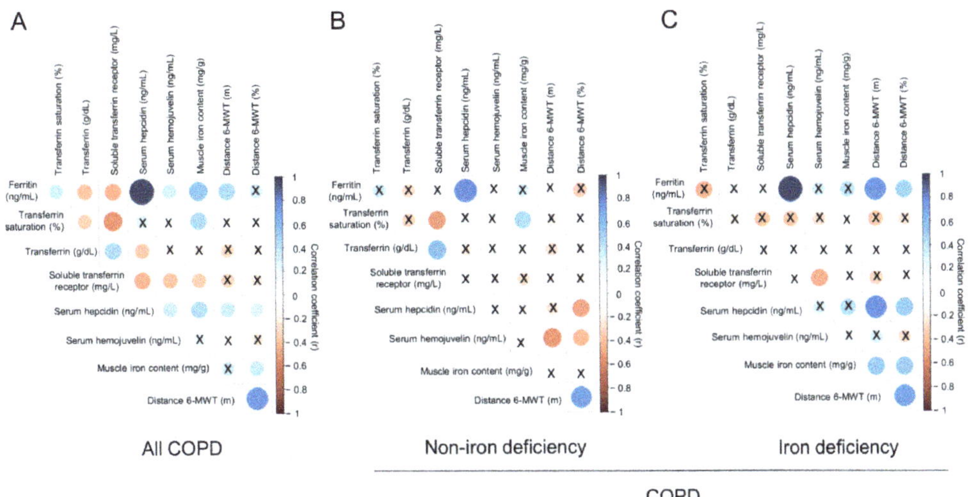

Figure 3. Correlation matrix of the iron metabolism parameters and six-minute walk test distance in (**A**) all COPD patients, (**B**) non-iron deficiency and (**C**) iron deficiency COPD patients. In matrix, blue color represents positive correlations, while negative correlations are indicated in red. Crosses inside the circle mean that the correlation is not statistically significant (*p* value > 0.05). The correlation coefficients (Y axis of the graph) are proportional to the circle color intensity and size. Abbreviations: 6-MWT, six-minute walk test. Definition of abbreviations: COPD, chronic obstructive pulmonary disease.

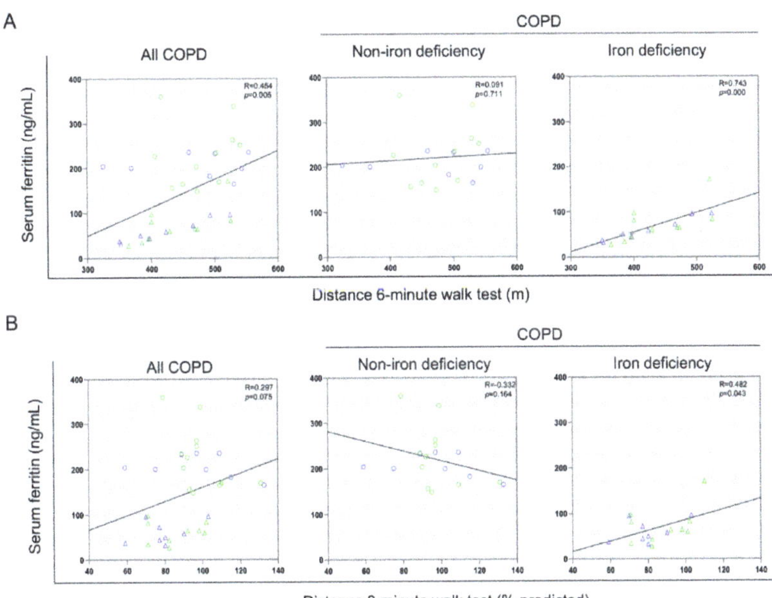

Figure 4. Scatter plots representation of correlations between serum ferritin levels and six-minute walk test distance (m) (**A**) and six-minute walk test distance (% predicted) (**B**) in all COPD patients (**left** panels), non-iron deficiency (**middle** panels) and iron deficiency (**right** panels) COPD patients. In the scatter plots, the green color dots or triangles indicate male patients in either study group, while the blue color dots or triangles indicate female patients in the same groups. Definition of abbreviations: COPD, chronic obstructive pulmonary disease.

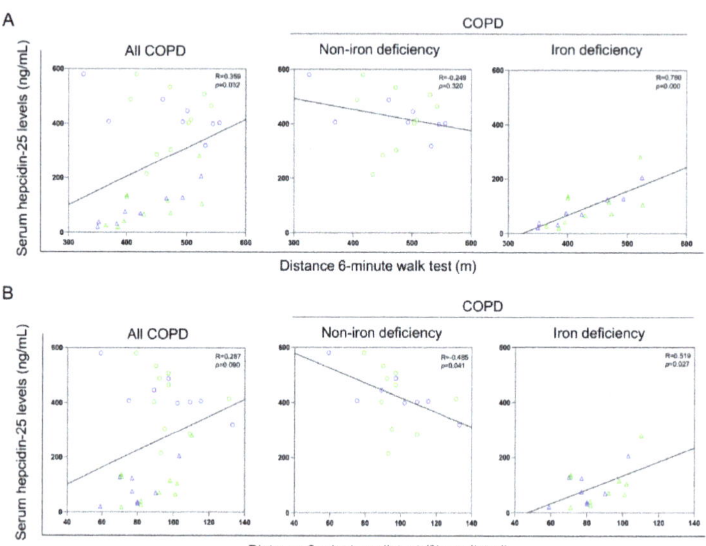

Figure 5. Scatter plots representation of correlations between serum hepcidin levels and distance six-minute walk test (m) (**A**) and distance six-minute walk test (% predicted) (**B**) in all COPD patients (**left** panels), non-iron deficiency (**middle** panels) and iron deficiency (**right** panels) COPD patients. In the scatter plots, the green color dots or triangles indicate male patients in either study group, while the blue color dots or triangles indicate female patients in the same groups. Definition of abbreviations: COPD, chronic obstructive pulmonary disease.

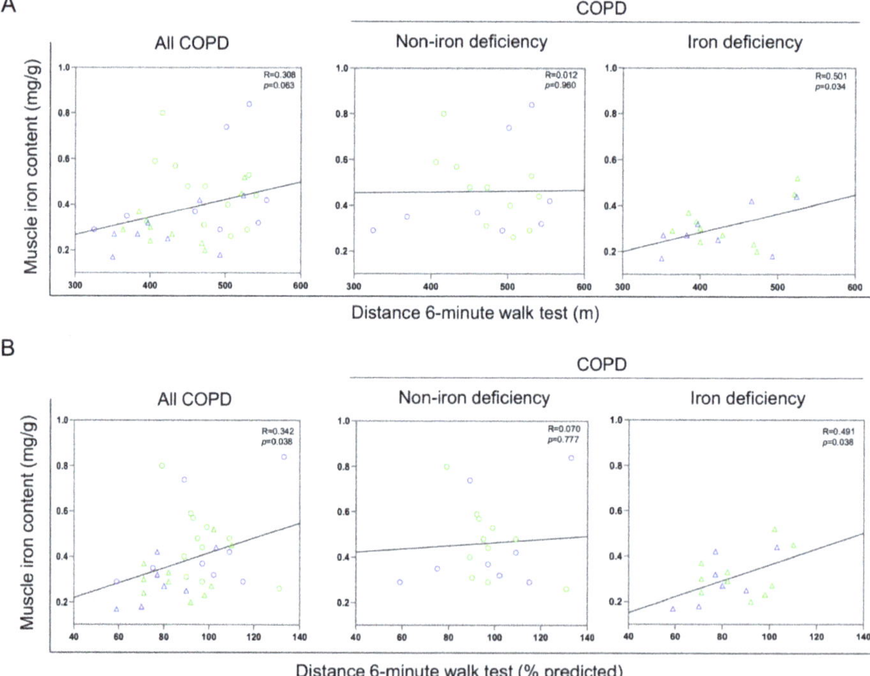

Figure 6. Scatter plots representation of correlations between muscle iron content and distance six-minute

walk test (m) (**A**) and distance six-minute walk test (% predicted) (**B**) in all COPD patients (**left** panels), non-iron deficiency (**middle** panels) and iron deficiency (**right** panels) COPD patients. In the scatter plots, the green color dots or triangles indicate male patients in either study group, while the blue color dots or triangles indicate female patients in the same groups. Definition of abbreviations: COPD, chronic obstructive pulmonary disease.

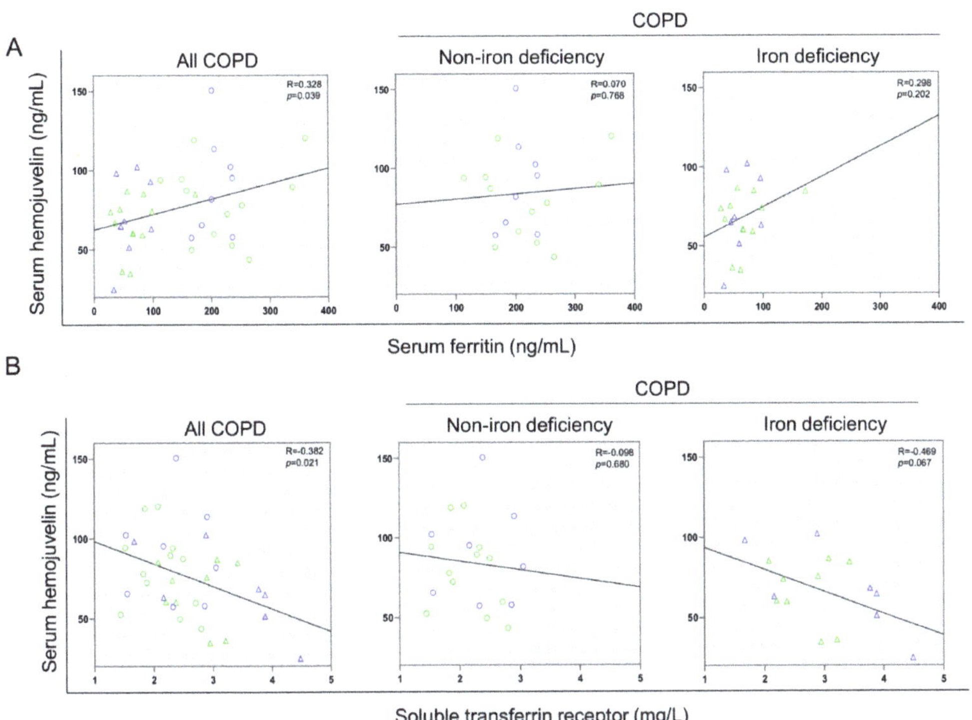

Figure 7. Scatter plots representation of correlations between serum hemojuvelin levels and serum ferritin (**A**) and soluble transferrin receptor (**B**) in all COPD patients (**left** panels), non-iron deficiency (**middle** panels) and iron deficiency (**right** panels) COPD patients. In the scatter plots, the green color dots or triangles indicate male patients in either study group, while the blue color dots or triangles indicate female patients in the same groups. Definition of abbreviations: COPD, chronic obstructive pulmonary disease.

3.3. Iron Metabolism in the Vastus Lateralis

No significant differences in muscle levels of myoglobin, transferrin receptor, and ferroportin-1 were observed between the two study groups (Figure 8A–D), whereas levels of ferritin were significantly lower in iron deficiency than in non-iron deficiency patients (Figure 8E). Protein content of IRP-1 and IRP-2 increased in the vastus lateralis of iron deficiency compared to non-iron deficiency patients (Figure 8F,G, respectively). Significant negative correlations were observed between either serum ferritin or serum hepcidin-25 levels and those of muscle IRP-2 among all COPD patients, while in iron deficiency patients positive correlations were found between the same variables (Figure 9A–C). In non-iron deficiency patients, no significant correlations were observed between either serum ferritin or serum hepcidin, and muscle IRP-2 content (Figure 9A–C).

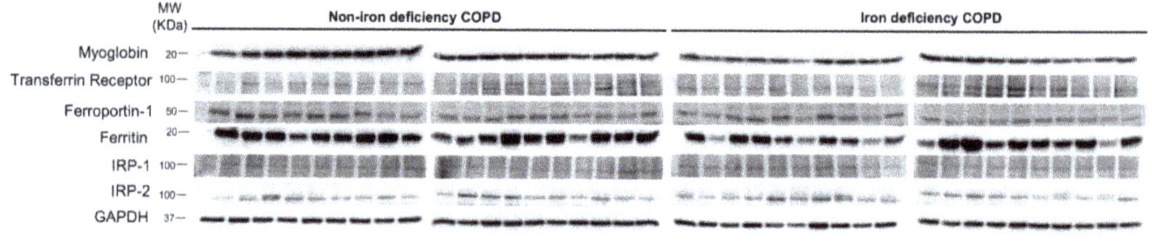

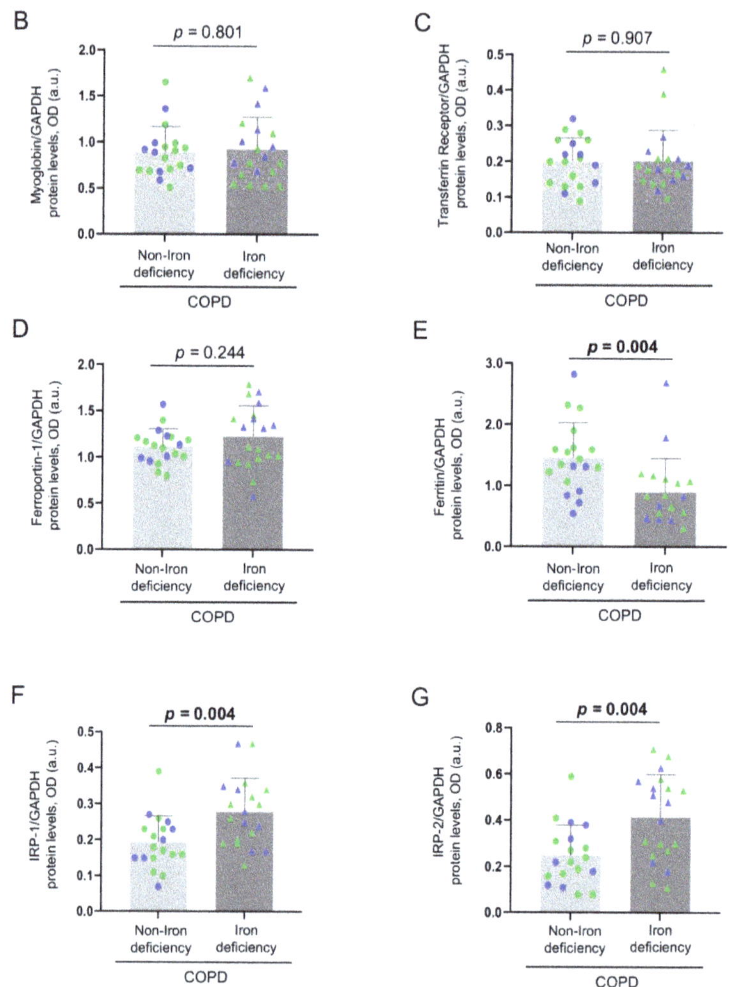

Figure 8. Representative immunoblots of myoglobin, transferrin receptor, ferroportin-1, ferritin, IRP-1, IRP-2 and GAPDH proteins in the vastus lateralis of all COPD patients (**A**). Mean values and standard deviation of myoglobin (**B**), transferrin receptor (**C**), ferroportin-1 (**D**), ferritin (**E**), IRP-1 (**F**)

and IRP-2 (**G**) protein content as measured by optical densities in arbitrary units (OD, a.u.). In the graphs, the green color dots or triangles indicate male patients in either study group, while the blue color dots or triangles indicate female patients in the same groups. Definition of abbreviations: IRP, Iron regulatory protein; GAPDH, glyceraldehyde-3-phosphate dehydrogenase; MW, molecular weight; kDa, kilodalton. Adjusted *p* values for gender differences are shown in the graphs.

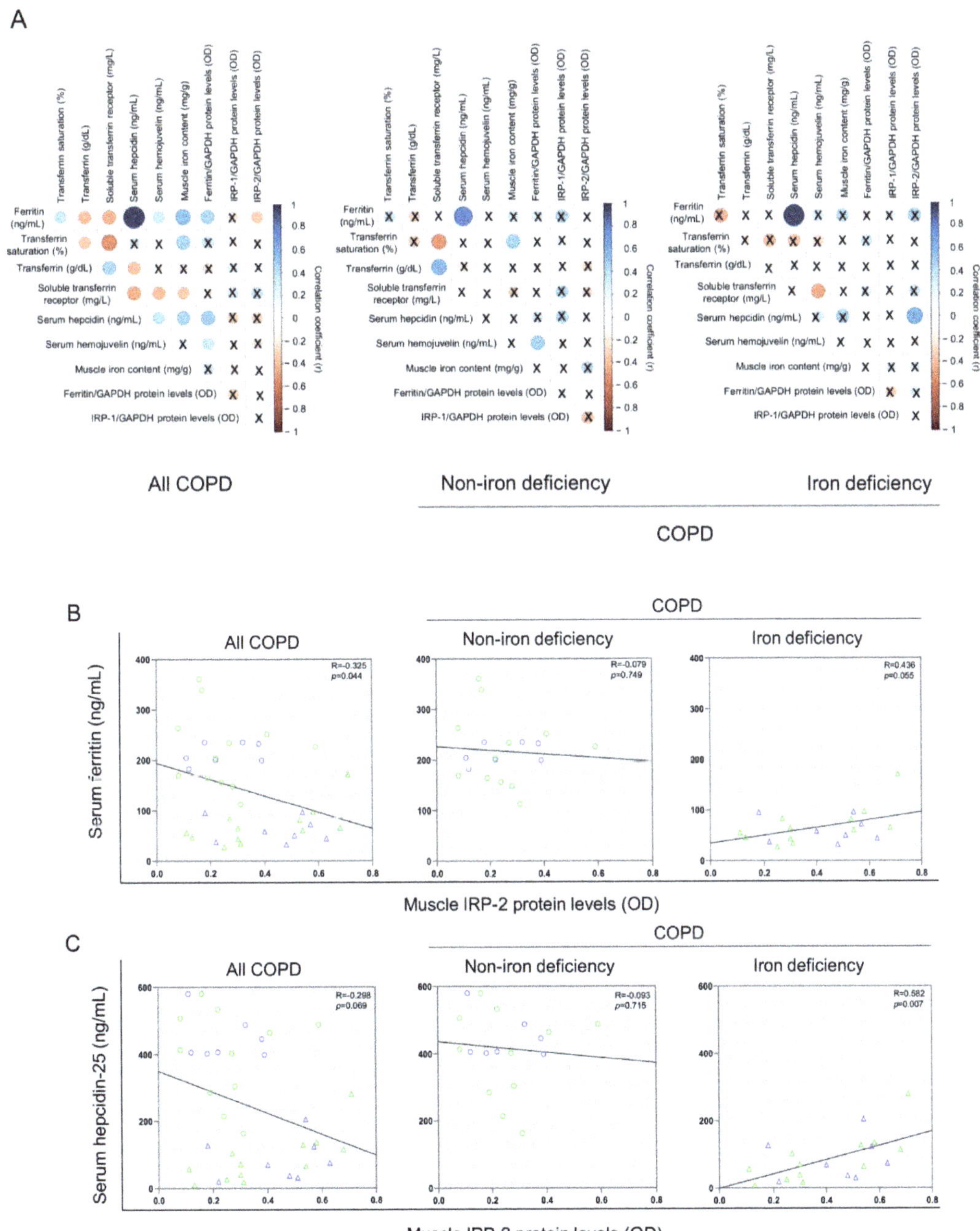

Figure 9. Correlation matrix of the blood and muscle iron metabolism parameters in all COPD patients

(**left** panels), non-iron deficiency (**middle** panel) and iron deficiency (**right** panel) COPD patients (**A**). In matrix, blue color represents positive correlations, while negative correlations are indicated in red. Crosses inside the circle mean that the correlation is not statistically significant (p value > 0.05). The correlation coefficients (Y axis of the graph) are proportional to the circle color intensity and size. Scatter plots representation of correlations between serum ferritin and muscle IRP-2 protein levels (**B**), and serum hepcidin and muscle IRP-2 protein levels (**C**), in all COPD patients (left panels), non-iron deficiency (middle panel) and iron deficiency (right panel) COPD patients. In the scatter plots, the green color dots or triangles indicate male patients in either study group, while the blue color dots or triangles indicate female patients in the same groups. Definition of abbreviations: IRP, Iron regulatory protein; GAPDH, glyceraldehyde-3-phosphate dehydrogenase; MW, molecular weight.

4. Discussion

The most relevant findings encountered in the investigation are that in severe COPD with iron deficiency, iron content and its regulators are differentially expressed compared to patients with identical degree of disease severity and muscle weakness with normal iron content. For the same degree of the airway obstruction, exercise capacity as measured by the six-minute walking test was reduced in the iron-deficient patients. Moreover, in the latter patients, levels of serum and muscle iron and ferritin content, and serum transferrin saturation, were also lower than in the non-iron deficient patients, whereas the content of transferrin and that of the soluble receptor of transferrin were greater. Airway obstruction as measured by FEV_1 positively correlated with serum transferrin saturation among all the study patients. Significant positive correlations were also detected between serum ferritin, hepcidin, and muscle iron content and exercise tolerance among the iron-deficient patients. In these patients compared to the non-iron deficient group, serum levels of hepcidin-25 and hemojuvelin decreased, while those of IRP-1 and IRP-2 increased. Positive associations were also detected between serum hepcidin and muscle IRP-2 levels among the patients with iron deficiency. These results remained equal when adjusted by gender. Iron deficiency defined a specific profile of patients with severe COPD and muscle weakness regardless of gender in this study. These are relevant novel results that shed light onto the potential implications of iron content and its regulation in the exercise capacity and the systemic component of patients with severe COPD.

Non-anemic iron deficiency can be encountered in up to 50% of COPD patients [3–5]. Iron deficient COPD patients showed a dysregulation of the iron metabolism parameters that were assessed specifically in their systemic compartment. Nonetheless, serum iron levels did not significantly differ between the two groups of COPD patients. As serum iron levels may fluctuate throughout the day, this parameter may not be all that reliable to properly diagnose the iron deficiency in patients [38–40]. At initial stages, iron deficiency can be defined as a result of reduced iron storage, which may translate into decreased serum ferritin levels [38–40]. Interestingly, serum iron concentration may remain within normal ranges in those early phases of iron depletion. As no significant differences in serum iron levels were observed between the two study groups, it may be possibly concluded that COPD iron deficient patients were probably at their initial stages of the iron depletion process [38,41]. Exercise capacity, however, was indeed reduced in the iron deficient patients. This finding suggests that decreased exercise capacity may be a clinical surrogate of iron depletion in chronic disease such as in COPD.

A positive strong correlation between ferritin levels and the six-minute walk distance was also seen in the iron deficient COPD patients. These results are consistent with previous findings. As such, the existence of a correlation between submaximal exercise capacity and the ferritin index was also detected in patients with chronic heart failure [42]. A positive correlation between the six-minute walk test distance and ferritin levels was also reported in iron deficient patients with heart failure [43]. Furthermore, other studies conducted on iron deficient COPD patients also demonstrated a reduction in the walked distance [6,44], reinforcing again this parameter as a hallmark in the overall assessment of severe COPD patients.

Iron homeostasis disruption has been proposed to influence the clinical course of COPD. Specifically, previous studies revealed a relationship between lung function parameters such as FEV_1 and FVC with different iron metabolism regulators in iron deficient COPD patients [45,46]. In line with these studies, in the present investigation a strong correlation was also found between FEV_1 and transferrin saturation, suggesting that disruptions in iron metabolism in these patients might exert deleterious effects on their lung function status [45,46]. Further studies are needed to confirm if iron homeostasis may influence lung function and disease severity in COPD.

Hepcidin is a peptide hormone that play an important role in the systemic iron regulation through its interaction with the major iron export protein, the ferroportin [12,14]. In patients with chronic diseases such as chronic hepatitis C, hepatocellular carcinoma, and inflammatory bowel disease, levels of hepcidin were significantly reduced compared to control subjects [47–49]. In COPD patients with normal iron metabolism, levels of hepcidin were also reduced [50,51]. In the current study, the iron deficient severe COPD patients were the ones showing a significant decrease in serum hepcidin levels. Furthermore, in the study, hepcidin levels positively correlated with ferritin, suggesting the potential implications of hepcidin in the systemic regulation of iron metabolism. Other studies also showed a positive correlation between those two biological parameters [52–56]. Moreover, serum hepcidin concentration also positively correlated with the six-minute walk distance in the iron deficient group of patients. Collectively, these results put the line forward the potential value of hepcidin as a surrogate marker of iron status and exercise capacity in COPD patients.

Hemojuvelin is a membrane molecule that acts as a co-receptor of bone morphogenetic protein (BMP). Hemojuvelin can negatively regulate hepcidin production through its soluble form [57]. Iron deficient COPD patients exhibited a decline in serum hemojuvelin levels in the current study. To the best of our knowledge, this is the first study that aimed to assess potential differences in serum hemojuvelin levels between iron deficient and normal iron status severe COPD patients. In general, hemojuvelin levels were shown to be increased in anemic patients compared to healthy subjects [58,59]. Among the COPD patients in this study, a significant positive correlation was also found between hemojuvelin and ferritin serum levels, while a negative association was observed between hemojuvelin and the soluble transferrin receptor. Other investigations also demonstrated the existence of a positive correlation between hemojuvelin and ferritin in iron-refractory anemia and in chronic kidney disease patients [59–61]. A significant positive correlation between hemojuvelin and hepcidin among all the COPD patients was also observed in the present study. Iron deficient patients exhibited a significant decline in ferritin along with a rise in IRP-1 and IRP-2 levels compared to non-iron deficient COPD patients. Moreover, significant positive correlations were observed between hepcidin and muscle IRP-2 in iron deficient patients. These results reveal the potential crosstalk between muscle and systemic compartment to target iron metabolism regulation. Collectively, the findings reported herein warrant further attention in the study of severe COPD as hemojuvelin and ferritin production may also be influenced by other stimuli such as inflammation [59,62,63], which is a major trigger of organ dysfunction and failure. Future investigations are needed to figure out which parameters can influence the control of the iron-regulatory proteins and to what extent they can affect patients with chronic diseases, in whom the inflammatory component plays a prominent role such as in COPD.

Study Limitations

Other parameters such as dynamic hyperinflation could have been measured in the study COPD patients. Nonetheless, due to the COVID-19 pandemic, only a few tests were possibly carried out in the lung function testing laboratory of our hospital. The measurements of dynamic hyperinflation were not included as patient recruitment took place to a great extent during the pandemic. It should be mentioned that despite results are shown for both female and male patients, the investigation was not aimed to analyze potential differences between them in each study group. Another potential limitation is related to the cross-sectional design of the study. However, the findings reported herein will help design future intervention studies with a follow-up component.

5. Conclusions

In severe iron deficient COPD patients, iron content and its regulators are differentially expressed compared to patients with normal iron content and identical degree of disease severity and muscle weakness. A potential crosstalk between systemic and muscle iron homeostasis regulation was revealed in severe COPD, particularly in the iron deficient patients. Lung function and exercise capacity were directly related to several markers of iron metabolism regulation, thus suggesting a potential role of the iron element in the disease status and severity. Iron regulation status should be included in the overall assessment of COPD patients due to its potential implications on their performance and disease progression.

Supplementary Materials: The following supporting information can be downloaded at: https://www.mdpi.com/article/10.3390/nu14193929/s1, Figure S1: Representation of scatter plots of correlations between serum ferritin (A), transferrin saturation (B), serum iron (C), serum soluble transferrin receptor, and correlations between serum ferritin (D) and serum iron (E) with transferrin saturation. Representation of scatter plots of correlations between transferrin and soluble transferrin receptor (F), ferritin (G) and transferrin saturation (H).

Author Contributions: Study conception and design: E.B., D.A.R.-C. and M.P.-P.; Patients recruitment: M.A. and C.M.-O.; Assessment of functional parameters: M.A. and M.P.-P.; Laboratory experiments: M.P.-P.; Statistical analyses and data interpretation: M.P.-P. and E.B.; manuscript drafting and intellectual input: M.P.-P., X.D., D.A.R.-C. and E.B.; manuscript writing final version: E.B. All authors have read and agreed to the published version of the manuscript.

Funding: María Pérez-Peiró was a recipient of a predoctoral fellowship from FIS Contratos Predoctorales de Formación en Investigación en Salud del Instituto de Salud Carlos III (ISC-III, FI19/00001) (Co-funded by European Social Fund "Investing in your future"). Mariela Alvarado was a recipient of a Contrato de Intensificación from FIS (ISC-III, INT19/00002). The research conducted in the study has been funded by Project FIS 21/00215 funded by ISC-III and co-funded by the European Union, CIBER—Consorcio Centro de Investigación Biomédica en Red—(CIBERES), Instituto de Salud Carlos III, Spanish Ministry of Science and Innovation and European Union, Spanish Respiratory Society (SEPAR), grant SEPAR 2020, and an unrestricted grant from Vifor Pharma 2018. Vifor Pharma did not have any role in the study or statistical analysis.

Institutional Review Board Statement: The research followed the guidelines of the World Medical Association for Research in Humans (Seventh revision of the Declaration of Helsinki, Fortaleza, Brazil, 2013) and the guidelines established by the International Committee of Medical Journal Editors (ICMJE) The editorial group MDPI strictly follows the guidelines recommended by ICMJE. The study was approved on 17 January 2018 by the local Ethics Committee at Hospital del Mar (CEIm Parc de Salut Mar, registration # 2017/7691/I).

Informed Consent Statement: All patients signed the informed written consent to participate in the study.

Data Availability Statement: The datasets generated and analyzed during the current study are available from the corresponding author on reasonable request.

Conflicts of Interest: The authors have none to disclose regarding this study.

References

1. Machado, A.; Marques, A.; Burtin, C. Extra-pulmonary manifestations of COPD and the role of pulmonary rehabilitation: A symptom-centered approach. *Expert Rev. Respir. Med.* **2020**, *15*, 131–142. [CrossRef]
2. Kaluźniak-Szymanowska, A.; Krzymińska-Siemaszko, R.; Deskur-śmielecka, E.; Lewandowicz, M.; Kaczmarek, B.; Wieczorowska-Tobis, K. Malnutrition, Sarcopenia, and Malnutrition-Sarcopenia Syndrome in Older Adults with COPD. *Nutrients* **2021**, *14*, 44. [CrossRef]

3. Cloonan, S.M.; Mumby, S.; Adcock, I.M.; Choi, A.M.K.; Chung, K.F.; Quinlan, G.J. The iron-y of iron overload and iron deficiency in chronic obstructive pulmonary disease. *Am. J. Respir. Crit. Care Med.* **2017**, *196*, 1103–1112. [CrossRef]

4. Silverberg, D.S.; Mor, R.; Weu, M.T.; Schwartz, D.; Schwartz, I.F.; Chernin, G. Anemia and iron deficiency in COPD patients: Prevalence and the effects of correction of the anemia with erythropoiesis stimulating agents and intravenous iron. *BMC Pulm. Med.* **2014**, *14*, 24. [CrossRef] [PubMed]

5. Nickol, A.H.; Frise, M.C.; Cheng, H.Y.; McGahey, A.; McFadyen, B.M.; Harris-Wright, T.; Bart, N.K.; Curtis, M.K.; Khandwala, S.; O'Neill, D.P.; et al. A cross-sectional study of the prevalence and associations of iron deficiency in a cohort of patients with chronic obstructive pulmonary disease. *BMJ Open* **2015**, *5*, e007911. [CrossRef]

6. Rathi, V.; Ish, P.; Singh, G.; Tiwari, M.; Goel, N.; Gaur, S.N. Iron deficiency in non-anemic chronic obstructive pulmonary disease in a predominantly male population: An ignored entity. *Arch. Monaldi Per Le Mal. Del Torace* **2020**, *90*, 38–42. [CrossRef]

7. Boldt, D.H. New perspectives on iron: An introduction. *Am. J. Med. Sci.* **1999**, *318*, 207. [CrossRef]

8. Winter, W.E.; Bazydlo, L.A.L.; Harris, N.S. The molecular biology of human iron metabolism. *Lab. Med.* **2014**, *45*, 92–102. [CrossRef]

9. Vogt, A.C.S.; Arsiwala, T.; Mohsen, M.; Vogel, M.; Manolova, V.; Bachmann, M.F. On Iron Metabolism and Its Regulation. *Int. J. Mol. Sci.* **2021**, *22*, 4591. [CrossRef]

10. Roemhild, K.; von Maltzahn, F.; Weiskirchen, R.; Knüchel, R.; von Stillfried, S.; Lammers, T. Iron metabolism: Pathophysiology and pharmacology. *Trends Pharmacol. Sci.* **2021**, *42*, 640–656. [CrossRef]

11. Mleczko-Sanecka, K.; Silvestri, L. Cell-type-specific insights into iron regulatory processes. *Am. J. Hematol.* **2021**, *96*, 110–127. [CrossRef]

12. Malyszko, J. Hemojuvelin: The hepcidin story continues. *Kidney Blood Press. Res.* **2009**, *32*, 71–76. [CrossRef] [PubMed]

13. Nemeth, E.; Ganz, T. Hepcidin-Ferroportin Interaction Controls Systemic Iron Homeostasis. *Int. J. Mol. Sci.* **2021**, *22*, 6493. [CrossRef]

14. Mackay, A.; Loza, M.; Branigan, P.; George, S.; Donaldson, G.; Baribaud, F.; Wedzicha, J. Inflammatory serum profiles are consistent across independent exacerbations in COPD patients with a history of frequent exacerbations. *Eur. Respir. J.* **2015**, *46*, PA407. [CrossRef]

15. Tandara, L.; Grubisic, T.Z.; Ivan, G.; Jurisic, Z.; Tandara, M.; Gugo, K.; Mladinov, S.; Salamunic, I. Systemic inflammation up-regulates serum hepcidin in exacerbations and stabile chronic obstructive pulmonary disease. *Clin. Biochem.* **2015**, *48*, 1252–1257. [CrossRef]

16. Wang, C.Y.; Babitt, J.L. Hepcidin Regulation in the Anemia of Inflammation. *Curr. Opin. Hematol.* **2016**, *23*, 189. [CrossRef] [PubMed]

17. Wyart, E.; Hsu, M.Y.; Sartori, R.; Mina, E.; Rausch, V.; Pierobon, E.S.; Mezzanotte, M.; Pezzini, C.; Bindels, L.B.; Lauria, A.; et al. Iron supplementation is sufficient to rescue skeletal muscle mass and function in cancer cachexia. *EMBO Rep.* **2022**, *23*, e53746. [CrossRef]

18. Zhou, D.; Zhang, Y.; Mamtawla, G.; Wan, S.; Gao, X.; Zhang, L.; Li, G.; Wang, X. Iron overload is related to muscle wasting in patients with cachexia of gastric cancer: Using quantitative proteome analysis. *Med. Oncol.* **2020**, *37*, 1–11. [CrossRef]

19. Wilson, A.C.; Kumar, P.L.; Lee, S.; Parker, M.M.; Arora, I.; Morrow, J.D.; Wouters, E.F.M.; Casaburi, R.; Rennard, S.I.; Lomas, D.A.; et al. Heme metabolism genes Downregulated in COPD Cachexia. *Respir. Res.* **2020**, *21*, 100. [CrossRef]

20. Leermakers, P.A.; Schols, A.M.W.J.; Kneppers, A.E.M.; Kelders, M.C.J.M.; de Theije, C.C.; Lainscak, M.; Gosker, H.R. Molecular signalling towards mitochondrial breakdown is enhanced in skeletal muscle of patients with chronic obstructive pulmonary disease (COPD). *Sci. Rep.* **2018**, *8*, 15007. [CrossRef]

21. Pérez-Peiró, M.; Martín-Ontiyuelo, C.; Rodó-Pi, A.; Piccari, L.; Admetlló, M.; Durán, X.; Rodríguez-Chiaradía, D.A.; Barreiro, E. Iron Replacement and Redox Balance in Non-Anemic and Mildly Anemic Iron Deficiency COPD Patients: Insights from a Clinical Trial. *Biomedicines* **2021**, *9*, 1191. [CrossRef] [PubMed]

22. Martín-Ontiyuelo, C.; Rodó-Pin, A.; Sancho-Muñoz, A.; Martinez-Llorens, J.M.; Admetlló, M.; Molina, L.; Gea, J.; Barreiro, E.; Chiaradía, D.A.R. Is iron deficiency modulating physical activity in COPD? *Int. J. Chron. Obstruct. Pulmon. Dis.* **2019**, *14*, 211–214. [CrossRef]

23. Martín-Ontiyuelo, C.; Rodó-Pin, A.; Echeverría-Esnal, D.; Admetlló, M.; Duran-Jordà, X.; Alvarado, M.; Gea, J.; Barreiro, E.; Rodríguez-Chiaradía, D.A. Intravenous Iron Replacement Improves Exercise Tolerance in COPD: A Single-Blind Randomized Trial. *Arch. Bronconeumol.* **2021**, *10*, 2903. [CrossRef]

24. Venkatesan, P. GOLD report: 2022 update. *Lancet Respir. Med.* **2022**, *10*, e20. [CrossRef]

25. Barberan-Garcia, A.; Rodríguez, D.A.; Blanco, I.; Gea, J.; Torralba, Y.; Arbillaga-Etxarri, A.; Barberà, J.A.; Vilaró, J.; Roca, J.; Orozco-Levi, M. Non-anaemic iron deficiency impairs response to pulmonary rehabilitation in COPD. *Respirology* **2015**, *20*, 1089–1095. [CrossRef] [PubMed]

26. Anker, S.D.; Comin Colet, J.; Filippatos, G.; Willenheimer, R.; Dickstein, K.; Drexler, H.; Lüscher, T.F.; Bart, B.; Banasiak, W.; Niegowska, J.; et al. Ferric Carboxymaltose in Patients with Heart Failure and Iron Deficiency. *N. Engl. J. Med.* **2009**, *361*, 2436–2448. [CrossRef] [PubMed]

27. Sancho-Muñoz, A.; Guitart, M.; Rodríguez, D.A.; Gea, J.; Martínez-Llorens, J.; Barreiro, E. Deficient muscle regeneration potential in sarcopenic COPD patients: Role of satellite cells. *J. Cell. Physiol.* **2021**, *236*, 3083–3098. [CrossRef]

28. Barreiro, E.; Salazar-Degracia, A.; Sancho-Muñoz, A.; Gea, J. Endoplasmic reticulum stress and unfolded protein response profile in quadriceps of sarcopenic patients with respiratory diseases. *J. Cell. Physiol.* **2019**, *234*, 11315–11329. [CrossRef] [PubMed]

29. Puig-Vilanova, E.; Martínez-Llorens, J.; Ausin, P.; Roca, J.; Gea, J.; Barreiro, E. Quadriceps muscle weakness and atrophy are associated with a differential epigenetic profile in advanced COPD. *Clin. Sci.* **2015**, *128*, 905–921. [CrossRef]

30. Roca, J.; Burgos, F.; Barberà, J.A.; Sunyer, J.; Rodriguez-Roisin, R.; Castellsagué, J.; Sanchis, J.; Antóo, J.M.; Casan, P.; Clausen, J.L. Prediction equations for plethysmographic lung volumes. *Respir. Med.* **1998**, *92*, 454–460. [CrossRef]

31. Roca, J.; Burgos, F.; Sunyer, J.; Saez, M.; Chinn, S.; Anto, J.; Rodriguez-Roisin, R.; Quanjer, P.; Nowak, D.; Burney, P. References values for forced spirometry. Group of the European Community Respiratory Health Survey. *Eur. Respir. J.* **1998**, *11*, 1354–1362. [CrossRef]

32. Roca, J.; Vargas, C.; Cano, I.; Selivanov, V.; Barreiro, E.; Maier, D.; Falciani, F.; Wagner, P.; Cascante, M.; Garcia-Aymerich, J.; et al. Chronic Obstructive Pulmonary Disease heterogeneity: Challenges for health risk assessment, stratification and management. *J. Transl. Med.* **2014**, *12*, S3. [CrossRef]

33. Rodriguez, D.A.; Kalko, S.; Puig-Vilanova, E.; Perez-Olabarría, M.; Falciani, F.; Gea, J.; Cascante, M.; Barreiro, E.; Roca, J. Muscle and blood redox status after exercise training in severe COPD patients. *Free Radic. Biol. Med.* **2012**, *52*, 88–94. [CrossRef]

34. Luna-Heredia, E.; Martín-Peña, G.; Ruiz-Galiana, J. Handgrip dynamometry in healthy adults. *Clin. Nutr.* **2005**, *24*, 250–258. [CrossRef]

35. Bui, K.L.; Mathur, S.; Dechman, G.; Maltais, F.; Camp, P.; Saey, D. Fixed Handheld Dynamometry Provides Reliable and Valid Values for Quadriceps Isometric Strength in People With Chronic Obstructive Pulmonary Disease: A Multicenter Study. *Phys. Ther.* **2019**, *99*, 1255–1267. [CrossRef] [PubMed]

36. Seymour, J.M.; Spruit, M.A.; Hopkinson, N.S.; Natanek, S.A.; Man, W.D.C.; Jackson, A.; Gosker, H.R.; Schols, A.M.W.J.; Moxham, J.; Polkey, M.I.; et al. The prevalence of quadriceps weakness in COPD and the relationship with disease severity. *Eur. Respir. J.* **2010**, *36*, 81–88. [CrossRef]

37. Puig-Vilanova, E.; Rodriguez, D.A.; Lloreta, J.; Ausin, P.; Pascual-Guardia, S.; Broquetas, J.; Roca, J.; Gea, J.; Barreiro, E. Oxidative stress, redox signaling pathways, and autophagy in cachectic muscles of male patients with advanced COPD and lung cancer. *Free Radic. Biol. Med.* **2015**, *79*, 91–108. [CrossRef]

38. Firkin, F.; Rush, B. Interpretation of biochemical tests for iron deficiency: Diagnostic difficulties related to limitations of individual tests. *Aust. Prescr.* **1997**, *20*, 74–76. [CrossRef]

39. Cohen-Solal, A.; Leclercq, C.; Deray, G.; Lasocki, S.; Zambrowski, J.J.; Mebazaa, A.; De Groote, P.; Dam, T.; Galinier, M. Iron deficiency: An emerging therapeutic target in heart failure. *Heart* **2014**, *100*, 1414–1420. [CrossRef]

40. Buttarello, M.; Pajola, R.; Novello, E.; Mezzapelle, G.; Plebani, M. Evaluation of the hypochromic erythrocyte and reticulocyte hemoglobin content provided by the Sysmex XE-5000 analyzer in diagnosis of iron deficiency erythropoiesis. *Clin. Chem. Lab. Med.* **2016**, *54*, 1939–1945. [CrossRef]

41. Özdemir, N. Iron deficiency anemia from diagnosis to treatment in children. *Turk. Arch. Pediatrics Pediatri Arşivi* **2015**, *50*, 11. [CrossRef] [PubMed]

42. Enjuanes, C.; Bruguera, J.; Grau, M.; Cladellas, M.; Gonzalez, G.; Meroño, O.; Moliner-Borja, P.; Verdú, J.M.; Farré, N.; Comín-Colet, J. Estado del hierro en la insuficiencia cardiaca crónica: Impacto en síntomas, clase funcional y capacidad de ejercicio submáxima. *Rev. Española Cardiol.* **2016**, *69*, 247–255. [CrossRef]

43. Barandiarán Aizpurua, A.; Sanders-van Wijk, S.; Brunner-La Rocca, H.P.; Henkens, M.T.H.M.; Weerts, J.; Spanjers, M.H.A.; Knackstedt, C.; van Empel, V.P.M. Iron deficiency impacts prognosis but less exercise capacity in heart failure with preserved ejection fraction. *ESC Heart Fail.* **2021**, *8*, 1304–1313. [CrossRef]

44. Cote, C.; Zilberberg, M.D.; Mody, S.H.; Dordelly, L.J.; Celli, B. Haemoglobin level and its clinical impact in a cohort of patients with COPD. *Eur. Respir. J.* **2007**, *29*, 923–929. [CrossRef]

45. Pizzini, A.; Aichner, M.; Sonnweber, T.; Tancevski, I.; Weiss, G.; Löffler-Ragg, J. The Significance of iron deficiency and anemia in a real-life COPD cohort. *Int. J. Med. Sci.* **2020**, *17*, 2232. [CrossRef] [PubMed]

46. Ghio, A.J.; Hilborn, E.D. Indices of iron homeostasis correlate with airway obstruction in an NHANES III cohort. *Int. J. Chron. Obstruct. Pulmon. Dis.* **2017**, *12*, 2075. [CrossRef]

47. Sharma, S.; Nemeth, E.; Chen, Y.H.; Goodnough, J.; Huston, A.; Roodman, G.D.; Ganz, T.; Lichtenstein, A. Involvement of hepcidin in the anemia of multiple myeloma. *Clin. Cancer Res.* **2008**, *14*, 3262–3267. [CrossRef]

48. Joachim, J.H.; Mehta, K.J. Hepcidin in hepatocellular carcinoma. *Br. J. Cancer* **2022**, *127*, 185–192. [CrossRef] [PubMed]

49. Krawiec, P.; Mroczkowska-Juchkiewicz, A.; Pac-Kozuchowska, E. Serum Hepcidin in Children with Inflammatory Bowel Disease. *Inflamm. Bowel Dis.* **2017**, *23*, 2165–2171. [CrossRef]

50. Attia, A.; Mohammed, T.; Abd Al Aziz, U. The relationship between serum hepcidin level and hypoxemia in the COPD patients. *Egypt. J. Chest Dis. Tuberc.* **2015**, *64*, 57–61. [CrossRef]

51. Duru, S.; Bilgin, E.; Ardiç, S. Hepcidin: A useful marker in chronic obstructive pulmonary disease. *Ann. Thorac. Med.* **2012**, *7*, 31. [CrossRef] [PubMed]

52. Wisaksana, R.; De Mast, Q.; Alisjahbana, B.; Jusuf, H.; Sudjana, P.; Indrati, A.R.; Sumantri, R.; Swinkels, D.; Van Crevel, R.; Van Der Ven, A. Inverse Relationship of Serum Hepcidin Levels with CD4 Cell Counts in HIV-Infected Patients Selected from an Indonesian Prospective Cohort Study. *PLoS ONE* **2013**, *8*, e79904. [CrossRef] [PubMed]

53. Gaillard, C.A.; Bock, A.H.; Carrera, F.; Eckardt, K.U.; Van Wyck, D.B.; Bansal, S.S.; Cronin, M.; Meier, Y.; Larroque, S.; Roger, S.D.; et al. Hepcidin Response to Iron Therapy in Patients with Non-Dialysis Dependent CKD: An Analysis of the FIND-CKD Trial. *PLoS ONE* **2016**, *11*, e0157063. [CrossRef] [PubMed]
54. Berglund, S.; Lönnerdal, B.; Westrup, B.; Domellöf, M. Effects of iron supplementation on serum hepcidin and serum erythropoietin in low-birth-weight infants. *Am. J. Clin. Nutr.* **2011**, *94*, 1553–1561. [CrossRef] [PubMed]
55. Lasocki, S.; Lefebvre, T.; Mayeur, C.; Puy, H.; Mebazaa, A.; Gayat, E.; Deye, N.; Fauvaux, C.; Mebazaa, A.; Damoisel, C.; et al. Iron deficiency diagnosed using hepcidin on critical care discharge is an independent risk factor for death and poor quality of life at one year: An observational prospective study on 1161 patients. *Crit. Care* **2018**, *22*, 1–8. [CrossRef]
56. Atkinson, S.H.; Armitage, A.E.; Khandwala, S.; Mwangi, T.W.; Uyoga, S.; Bejon, P.A.; Williams, T.N.; Prentice, A.M.; Drakesmith, H. Combinatorial effects of malaria season, iron deficiency, and inflammation determine plasma hepcidin concentration in African children. *Blood* **2014**, *123*, 3221–3229. [CrossRef]
57. Core, A.B.; Canali, S.; Babitt, J.L. Hemojuvelin and bone morphogenetic protein (BMP) signaling in iron homeostasis. *Front. Pharmacol.* **2014**, *5*, 104. [CrossRef]
58. Shalev, H.; Perez-Avraham, G.; Kapelushnik, J.; Levi, I.; Rabinovich, A.; Swinkels, D.W.; Brasse-Lagnel, C.; Tamary, H. High levels of soluble serum hemojuvelin in patients with congenital dyserythropoietic anemia type I. *Eur. J. Haematol.* **2013**, *90*, 31–36. [CrossRef]
59. Brasse-Lagnel, C.; Poli, M.; Lesueur, C.; Grandchamp, B.; Lavoinne, A.; Beaumont, C.; Bekri, S. Immunoassay for human serum hemojuvelin. *Haematologica* **2010**, *95*, 2031–2037. [CrossRef]
60. Malyszko, J.; Malyszko, J.S.; Levin-Iaina, N.; Koc-Zorawska, E.; Kozminski, P.; Mysliwiec, M. Is hemojuvelin a possible new player in iron metabolism in hemodialysis patients? *Int. Urol. Nephrol.* **2012**, *44*, 1805–1811. [CrossRef]
61. Rumjon, A.; Sarafidis, P.; Brincat, S.; Musto, R.; Malyszko, J.; Bansal, S.S.; Macdougall, I.C. Serum hemojuvelin and hepcidin levels in chronic kidney disease. *Am. J. Nephrol.* **2012**, *35*, 295–304. [CrossRef] [PubMed]
62. Constante, M.; Wang, D.; Raymond, V.A.; Bilodeau, M.; Santos, M.M. Repression of repulsive guidance molecule C during inflammation is independent of Hfe and involves tumor necrosis factor-α. *Am. J. Pathol.* **2007**, *170*, 497–504. [CrossRef] [PubMed]
63. Kernan, K.F.; Carcillo, J.A. Hyperferritinemia and inflammation. *Int. Immunol.* **2017**, *29*, 401. [CrossRef] [PubMed]

Article

Positive or U-Shaped Association of Elevated Hemoglobin Concentration Levels with Metabolic Syndrome and Metabolic Components: Findings from Taiwan Biobank and UK Biobank

Vanessa Joy Timoteo [1,2], Kuang-Mao Chiang [2] and Wen-Harn Pan [2,*]

[1] Taiwan International Graduate Program in Molecular Medicine, National Yang Ming Chiao Tung University and Academia Sinica, Taipei City 115, Taiwan
[2] Institute of Biomedical Sciences, Academia Sinica, Taipei City 115, Taiwan
* Correspondence: pan@ibms.sinica.edu.tw; Tel.: +886-2-27899121; Fax: +886-2-27823047

Abstract: Iron overnutrition has been implicated with a higher risk of developing metabolic and cardiovascular diseases, including metabolic syndrome (MetS), whereas iron deficiency anemia exacerbates many underlying chronic conditions. Hemoglobin (Hb) concentration in the blood, which reflects a major functional iron (i.e., heme iron) in the body, may serve as a surrogate of the nutritional status of iron. We conducted sex-specific observational association studies in which we carefully titrated the association between Hb deciles and MetS and its components among the Taiwanese Han Chinese (HC) from the Taiwan Biobank and Europeans of White ancestry from the UK Biobank, representing two large ethnicities. Our data show that at higher-than-normal levels of Hb, increasing deciles of Hb concentration were significantly associated with MetS across all sex subgroups in both ethnicities, with the highest deciles resulting in up to three times greater risk than the reference group [Taiwanese HC: OR = 3.17 (95% CI, 2.75–3.67) for Hb $\geq$ 16.5 g/dL in men, OR = 3.11 (2.78–3.47) for Hb $\geq$ 14.5 g/dL in women; European Whites: OR = 1.89 (1.80–1.98) for Hb $\geq$ 16.24 g/dL in men, OR = 2.35 (2.24–2.47) for Hb $\geq$ 14.68 g/dL in women]. The association between stronger risks and increasing Hb deciles was similarly observed with all metabolic components except diabetes. Here we found that both the highest Hb decile groups and contrarily the lowest ones, with respect to the reference, were associated with higher odds of diabetes in both ethnic groups [e.g., Taiwanese HC men: OR = 1.64 (1.33–2.02) for Hb $\geq$ 16.5 g/dL, OR = 1.71 (1.39–2.10) for Hb $\leq$ 13.5 g/dL; European Whites women: OR = 1.39 (1.26–1.45) for Hb $\geq$ 14.68 g/dL, OR = 1.81 (1.63–2.01) for Hb $\leq$ 12.39 g/dL]. These findings confirm that elevated Hb concentrations, a potential indicator of iron overnutrition, may play a role in the pathophysiology of MetS and metabolic components.

Keywords: hemoglobin; iron nutrition status; metabolic syndrome; metabolic disorders; observational study; Taiwanese Han Chinese; European White

Citation: Timoteo, V.J.; Chiang, K.-M.; Pan, W.-H. Positive or U-Shaped Association of Elevated Hemoglobin Concentration Levels with Metabolic Syndrome and Metabolic Components: Findings from Taiwan Biobank and UK Biobank. *Nutrients* 2022, 14, 4007. https://doi.org/10.3390/nu14194007

Academic Editor: Dietmar Enko

Received: 10 August 2022
Accepted: 22 September 2022
Published: 27 September 2022

Publisher's Note: MDPI stays neutral with regard to jurisdictional claims in published maps and institutional affiliations.

1. Introduction

Iron is an essential mineral that plays crucial roles in a wide variety of metabolic processes in the human body [1]. It primarily functions in the binding and transporting of oxygen molecules in circulation. It further takes part in immune function, DNA synthesis and cell division, electron transport within cells, and forms an integral part of the vital enzyme systems found in various tissues.

Among the most common diseases in humans are those linked to iron metabolism and homeostasis, ranging from anemia (i.e., commonly caused by iron deficiency) to hemochromatosis (i.e., iron overload) [2]. Anemia, clinically described as a condition wherein the number and size of red blood cells (RBCs) and their oxygen-carrying capacity become insufficient to meet physiological needs, remains to be a global public health problem [3]. It afflicted 1.62 billion individuals worldwide based on the published data of the World Health Organization (WHO) Global Database in 2008 [4]. More recently, in

2019, anemia was reported to globally affect 39.8% of children and 29.9% of women at reproductive age [5].

Meanwhile, epidemiological studies have implicated that iron overnutrition and elevated iron levels are associated with higher risk of adverse metabolic and cardiovascular outcomes [6]. Metabolic syndrome (MetS) is a disease entity characterized by the clustering of insulin resistance, impaired glucose tolerance, dyslipidemia, hypertension, and central obesity–all of which can lead to major cardiovascular events [7]. A global epidemic of MetS has been recognized for some time as a result of genetic susceptibility and lifestyle changes associated with modernization and urbanization (e.g., poor dietary quality, sedentary lifestyle, and physical inactivity). There has been a continuous increase in the prevalence of MetS in both men and women in all age groups on a global scale [7]. A 25.5% prevalence of MetS among Taiwanese Han Chinese (HC) was reported in the Nutrition and Health Survey in Taiwan (NAHSIT) from 2005 to 2008, with a lower proportion of women affected than men aged below 45 years, but with a higher proportion of women affected than men for those aged 45 years and over [8,9]. Similarly, in the United Kingdom, a MetS prevalence of 24.3% (i.e., one in four adults) was reported across several European cohorts [10].

Oxidative stress is known to be involved in the pathophysiology of many chronic non-communicable diseases, including MetS. Although there are other processes contributing to their onset, oxidative stress leads to the development of prolonged inflammatory state and further complications [11]. Iron deficiency and overload both influence the redox states. Nonetheless, the exact mechanisms as to how iron contributes to metabolic derangements are yet to be fully elucidated. It was proposed long before, through the 'iron hypothesis', that iron depletion protects against heart disease whereas high levels of body iron stores promote cardiovascular and metabolic diseases [12]. To date, most evidence supports that iron is a powerful pro-oxidant that in excess can cause cellular damage by producing reactive oxygen species in different tissues of the body, which may eventually lead to atherosclerotic events [13].

Iron, being an essential micronutrient, comes entirely from the diet. However, the bioavailability of iron depends on the dietary matrix, which, in turn, is constituted by the major dietary patterns found in a population [1,14]. Heme iron from meat, poultry, and fish is a superb source of iron compared to the non-heme iron from plant sources. Red meat consumption is known to contribute to the storage of excess iron in the body. On the contrary, non-heme iron, despite its lower bioavailability, was shown to contribute more to iron nutrition due to its abundant quantity in the diet [15].

Approximately two-thirds (65%) of iron in the human body are integrated in the protein hemoglobin (Hb) in circulating RBCs. The four heme units of Hb contain an iron cation (Fe^{2+}) that switches redox states upon the binding and release of oxygen from the lungs to cells and of carbon dioxide from the cells to the lungs [16]. Therefore, Hb concentration in the blood may serve as a surrogate of iron nutrition status as it reflects a major functional iron in the body (i.e., heme iron) [17]. Nonetheless, Hb concentration may be affected by other prevailing micronutrient deficiencies, acute or chronic infections, inflammation, and disorders that alter RBC metabolism [18]. Hb is a good iron marker when the body iron store is low, whereas at the higher end of iron spectrum, variations in Hb concentration may be due to other nutritional or non-nutritional factors including autoimmune disease [19]. It is further important to note that other biomarkers (i.e., serum ferritin, transferrin saturation, etc.) should be considered with respect to the storage and transport of iron in humans.

There have been on-going controversies in human epidemiological studies as to whether the disruption of iron homeostasis that leads to elevated Hb levels may result in higher risks of metabolic disorders. Hence, we conducted sex-specific association studies to determine and validate the relationship of iron in excess with the risks of MetS and its individual components. We carefully titrated the association between Hb deciles and MetS, central obesity, hypertension (HTN), diabetes (i.e., type 2 diabetes or T2D), dyslipidemia (DLP), and gout. Gout was additionally included as an outcome due to the emerging evidence that suggests its role in the pathophysiology of MetS [20]. We utilized two, large population-based cohorts, the Taiwan Biobank (TWB) and UK Biobank (UKB), to represent the distinct ethnicities of East Asians and Caucasians, respectively. Our findings then confirm the role of elevated iron status, as demonstrated by increasing Hb concentrations, in the pathophysiology of MetS and metabolic outcomes in both the Taiwanese HC and European Whites. Intriguingly, we observed that both low and high levels of Hb contribute to the pathogenesis of T2D, whereas within normal levels, Hb confers protective effects. These findings will help to further our understanding on the role of iron dysregulation in human health and better identify at-risk individuals; hence, preventing further complications associated with the MetS and other cardiometabolic outcomes–ultimately leading to the improvement of public health in the future.

2. Materials and Methods

2.1. Study Design, Study Populations, and Ethical Considerations

We conducted cross-sectional observational studies utilizing two population-based cohorts, the Taiwan Biobank and the UK Biobank. The TWB is a large-scale collection of data on the genomic profiles, lifestyle, and environmental exposure history, and long-term health outcomes of the Han Chinese population in Taiwan. It has been established to elucidate the interrelationship between genetic and environmental factors in disease onset and progression [21]. The UKB, on the other hand, is a large prospective study of more than 500,000 individuals residing in the UK, whose extensive phenotype information, genotypes, and details on assessment and follow-ups have been reported in detail elsewhere [22]. Information on the TWB and UKB Projects can be accessed at https://www.twbiobank.org.tw/ (accessed on 6 June 2022) and https://www.ukbiobank.ac.uk/ (accessed on 6 June 2022).

The TWB cohort initially comprised 67,515 participants (20,764 men and 46,751 women), aged from 30 to 70 years old at the time of recruitment. These were community-dwelling and non-handicapped individuals during assessment. In order to maximize the power of the TWB study, we included all subjects at first, regardless of age. We excluded subjects who had self-reported renal/kidney failure, been diagnosed with cancer, or had missing data on Hb concentration.

From a total of 502,536 UKB participants (229,134 men; 273,402 women) aged from 40 to 70 years old during assessment (UKB data field 21003), we first selected participants who identified themselves as White (i.e., British, Irish, or any other White background) (field 21000) and were born in the UK (field 1647). We removed subjects who had inconsistent reported and genetic sex (fields 31 and 22001), were pregnant during data collection (field 3140), had been diagnosed with any form of cancer (fields 134 and 20001), and had self-reported renal/kidney failure, hereditary/genetic hematological disorder, clotting disorder/excessive bleeding, acquired immunodeficiency syndrome (HIV/AIDS), tuberculosis, or tropical and travel-related infections (field 135). These conditions are known to affect erythropoiesis and, thus, may influence the Hb levels of an individual [18]. We further excluded subjects with fasting time of 0 or beyond 24 h (field 74) and those with missing Hb data (field 30020). File S1 provides the list of variables and data fields used in the UKB study.

Upon further exclusion of subjects with extreme Hb values, a total of 67,237 Taiwanese HC (20,670 men and 46,567 women) were included in the analyses. On the other hand, a final population of 386,477 Europeans of White ancestry (182,048 men and 204,429 women) were analyzed.

This study has been carried out in accordance with The Code of Ethics of the World Medical Association (Declaration of Helsinki). Study protocols were evaluated and approved by the Institutional Review Board of Academia Sinica, with reference number AS-IRB 02. The UKB Project has been given ethical clearance by the UK National Health Service's National Research Ethics Committee with reference number 11/NW/TWB and

UKB participants gave duly signed written informed consents prior to the conduct of data collection. All data obtained were treated with the utmost confidentiality.

2.2. Blood and Data Collection

Venous blood samples were drawn from the TWB and UKB participants by qualified and trained medical researchers for hematological and biochemical assessments. Blood samples collected in EDTA (Ethylenediaminetetraacetic acid) vacutainer tubes were analyzed within 24–48 h after blood draw using automated, quality control-checked clinical hematology analyzers. Hemoglobin is one of the hematological traits measured at baseline recruitment, where anemia is defined by a Hb level <13.0 g/dL in men and <12.0 g/dL in women [23]. Serum and plasma were then separated from the whole blood samples for the measurement of a wide range of biochemical markers in the TWB and UKB. Glucose (GLU), triglyceride (TG), cholesterol (TC), high-density (HDL-C), and low-density (LDL-C) lipoprotein-associated cholesterol, and uric acid (UA) were measured via enzymatic assays using routinely calibrated clinical chemistry analyzers, whereas glycated hemoglobin (HbA1c) was determined by a high-performance liquid chromatography method. In the UKB, however, blood biochemistry was based on a random and non-fasting state of participants.

Standard procedures were followed in collecting anthropometric and blood pressure readings. Briefly, weight was recorded using a platform weighing scale whereas standing height was measured using a microtoise or medical measuring rod. Body mass index (BMI) was calculated as weight in kg divided by the square of height in m (kg/m^2). Waist (WC) and hip circumferences were measured at the approximate midpoint between the lowest rib bone and super iliac point and the widest part of the hips, respectively, with a tape measure placed parallel to the floor at the end of a relaxed expiration of participants while standing. Waist-hip ratio (WHR) was simply calculated as waist circumference divided by the hip circumference. Systolic (SBP) and diastolic blood pressure (DBP) levels were obtained using automated sphygmomanometer following stringent protocol. All measurements were taken twice.

The socio-demographic profile, behavioral risk factors such as smoking and alcohol consumption, physical activity, and other health-related information about the participants were obtained through a face-to-face interview in the TWB. On the other hand, an extensive array of lifestyle, environmental, health, and medical data among UKB participants were collected via a touchscreen questionnaire, followed by a verbal interview for some variables that required verification (i.e., type of prescription medications).

2.3. Definition and Ascertainment of Metabolic Outcomes

Metabolic syndrome and its associated components (i.e., central obesity, hypertension, diabetes, dyslipidemia, and gout) are the metabolic outcomes, as dichotomous variables, in this study. We defined outcomes based on self-reports of physician-diagnosed health conditions by the participant and the widely used cut-offs for WC, SBP/DBP, GLU, TG, HDL-C, and UA. In the UKB, outcomes were additionally classified using data on medications and hospital episode statistics following the International Classification of Diseases 9th (ICD9) and 10th (ICD10) revisions. File S2 summarizes the disease definitions based on ICD and self-reported fields, whereas File S3 lists the prescription medications in the ascertainment of HTN, T2D, DLP, and gout in the UKB.

The definition of central or abdominal obesity was adopted from the International Diabetes Federation (IDF) criteria: a WC > 90 cm in men or >80 cm in women for East Asians and a WC > 94 cm in men or >80 cm in women for Europeans [24,25]. An SBP/DBP reading of $\geq$140/$\geq$90 mm Hg classified hypertension in the TWB and UKB following the Taiwan Society of Cardiology/Taiwan Hypertension Society and the European Society of Cardiology/European Society of Hypertension criteria, respectively [26,27]. Type 2 diabetes in TWB was defined by a high fasting blood glucose (FBG) level $\geq$ 126 mg/dL [28] or a glycated hemoglobin $\geq$ 6.5% [29]. To convert FBG in mg/dL to SI units, multiply by

18 mmol/L [30]. HbA1c in mmol/mol is calculated as 10.93 × (HbA1c in %) − 23.50 [31]. In the ascertainment of T2D in UKB subjects, we followed the algorithms proposed by Eastwood and team, where GLU ≥ 11.1 mmol/L or HbA1c ≥ 48 mmol/mol (6.5%) captured hyperglycemia from blood tests even on a non-fasting state [32]. Hypertriglyceridemia, hypercholesterolemia, high LDL-cholesterol levels, and low HDL-cholesterol levels were classified in the TWB and UKB using the following cut-offs: TG ≥ 200 mg/dL (≥2.30 mmol/L), TC ≥ 240 mg/dL (≥6.20 mmol/L), LDL-C ≥ 160 mg/dL (≥4.10 mmol/L), and HDL-C < 40 mg/dL (<1.00 mmol/L) in men or <50 mg/dL (<1.30 mmol/L) in women [33,34]. Dyslipidemia in this study is described as a combination of either hypertriglyceridemia or low levels of HDL-C. TG in mg/dL is converted to SI units by multiplying by 0.01129 mmol/L, whereas TC, LDL-, and HDL-C are multiplied by 0.02586 mmol/L [35]. Lastly, gout was ascertained from a hyperuricemic level of >7.0 mg/dL (>416.0 µmol/L) in men or >6.0 mg/dL (>357.0 µmol/L) in women [36,37]. UA from mg/dL to µmol/L is computed using a conversion factor of 1 mg/dL = 59.48 µmol/L [38].

The definition of metabolic syndrome was mainly based on the NCEP ATP III criteria [39], whereas the WC cut-off point for Asians from IDF [24] was used in analyzing the Taiwanese HC data. In the NCEP ATP III classification, MetS was present upon meeting at least three of the following component risk factors: (1) WC > 102 cm in men or >88 cm in women; (2) TG ≥ 150 mg/dL (≥1.70 mmol/L) or taking triglyceride-lowering drugs; (3) HDL-C < 40 mg/dL in men or <50 mg/dL in women or taking statins or other medicines for high cholesterol; (4) blood pressure ≥130/≥85 mmHg or current use of anti-hypertensive drugs; and (5) FBG ≥ 100 mg/dL (≥5.6 mmol/L) or current use of anti-hyperglycemic drugs. The ethnic-specific criteria for central obesity in IDF (i.e., for Taiwanese HC) is WC > 90 cm in men and >80 cm in women.

2.4. Data Processing and Statistical Analyses

Data were analyzed using SAS v.9.4. The criteria for statistical significance were a *p*-value of <0.05.

For describing the characteristics of participants, continuous variables were expressed as mean ± S.D., whereas categorical characteristics were in counts and percentage values. The normal distribution of quantitative parameters was assessed, and extreme outliers were removed. We excluded subjects who were more than ± 5 S.D. from the sex-specific Hb means per ethnic group.

For comparing the risks of MetS and associated components, we categorized subjects according to deciles of increasing hemoglobin concentration. The decile with the lowest outcome prevalence served as the reference. Men and women were analyzed separately. Trends in quantitative characteristics and prevalence across Hb deciles were respectively tested by linear regression and the Cochran–Armitage trend test.

Multivariate logistic regression models were used to estimate the ORs and CIs of MetS and its individual components across Hb deciles. Age and age-squared were adjusted in the first model. Smoking status and alcohol consumption were further adjusted as covariates in the second model. The final model was fully adjusted for physical activity level, highest educational attainment, and comorbidities (i.e., for individual components of MetS). Fasting time was included in the UKB analyses to adjust for random/non-fasting blood biochemistry collection. Menopausal status was also added as a covariate in the women subgroups, wherein subjects were categorized as either non-menstruating, unsure of their menopausal status (i.e., had hysterectomy), or had age-at-menopause <44, 45–49, 50–54 (reference), or >55 years. We further adjusted for ethnic-specific BMI category groups [40,41] in all models when testing the associations between Hb and HTN, T2D, DLP, and gout.

In the UKB, the final categorization of smoking status (i.e., never smoked, stopped smoking, occasionally smoking, and currently smoking) was derived from smoking status (field 20116) and current tobacco smoking (field 1239). Alcohol drinking status (i.e., never drank, stopped drinking, occasionally drinking, and currently drinking) was similarly recoded from alcohol drinker status (field 20117) and alcohol intake frequency (1558). The highest educational qualification in the UK (field 6138) was recategorized into four levels as none; O-levels, CSEs, or equivalent; A-levels, NVQ/HND/HNC or equivalent, or other professional qualifications; and college or university degree holder.

3. Results

3.1. Characteristics of Taiwanese Han Chinese and European White Cohorts

The characteristics of TWB and UKB cohorts by sex and increasing Hb deciles were presented in detail in Tables S1 and S2, respectively. A total of 67,237 Taiwanese Han Chinese (20,670 men; 46,567 women) from the TWB and 386,477 European Whites (182,048 men; 204,429 women) from the UKB were included in the analyses. The mean ages were around 50 years old for male and female TWB participants and around 56–57 years for UKB counterparts, respectively. Around half of the Taiwanese HC women had already entered menopause (51.3%) during assessment, with a mean age-at-menopause of 49.4 ± 5.0 years. A large proportion of European women subjects also had menopause (60.6%), with a mean menopausal age of 49.8 ± 5.1 years.

The mean Hb levels of men subgroups were found to be similar in both populations (15.0 g/dL), whereas European women had a slightly higher mean Hb (13.5 ± 0.93 g/dL) than Taiwanese HC women (13.0 ± 1.3 g/dL). The interquartile ranges of Hb concentration were wider in the Taiwanese HC (14.4–15.8 g/dL in men; 12.5–13.8 g/dL in women) than in Europeans (14.4–15.7 g/dL in men; 12.9–14.1 g/dL in women). In both ethnic groups, men had higher mean measurements in all anthropometric, biochemical, and clinical parameters than women, except TC and LDL-C.

Majority of the subjects were never-smokers. In particular, nearly all of the Taiwanese HC women had never smoked a cigarette. Taiwanese HC were mostly occasional-drinkers whereas larger percentages of Europeans were current-drinkers. More than half of the Taiwanese HC did not regularly exercise. Meanwhile, majority of European men and women engaged in high and moderate physical activities, respectively. For both ethnic groups, however, a general increase in the proportion of current-smokers, current-drinkers, and those with no regular exercise (or had low physical activity levels) were observed with increasing Hb deciles. For instance, 32.9% of Taiwanese HC men and 14.1% of European men were current-smokers in the highest Hb deciles as compared to the 16.3% and 7.4%, respectively, in the lowest deciles. Lastly, at least half of the Taiwanese HC reached university or post-graduate studies, whereas relatively lower percentages of Europeans obtained a degree. Across increasing Hb concentration, we also observed a general increase in percentages of those with no formal education or a decrease in percentages of degree-holders.

3.2. Prevalence Rates of Metabolic Syndrome and Metabolic Components and Associations with Increasing Hemoglobin Deciles among Taiwanese Han Chinese and European Whites

There were significant differences in the prevalence rates of MetS and metabolic components by ethnic group (Tables S3 and S4 show by sex and Hb deciles the frequency distributions of outcomes based on measured health indicators; Tables S5 and S6 present the frequencies of cases based on self-reported data; Table S7 provides the frequencies from hospital in-patient records in the UKB).

Europeans were observed to have a higher mean weight, BMI, WC, SBP, and DBP than the Taiwanese HC counterparts. The mean TG, TC, and LDL-C of Europeans were also higher than the Taiwanese HC (i.e., TG: 1.98 mmol/L in European men versus 1.57 mmol/L in Taiwanese HC men; 1.55 mmol/L in European women versus 1.17 mmol/L in Taiwanese HC women). Conversely, the mean glucose, HbA1c, and UA levels of Taiwanese HC were found larger than the Europeans, despite a non-fasting blood collection among the latter (i.e., glucose: 5.5 mmol/L in Taiwanese HC men versus 5.1 mmol/L in European men; 5.2 mmol/L in Taiwanese HC women versus 5.1 mmol/L in European women).

Overall, we present that among the MetS components, central obesity was consistently highly prevalent among the Taiwanese HC (44.8%) and European Whites (56.7%), followed by dyslipidemia (i.e., combination of either hypertriglyceridemia or low levels of HDL-C) (34.3%; 51.9%), hypertension (23.9%; 56.6%), gout (20.2%; 14.4%), and type 2 diabetes (10.2%; 7.2%) (Tables S8 and S9).

We noted the highly significant increasing trends in the prevalence rates of MetS and other metabolic components with increasing deciles of Hb concentration among Taiwanese HC and European Whites ($p < 0.0001$) (Tables S3–S7). However, on the other side of the Hb distribution, we also observed that the Hb decile groups which comprised mostly the anemic subjects (i.e., D1 in Taiwanese HC men and Europeans; D1 and D2 in Taiwanese HC women) had higher prevalence rates of MetS, central obesity, hyperglycemia, low HDL-C, and hyperuricemia than those in the reference deciles. The elevation of rates is strong for hyperglycemia but modest for others.

Consistent with the above observations, increasing Hb was found significantly associated with the increasing odds of MetS (Figure 1; Tables S10 and S11) and its metabolic components (Figure 2; Tables S12–S21) across all sex subgroups in both ethnicities, except for T2D. Higher deciles of Hb concentration further corresponded to stronger odds of central obesity (Figure 2; Tables S12 and S13), hypertension (Tables S14 and S15), dyslipidemia (Tables S16 and S17), and gout (Tables S20 and S21) in both ethnic groups. However, the slightly higher MetS prevalence observed among the mostly anemic subjects in the lower Hb deciles corresponded to ORs that were non-significant. Our findings on diabetes were different from those of other MetS components (Figure 3; Tables S22 and S23). Interestingly, the highest or lower Hb deciles, as compared to the reference, were associated with stronger odds of T2D.

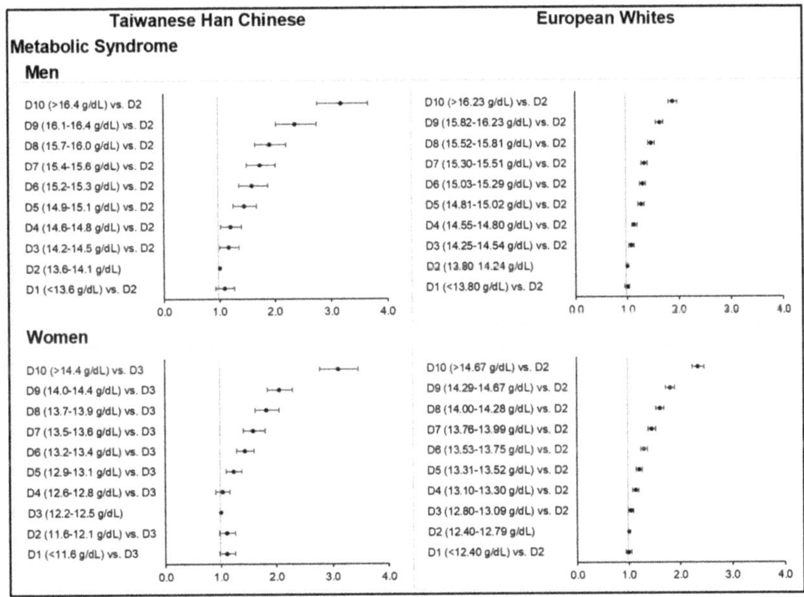

Figure 1. Odds ratios of metabolic syndrome across increasing deciles of hemoglobin concentration among Taiwanese Han Chinese and European Whites (top: men, bottom: women). Final models adjusted for age, age-squared, smoking status, alcohol consumption, physical activity level, and highest educational attainment, as well as menopausal status in women.

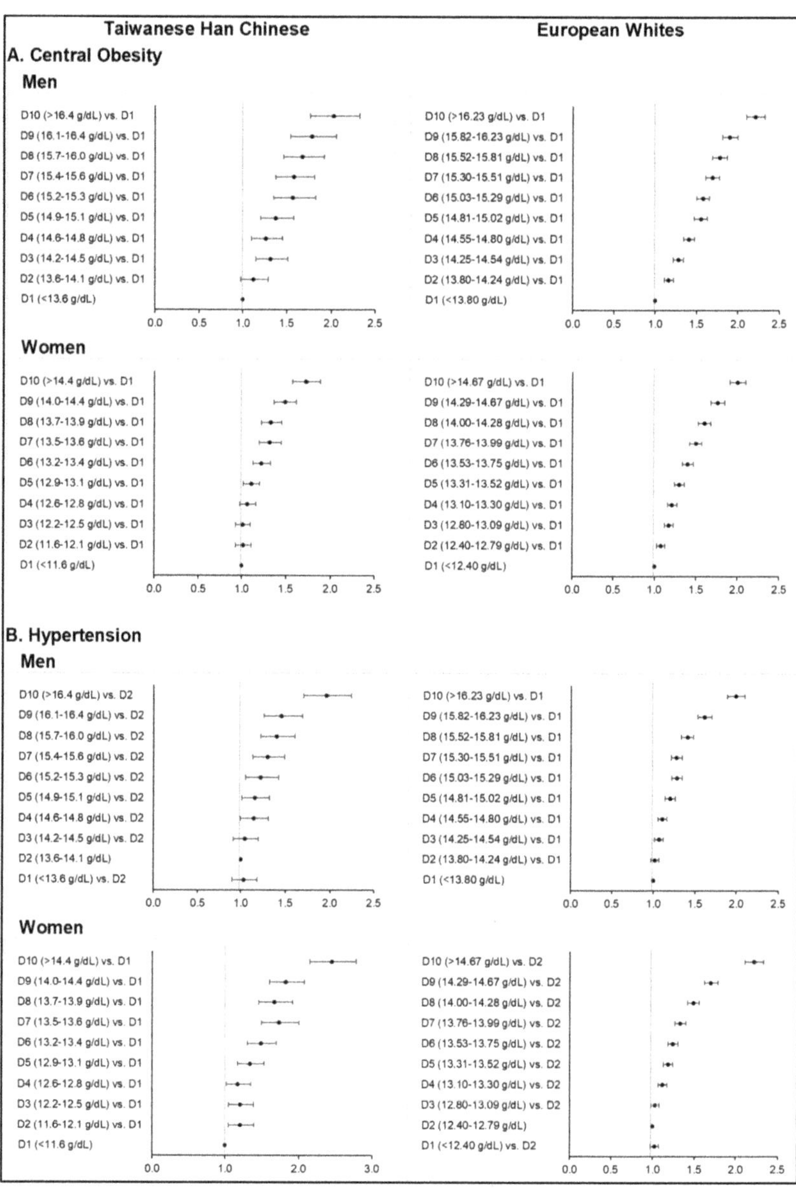

Figure 2. *Cont.*

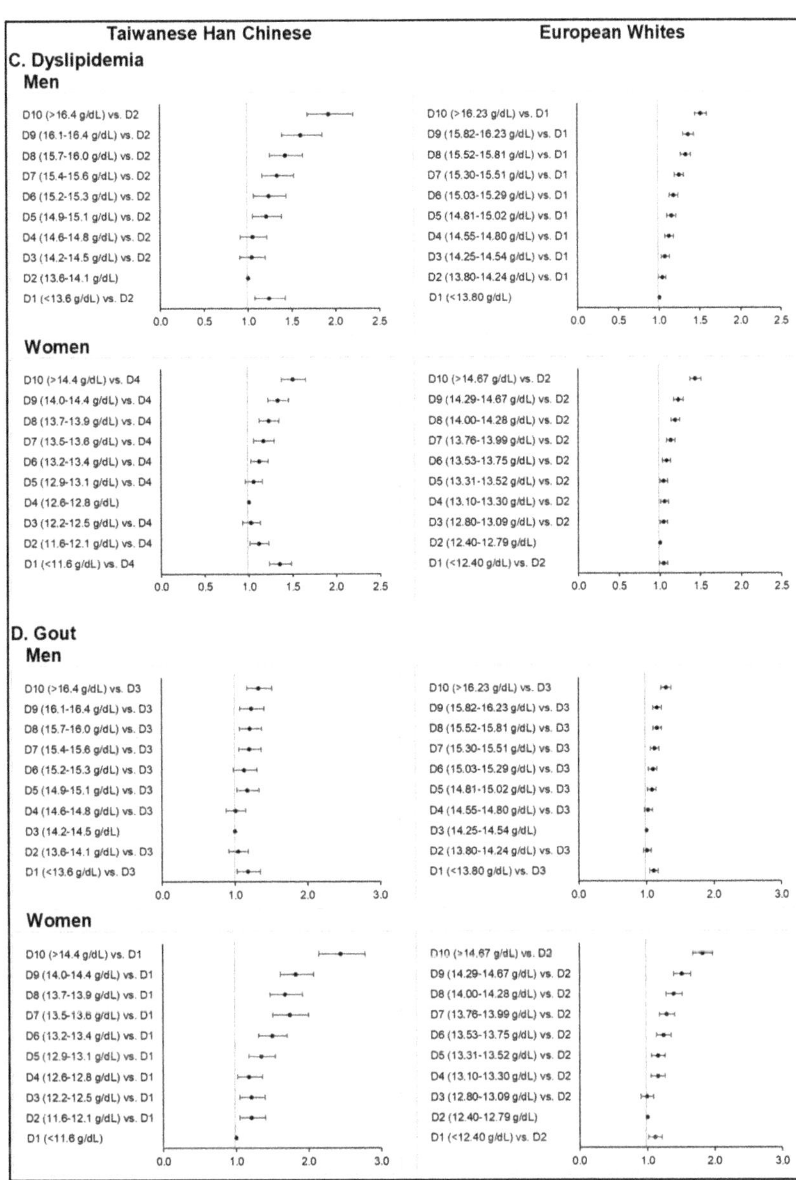

Figure 2. Odds ratios of central obesity, hypertension, dyslipidemia, and gout across increasing deciles of hemoglobin concentration among Taiwanese Han Chinese and European Whites (top: men, bottom: women). Final models adjusted for age, age-squared, smoking status, alcohol consumption, physical activity level, highest educational attainment, and comorbidities, as well as menopausal status in women.

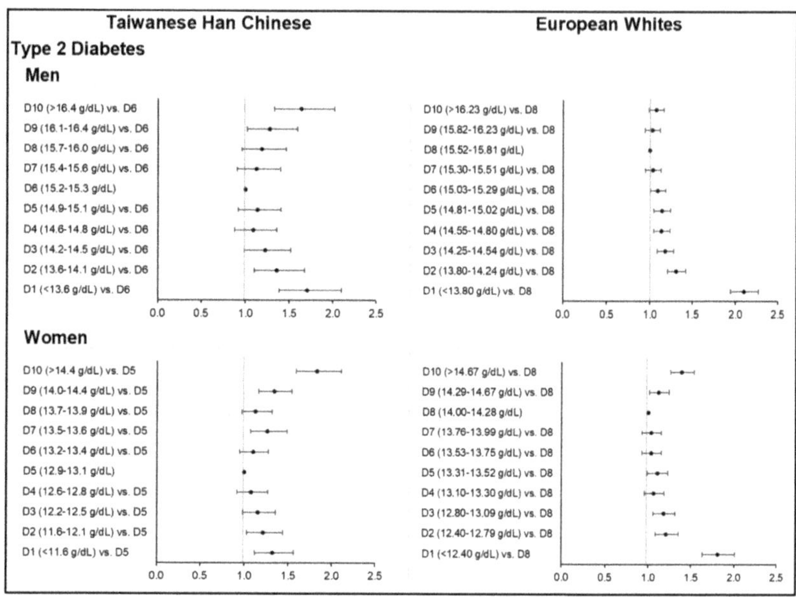

Figure 3. Odds ratios of diabetes across increasing deciles of hemoglobin concentration among Taiwanese Han Chinese and European Whites (top: men, bottom: women). Final models adjusted for age, age-squared, smoking status, alcohol consumption, physical activity level, highest educational attainment, and comorbidities, as well as menopausal status in women.

3.3. Metabolic Syndrome

The prevalence of MetS following the NCEP ATP III criteria was 25.7% among Taiwanese HC men and 18.9% among women at the time of assessment. MetS prevalence was slightly increased to 26.8% in men when using the Taiwan's modified criteria. Higher prevalence rates among European Whites were observed in men at 49.0% and in women at 37.5%. Subjects in the highest Hb deciles who were classified as having MetS were at 40.7% and 34.4% among Taiwanese HC men and women, respectively, and 58.4% and 53.0% among European men and women, respectively.

At higher-than-normal range of Hb concentration, increasing deciles of Hb were significantly associated with MetS across all sex subgroups in both ethnicities (Figure 1). The highest Hb deciles resulted in up to three times greater odds of having MetS than the reference group, upon fully adjusting for age, age-squared, smoking status, drinking status, physical activity level, and highest educational attainment (and menopausal status in women) (Tables S10 and S11). Particularly among Taiwanese HC men and women, ORs for MetS were 3.17 (2.75–3.67, $p < 0.0001$) for Hb $\geq$ 16.5 g/dL and 3.11 (2.78–3.47, $p < 0.0001$) for Hb $\geq$ 14.5 g/dL, respectively. Abrupt increases in ORs between D8 and D10 in this group were evident (i.e., from 1.90 (1.65–2.20) at 15.7–16.0 g/dL to 2.36 (2.02–2.74) at 16.1–16.4 g/dL to 3.17 (2.75–3.67) when Hb was $\geq$16.5 g/dL in men). Among Europeans, much lower ORs were obtained despite a larger sample size and higher prevalence rates of MetS across Hb deciles: 1.89 (1.80–1.98, $p < 0.0001$) at Hb $\geq$16.24 g/dL in men and 2.35 (2.24–2.47, $p < 0.0001$) at Hb $\geq$ 14.68 g/dL in women. ORs were generally attenuated after fully adjusting for covariates.

Central obesity was more prevalent among Europeans than Taiwanese HC when following the WHO's WC criteria and ethnic-specific cut-off points. Similar to MetS, more than half of the European subjects were with central obesity. It is also evident that among Europeans and Taiwanese HC in the highest deciles of Hb concentration, central obesity was highly prevalent.

Among the Taiwanese HC men and women, fully adjusted ORs from the highest Hb deciles were respectively 2.03 (1.77–2.33, $p < 0.0001$) and 1.72 (1.57–1.89, $p < 0.0001$) for central obesity (Table S12). On the other hand, ORs obtained from the highest deciles among European men and women were 2.21 (2.11–2.33, $p < 0.0001$) and 2.00 (1.91–2.11, $p < 0.0001$) (Table S13).

Monotonous risk elevations with increasing Hb deciles were observed in men and women for both ethnic groups. Among Europeans, starting from the lowest decile within normal Hb concentrations, highly significant positive associations with central obesity were noted.

3.4. Hypertension

Greater proportions of Europeans were found to be hypertensive than the Taiwanese HC, either based on blood pressure measurement, self-reported data, or hospital in-patient records. More than half of the European men (i.e., 52.8%) were hypertensive based on office blood pressure readings alone. Furthermore, the majority of European subjects in the 10th deciles were classified as hypertensive (i.e., 63.5% in men and 55.2% in women). On the other hand, hypertensive Taiwanese HC based on blood pressure readings were 24.5% among men and 13.0% among women, whereas only 18.4% of men and 10.6% of women had self-reported hypertension.

The fully adjusted ORs for hypertension from the highest Hb deciles in Taiwanese HC men and women were 1.95 (1.70–2.24, $p < 0.0001$) and 2.12 (1.89–2.39, $p < 0.0001$), respectively (Table S14). Among European men and women, ORs for hypertension were at 1.99 (1.89–2.10, $p < 0.0001$) and 2.22 (2.11–2.33, $p < 0.0001$), respectively (Table S15). Significant risk elevation starts from the 4th decile for both gender and ethnicities. However, ORs became non-significant if we further adjusted for BMI groups.

3.5. Dyslipidemia

Europeans with hypertriglyceridemia, hypercholesterolemia, high LDL-C, and low HDL-C levels based on blood lipid profile were all higher in prevalence rates as compared to the Taiwanese HC. Notably, among the Taiwanese HC, observed prevalence rates in the highest Hb deciles were as high as thrice of that in the reference deciles. Hypertriglyceridemia was 28.8% among the Taiwanese HC men whose Hb levels were ≥16.5 g/dL; whereas hypertriglyceridemia was only 10.9% among those whose Hb were 13.6–14.1 g/dL. Among Taiwanese HC women, the prevalence of high LDL-C was 19.8% when Hb was ≥14.5 g/dL, whereas only 7.7% were identified in the 12.2–12.5 g/dL Hb range.

The fully adjusted ORs for dyslipidemia (i.e., combination of either hypertriglyceridemia or low HDL-C levels) from the highest Hb deciles were 1.92 (1.68–2.20, $p < 0.0001$) among Taiwanese HC men and 1.50 (1.37–1.65, $p < 0.0001$) among women (Table S16); whereas ORs obtained in the highest deciles among European men and women were 1.51 (1.44–1.59, $p < 0.0001$) and 1.44 (1.37–1.51, $p < 0.0001$), respectively (Table S17). Specific to the Taiwanese HC group, we observed the lowest Hb deciles to be positively associated with dyslipidemia as compared to the reference: OR = 1.24 (1.07–1.42, $p = 0.0034$) at Hb ≤13.5 g/dL in men and OR = 1.35 (1.23–1.48, $p < 0.0001$) at Hb ≤ 11.5 g/dL in women.

We also observed higher risks at higher Hb deciles for hypercholesterolemia or raised levels of either total cholesterol or LDL-C (Tables S18 and S19). However, highly significant positive associations were only identified in the Taiwanese HC group, especially among women, whereas risk elevations started from the 4th decile in the European group. ORs from the highest Hb deciles were 1.92 (1.65–2.25) among Taiwanese HC men, 2.67 (2.37–3.02) among Taiwanese HC women, 1.23 (1.18–1.29) among European men, and 1.35 (1.29–1.42) among European women ($p < 0.0001$). As opposed to a combination of hypertriglyceridemia and low HDL-C levels, we did not find any significant positive association with the lowest Hb decile for elevated cholesterol levels.

3.6. Gout or Hyperuricemia

The percentages of Taiwanese HC men and women with hyperuricemia were almost twice of that in Europeans. In particular, 29.7% of Taiwanese HC men and 17.7% of European men had hyperuricemia whereas there were 14.1% of Taiwanese HC women and 9.2% of European women with hyperuricemia. Similar observations on self-reported gout were noted for both ethnic groups.

Among the Taiwanese HC men and women, fully adjusted ORs for gout from the highest Hb deciles were 1.33 (1.17–1.51, $p < 0.0001$) and 2.44 (2.15–2.78, $p < 0.0001$) (Table S20). A highly significant monotonous risk elevation was evident among the Taiwanese HC women. ORs obtained in the highest deciles among European men and women were 1.29 (1.22–1.37, $p < 0.0001$) and 1.84 (1.70–1.99, $p < 0.0001$), respectively (Table S21). Similar to dyslipidemia, significant positive associations with gout were also identified in the lowest Hb deciles: OR = 1.18 (1.03–1.35, $p = 0.0181$) at Hb $\leq$ 13.5 g/dL among Taiwanese HC men, OR = 1.11 (1.04–1.17, $p = 0.0010$) at Hb $\leq$ 13.79 g/dL among European men, and OR = 1.12 (1.03–1.23, $p = 0.0113$) at Hb $\leq$ 12.39 g/dL among European women.

3.7. Type 2 Diabetes

Type 2 diabetes based on HbA1c cut-off and self-reported data were observed to be higher among Taiwanese HC (i.e., 11.7% and 7.4%, respectively, in men) than Europeans (i.e., 4.4% and 6.5%, respectively, in men). However, the percentage of European men with diabetes increased (i.e., 17.4%) using hospital in-patient records. On the basis of FBG criteria, lower percentages of subjects with hyperglycemia were noted. There were even much lower counts of diabetic European subjects identified from an 11.1 mmol/L GLU threshold, which was set to exclude false positives from a non-fasting glucose reading in the UKB. Among Europeans, the mean GLU and HbA1c levels of men were in a down-and-up trend with increasing Hb concentration, although the upward trend was rather small. On the other hand, the mean GLU levels of women continuously decreased from the first to fourth decile, followed by a constant increase from the fifth to the last decile. Hyperglycemia prevalence across Hb deciles was not statistically significant in the Taiwanese HC men and European women subgroups.

A U-shaped curvilinear association between Hb and diabetes risk was evident (Tables S22 and S23). The ORs obtained for diabetes in the lowest or highest decile were as follows: Taiwanese HC men: OR = 1.71 (1.39–2.10, $p < 0.0001$) at $\leq$13.5 g/dL and OR = 1.64 (1.33–2.02, $p < 0.0001$) at $\geq$16.5 g/dL Hb; Taiwanese HC women: OR = 1.32 (1.12–1.56, $p = 0.0010$) at $\leq$11.5 g/dL and OR = 1.83 (1.59–2.10, $p < 0.0001$) at $\geq$14.5 g/dL Hb; European Whites men: OR = 2.10 (1.94–2.27, $p < 0.0001$) at $\leq$13.79 g/dL Hb; European Whites women: OR = 1.81 (1.63–2.01, $p < 0.0001$) at $\leq$12.39 g/dL and OR = 1.39 (1.26–1.54, $p < 0.0001$) at $\geq$14.68 g/dL Hb.

In both ethnic groups, the associations between increasing Hb and risks of individual metabolic components were found modest, as compared to the Hb-MetS associations. ORs were attenuated in the final regression models, wherein we additionally accounted for the presence of comorbidities. ORs were further attenuated upon adjusting for BMI groups in the models for hypertension, dyslipidemia, and gout, but the trends still remained (Tables S24.1–S24.8).

4. Discussion

Our sex-specific observational association studies determined and validated the relationship between iron excess, as depicted by elevated hemoglobin concentration levels, and the risks of metabolic syndrome and metabolic components (i.e., central obesity, hypertension, type 2 diabetes, dyslipidemia, and gout) in two major ethnic populations, the Han Chinese and Europeans. With the careful titration of the associations between Hb deciles and most MetS components, monotonous MetS risk elevations in both men and women were found to be evident with an increasing Hb concentration. We additionally discovered that the Hb-T2D association was different from other MetS components investigated, which showed more of an inverse association, but both the highest and lowest Hb deciles led to higher odds of T2D as compared to the middle deciles for both the Taiwanese HC and European cohorts. Few studies made comparison among all MetS components to provide insight to MetS etiology.

4.1. Elevated Hemoglobin Concentration Is Associated with Increased Risks of Metabolic Syndrome and Several Individual Metabolic Components

The positive association between Hb and prevalence of the MetS and metabolic components has been illustrated in a few previous studies [42–44]. Recently, in a cohort of Finnish men and women after a 20-year follow-up [45], higher Hb levels within the normal range were further shown to be associated with an increased incidence of MetS and key components, as well as higher risks of cardiovascular and total mortalities. A similar study conducted in a population of middle-aged and elderly Chinese also demonstrated high Hb levels as a potential predictor of the incidence of MetS and its components, including gout and non-alcoholic fatty liver disease, after years of follow-up [46]. Our present findings corroborated these previous results, while carefully depicting in two ethnicities the relationships for the whole Hb range.

The mechanisms underlying the reported associations between Hb levels and metabolic disorders (or markers) have been poorly understood and not clearly identified. Most evidence points to iron-mediated oxidative stress. Oxidative stress is defined as an imbalance between the overproduction of reactive oxygen species and insufficient antioxidant defenses, which may generally lead to negative health consequences due to free radical-mediated tissue damage [47]. Hb at higher end may also be regarded as a biomarker of the body's inflammatory state [48], aside from being a putative indicator of oxidative stress. Both inflammation and oxidative stress are well-known associated factors in the development of obesity [48,49] and hypertension [50,51]. In dyslipidemia, one of the possible consequences of free radical damage is on lipid peroxidation, which causes the depletion of the cellular content of reduced glutathione, an important player for preventing cell damage [52]. Another proposed mechanism of action of high Hb levels is by increasing blood viscosity (i.e., hyperviscosity) following a prolonged state of oxidative stress that eventually decreases the blood flow throughout the body [53], raising the blood pressure systemically [54]. Elevated blood viscosity might also induce insulin resistance, which is believed to be the root cause of MetS [7]. Other suggested mechanisms linking Hb and MetS include changes in plasma volume, endothelial cell dysfunction, etc. [45].

Meanwhile, hyperuricemia has been recently proposed to be an additional cause of the MetS. Epidemiological studies have likewise shown a positive association between hyperuricemia or gout and MetS [55,56]. The relationship between hyperuricemia and iron accumulation has been illustrated with a strong correlation between serum uric acid and serum ferritin levels, the most commonly deployed indicator for determining iron status [57]. For this reason, elevated uric acid was further proposed as a potential indicator of iron overload [58]. It was hypothesized that the elevation in uric acid levels is a response to counter an increased oxidative stress in the body (i.e., compensatory rather than protective) as urate assumes the coordination sites of iron, leading to a decline in electron transport and subsequent oxidant generation [57]. Although the mechanism remains unclear, looking into the etiology and pathology of hyperuricemia is highly important for Asians due to the relatively higher prevalence rates seen in this group [59]. Based on our data alone, MetS and gout were both found to be more prevalent among the Taiwanese HC than Europeans. However, our findings further demonstrated that although a relationship has been detected, the association between elevated Hb and hyperuricemia or gout was not as great as compared to the other MetS components. This requires further studies.

Taken together, future studies are highly needed to confirm the role of elevated Hb concentration as a risk factor of MetS and its individual components. MetS and all components are considered as major risk factors for cardiovascular diseases and total mortality. In fact, MetS was reported to increase the mortality risk by 46% among individuals with

MetS as compared to those without [60]. Therefore, if these associations are confirmed as causal, primary prevention and optimal treatment regimen employing measures to control iron overnutrition may eventually lead to decreased MetS incidence and morbidity and mortality.

4.2. U-Shaped Association between Hemoglobin Concentration and Type 2 Diabetes Risk

Epidemiological studies that determined the association between iron intake or body iron stores and T2D risk have produced conflicting results between sex and different ethnicities to date [61,62], probably due to failure to carefully titrate the association. However, few studies have demonstrated that the association of Hb concentration with cause-specific and total mortality rates is rather U-shaped and not linear [63,64], and that in general, both low- and high-end Hb levels within the normal variation have been considered beneficial for human health [65]. We are the first to carefully investigate the Hb–T2D relationship.

In diabetes, high levels and stores of iron in the body may result in a selective damage of the pancreatic beta-cells through excessive oxidative stress, which then leads to impaired insulin synthesis and release [66]. Hereditary or primary hemochromatosis frequently gives rise to T2D (i.e., referred to as secondary diabetes), and there are instances when diabetes is the only apparent manifestation of hemochromatosis in patients [67]. Primary hemochromatosis particularly affects the parenchymal cells; thus, producing significant tissue damage [68]. Secondary hemochromatosis or the acquired abnormalities of iron overload, on the other hand, initially affects the reticuloendothelial cells in many organ systems, but eventually will involve the parenchymal cells [69].

Lower-than-normal Hb concentrations were also implicated with the risk of T2D from our findings. In a study that utilized the Women's Health Initiative data of ~160,000 postmenopausal women in order to examine the associations of Hb levels with cancer mortality, coronary heart disease mortality, and total mortality, it showed that both low and high deciles of Hb were positively associated with all outcomes, and that a low mean Hb level specifically demonstrated a robust, positive association even after three years of follow-up [63]. Another study similarly reported the association of both low and high Hb concentrations with cardiovascular and all-cause mortality after an eight-year follow-up [64]. T2D was not investigated in the above two studies. The lower end risk elevation may represent an end-stage anemia phenomenon, whereas the low Hb-T2D association due to the chronic impairment of insulin action and low-grade inflammation, among others, that eventually lead to many long-term micro- and macrovascular complications [70], suggests more of a causal relationship specifically for the etiology of diabetes mellitus, since other MetS components did not show this clear negative association.

Similar inverse association between Hb concentrations and HbA1c levels for both men and women, across the fasting glucose quintiles within the non-diabetic range, was observed in a cross-sectional analysis among ~87,000 adult non-anemic and non-diabetic Korean adults [71,72]. The inverse relationship between Hb and HbA1c has been also observed in other populations, and anemia has been considered as a confounder of HbA1c values [73]. Nonetheless, the underlying mechanisms and clinical relevance of the effect of low Hb concentrations, iron deficiency, and iron deficiency anemia on glucose homeostasis are yet to be clearly elucidated.

We surmise that the observed increased odds of T2D at lower-than-normal Hb concentrations among Taiwanese HC and Europeans in the present study may be caused by anemia that reflects poor nutrition status, which results in poor blood circulation, decreased oxidative capacity, and pancreatic islet malfunction [74]. The possibility of a reverse causation should also be explored. Type 2 diabetes leads to chronic kidney disease that in turn causes anemia, contributing to the poor prognosis for T2D patients [75]. Whether it is reasonable to see a very modest effect of low Hb concentrations on the pathology of other MetS components, but primarily on T2D alone, warrants additional investigation.

In the context of iron metabolism, hepcidin has been recently discovered and is now widely accepted as the key systemic iron-regulatory hormone [76]. The expression of hepcidin in the liver is stimulated by inflammation or iron overload. During inflammation or infection, when hepcidin level is high, erythrocyte production becomes restricted through the inhibition of intestinal iron absorption and macrophage iron recycling, which reduces Hb concentration and ultimately causes anemia (i.e., anemia of inflammation) [77]. Hepcidin is also thought to be homeostatically regulated by the iron requirements of erythroid precursors for hemoglobin synthesis, so that during active erythropoiesis (i.e., during anemia), hepcidin production is suppressed to make more iron available for hemoglobin synthesis [78].

Serum hepcidin level was similarly shown to be increased among subjects with the MetS at a population level, wherein a linear increase in hepcidin levels in both sex groups was observed with increasing number of MetS components [79]. Nonetheless, there is still scarcity of studies linking hepcidin and MetS to date, which is probably due to a number of challenges related to hepcidin assay development [80]. Although the hepcidin-MetS association has been established for the first time, more studies are needed for validation and further elucidation of the mechanisms relating hepcidin to other iron biomarkers, including Hb, in the complex pathophysiology of the MetS.

In the context of iron nutrition, neither a too low nor too high iron status is desirable for human health. We have consistently shown that higher-than-normal levels of Hb increased risks of MetS and majority of the metabolic components, whereas too less or too much may confer risk against T2D. Such findings may be used in the future to support the rectification of the estimated average requirements, recommended dietary allowances, and tolerable upper intake levels for iron among Taiwanese HC and European Whites. Additionally, the utilization of a binary, single cut-off point to define optimal iron status may need to be revisited as it tends to obscure the more subtle disease-Hb associations [63,81].

4.3. Strengths and Limitations of the Study

The present study has several strengths. First, as we try to answer what levels of iron nutrition is optimal in order to prevent or delay the onset and progression of MetS and its metabolic components, we are the first to titrate Hb concentrations by decile categorization (i.e., investigating the effects of small increments in Hb concentration) against most components of MetS altogether. Second, our analyses employed relatively large populations of the Taiwanese HC and European Whites, which allowed us to further conduct sex-specific titration analyses. The large sample size and consistent results between the two ethnic groups strengthen the findings of the present study. To our knowledge, we are the first to explore the association of Hb levels with the risk of MetS among the Han Chinese of Taiwan. Our ethnic-specific analyses addressed the apparent differences in terms of iron status between the East Asians and Caucasians [82,83]. Hemochromatosis is common among Caucasians or populations of European origin. Iron overload among Asian populations, on the contrary, is rare and not well understood. This is partly due to the relatively higher prevalence of hemoglobinopathies (i.e., thalassemia) in Asians such that there is a possible masking by the coexistence and high prevalence of iron deficiency and anemia [84]. Lastly, our sex-specific analyses addressed the known differences between genders in terms of Hb concentration and distribution and other indicators of iron metabolism [85]. Although the underpinning mechanism has not been clearly depicted, sex hormones and menstruation are thought to play roles that might cause such sex-differences. Men had a consistently higher prevalence of MetS and metabolic outcomes than women in both ethnic groups, except for central obesity, yet we found that for most outcomes, women were likely to have higher relative risks of developing metabolic disorders with higher levels of Hb concentration. Interestingly, for central obesity, the relative risks in men were contrarily higher than women. Further studies of other factors, be it genetic, environmental, or lifestyle, will elucidate the subtle ethnic- and sex-differences in iron nutrition and metabolism and on how it affects an individual's susceptibility to the MetS and metabolic components. This will be needed in order to come up with more comprehensive population- and gender-specific recommendations and personalized nutrition interventions for iron-related disorders.

Our study has some limitations as well. Findings were derived from cross-sectional data and analyses, which means that causality cannot be inferred. Both the TWB and UKB were comprised datasets that are comprehensive enough to adjust for variables or parameters that may confound any Hb-outcome association observed. However, there is the possibility of unadjusted confounders. We were unable to consider the consumption of total dietary energy, saturated fat, and other nutrients, as well as the intake of iron supplements, iron absorption enhancers (i.e., vitamin C), or iron inhibitors (i.e., tannins) in the regression models because of some incomplete information collected during a personal interview in the TWB. We did not screen for vegetarians, subjects taking dietary supplements, or those who had consumed more red meat, which may have influenced the outcomes. Concurrent acute or chronic infections and autoimmune diseases of subjects and the confounding effect of inflammatory markers (i.e., C-reactive protein) may have been overlooked as well. Additionally, although our findings may have substantial impact on two major ethnicities, the East Asians and Caucasians, the generalizability of results to other groups (i.e., Africans, South Asians, etc.) is still not clear. Finally, as we anticipate the measurement and availability of other relevant iron status markers (i.e., serum hepcidin, serum ferritin, serum iron, transferrin saturation) in the TWB and UKB in the near future, exploration of the association of these markers with metabolic outcomes is highly warranted. Low and high Hb concentrations may be the results of other pathological conditions irrelevant to iron status. All these should be considered in future studies.

5. Conclusions

Hemoglobin reflects the major functional iron in the body and may serve as a surrogate of iron status. Elevated Hb concentration, a potential indicator of iron overnutrition, may play a role in the pathophysiology of MetS and its key components. It is highly warranted to further explore the observed protection against diabetes within normal Hb levels, the increased risks in both below normal and relatively higher levels, and the underlying mechanisms for such findings.

Supplementary Materials: The following supporting information can be downloaded at: https: //www.mdpi.com/article/10.3390/nu14194007/s1, File S1: List of variables and data field codes utilized in the UKB analyses, File S2: Disease definitions used in the UKB analyses, File S3: Prescription medication codes (data field 20003) in the UKB analyses, Table S1: Characteristics of subjects in the TWB across increasing deciles of Hb levels; Table S2: Characteristics of subjects in the UKB across increasing deciles of Hb levels; Table S3: Frequency distribution of metabolic outcomes among subjects in the TWB across increasing deciles of Hb levels; Table S4: Frequency distribution of metabolic outcomes among subjects in the UKB across increasing deciles of Hb levels; Table S5: Frequency distribution of self-reported metabolic conditions among subjects in the TWB across increasing deciles of Hb levels; Table S6: Frequency distribution of self-reported metabolic conditions among subjects in the UKB across increasing deciles of Hb levels; Table S7: Cases of metabolic disorders from hospital in-patient records of UKB subjects across increasing deciles of Hb levels; Table S8: Cross-tabulation of cases in the TWB based on self-reported metabolic disorders and high levels of biochemical markers; Table S9: Cross-tabulation of cases in the UKB based on high levels of anthropometric and biochemical markers, self-reported metabolic conditions, and hospital in-patient records of subjects; Table S10: Odds ratios of metabolic syndrome (NCEP ATP III criteria) across increasing deciles of Hb levels among Taiwanese Han Chinese; Table S10.1: Odds ratios of metabolic syndrome (Taiwan's criteria) across increasing deciles of Hb levels among Taiwanese Han Chinese; Table S11: Odds ratios of metabolic syndrome (NCEP ATP III criteria) across increasing deciles of Hb levels among European Whites; Table S12: Odds ratios of central obesity across increasing deciles of Hb levels among Taiwanese Han Chinese; Table S13: Odds ratios of central obesity across increasing deciles of Hb levels among European Whites; Table S14: Odds ratios of hypertension across increasing deciles of Hb levels among Taiwanese Han Chinese; Table S15: Odds ratios of hypertension across increasing deciles of Hb levels among European Whites; Table S16: Odds ratios of dyslipidemia across increasing deciles of Hb levels among Taiwanese Han Chinese; Table S17: Odds ratios of dyslipidemia across increasing deciles of Hb levels among European Whites; Table S18: Odds ratios of hypercholesterolemia across increasing deciles of Hb levels among Taiwanese Han Chinese; Table S19: Odds ratios of hypercholesterolemia across increasing deciles of Hb levels among European Whites; Table S20: Odds ratios of gout across increasing deciles of hemoglobin levels among Taiwanese Han Chinese; Table S21: Odds ratios of gout across increasing deciles of hemoglobin levels among European Whites; Table S22: Odds ratios of diabetes across increasing deciles of hemoglobin levels among Taiwanese Han Chinese; Table S23: Odds ratios of diabetes across increasing deciles of hemoglobin levels among European Whites; Tables S24.1–S24.8: With BMI as additional covariate in the models.

Author Contributions: Conceptualization, W.-H.P. and V.J.T.; methodology, V.J.T., K.-M.C. and W.-H.P.; software, V.J.T. and K.-M.C.; validation, K.-M.C.; formal analysis, V.J.T.; investigation, V.J.T.; resources, K.-M.C. and W.-H.P.; data curation, V.J.T.; writing–original draft preparation, V.J.T.; writing–review and editing, W.-H.P.; visualization, V.J.T.; supervision, K.-M.C. and W.-H.P.; project administration, K.-M.C. and W.-H.P.; funding acquisition, W.-H.P. All authors have read and agreed to the published version of the manuscript.

Funding: This research was funded by the Health Cloud Project of Academia Sinica with grant number AS-PH-109-02.

Institutional Review Board Statement: The UK National Health Service's National Research Ethics Committee gave ethical clearance to the UKB Project with reference number 11/NW/The Institutional Review Board of Academia Sinica approved the TWB study and protocols on June 2019 with reference number AS-IRB02-This study has been carried out in accordance with The Code of Ethics of the World Medical Association (Declaration of Helsinki).

Informed Consent Statement: Informed consent was obtained from all subjects involved in the study.

Data Availability Statement: Not applicable.

Acknowledgments: We thank all the participants and investigators of the Taiwan Biobank and UK Biobank. We thank the Institute of Statistical Science of Academia Sinica and the Multidisciplinary Health Cloud Research Program Project–the Statistical Science unit and the Humanities and Social Sciences unit for the facility and financial support. This study has been undertaken using the UK Biobank resource under Application Number 48272.

Conflicts of Interest: The authors declare no conflict of interest.

References

1. Abbaspour, N.; Hurrell, R.; Kelishadi, R. Review on iron and its importance for human health. *J. Res. Med. Sci.* **2014**, *19*, 164–174. [PubMed]
2. Dev, S.; Babitt, J.L. Overview of iron metabolism in health and disease. *Hemodial. Int.* **2017**, *21* (Suppl. 1), S6–S20. [CrossRef] [PubMed]
3. Kassebaum, N.J.; GBD 2013 Anemia Collaborators. The Global Burden of Anemia. *Hematol. Oncol. Clin. North Am.* **2016**, *30*, 247–308. [CrossRef] [PubMed]
4. World Health Organization. *Worldwide Prevalence of Anaemia 1993–2005: WHO Global Database on Anaemia*; De Benoist, B., McLean, E., Egli, I., Cogswell, M., Eds.; World Health Organization: Geneva, Switzerland, 2008; Available online: https://apps.who.int/iris/handle/10665/43894 (accessed on 6 June 2022).
5. WHO Global Anaemia Estimates, 2021 Edition. Available online: https://www.who.int/data/gho/data/themes/topics/anaemia_in_women_and_children (accessed on 6 June 2022).
6. Basuli, D.; Stevens, R.G.; Torti, F.M.; Torti, S.V. Epidemiological associations between iron and cardiovascular disease and diabetes. *Front. Pharmacol.* **2014**, *5*, 117. [CrossRef]
7. Cornier, M.A.; Dabelea, D.; Hernandez, T.L.; Lindstrom, R.C.; Steig, A.J.; Stob, N.R.; Van Pelt, R.E.; Wang, H.; Eckel, R.H. The metabolic syndrome. *Endocr. Rev.* **2008**, *29*, 777–822. [CrossRef]
8. Pan, W.H.; Wu, H.J.; Yeh, C.J.; Chuang, S.Y.; Chang, H.Y.; Yeh, N.H.; Hsieh, Y.T. Diet and health trends in Taiwan: Comparison of two nutrition and health surveys from 1993–1996 and 2005–2008. *Asia Pac. J. Clin. Nutr.* **2011**, *20*, 238–250.
9. Yeh, C.J.; Chang, H.Y.; Pan, W.H. Time trend of obesity, the metabolic syndrome and related dietary pattern in Taiwan: From NAHSIT 1993–1996 to NAHSIT 2005–2008. *Asia Pac. J. Clin. Nutr.* **2011**, *20*, 292–300.
10. Scuteri, A.; Laurent, S.; Cucca, F.; Cockcroft, J.; Cunha, P.G.; Mañas, L.R.; Mattace Raso, F.U.; Muiesan, M.L.; Ryliškytė, L.; Metabolic Syndrome and Arteries Research (MARE) Consortium; et al. Metabolic syndrome across Europe: Different clusters of risk factors. *Eur. J. Prev. Cardiol.* **2015**, *22*, 486–491. [CrossRef]
11. Imam, M.U.; Zhang, S.; Ma, J.; Wang, H.; Wang, F. Antioxidants Mediate Both Iron Homeostasis and Oxidative Stress. *Nutrients* **2017**, *9*, 671. [CrossRef]

12. Sullivan, J.L. Iron and the sex difference in heart disease risk. *Lancet* **1981**, *1*, 1293–1294. [CrossRef]
13. Reddy, M.B.; Clark, L. Iron, Oxidative Stress, and Disease Risk. *Nutr. Rev.* **2004**, *62*, 120–124. [CrossRef]
14. World Health Organization. *Vitamin and Mineral Requirements in Human Nutrition*, 2nd ed.; World Health Organization: Bangkok, Thailand, 1998; Available online: https://apps.who.int/iris/handle/10665/42716 (accessed on 6 June 2022).
15. Monsen, E.R.; Hallberg, L.; Layrisse, M.; Hegsted, D.M.; Cook, J.D.; Mertz, W.; Finch, C.A. Estimation of available dietary iron. *Am. J. Clin. Nutr.* **1978**, *31*, 134–141. [CrossRef]
16. Marengo-Rowe, A.J. Structure-Function Relations of Human Hemoglobins. *Bayl. Univ. Med Cent. Proc.* **2006**, *19*, 239–245. [CrossRef]
17. Pfeiffer, C.M.; Looker, A.C. Laboratory methodologies for indicators of iron status: Strengths, limitations, and analytical challenges. *Am. J. Clin. Nutr.* **2016**, *106*, 1606S–1614S. [CrossRef]
18. World Health Organization. *Nutritional Anaemias: Tools for Effective Prevention and Control*; World Health Organization: Geneva, Switzerland, 2017; Available online: https://apps.who.int/iris/handle/10665/259425 (accessed on 6 June 2022).
19. Jurado, R.L. Iron, Infections, and Anemia of Inflammation. *Clin. Infect. Dis.* **1997**, *25*, 888–895. [CrossRef]
20. Thottam, G.E.; Krasnokutsky, S.; Pillinger, M.H. Gout and Metabolic Syndrome: A Tangled Web. *Current Rheumatol. Rep.* **2017**, *19*, 60. [CrossRef]
21. Fan, C.T.; Lin, J.C.; Lee, C.H. Taiwan Biobank: A project aiming to aid Taiwan's transition into a biomedical island. *Pharmacogenomics* **2008**, *9*, 235–246. [CrossRef]
22. Sudlow, C.; Gallacher, J.; Allen, N.; Beral, V.; Burton, P.; Danesh, J.; Downey, P.; Elliott, P.; Green, J.; Landray, M.; et al. UK Biobank: An Open Access Resource for Identifying the Causes of a Wide Range of Complex Diseases of Middle and Old Age. *PLoS Med.* **2015**, *12*, e1001779. [CrossRef]
23. World Health Organization. *Haemoglobin Concentrations for the Diagnosis of Anaemia and Assessment of Severity*; World Health Organization: Geneva, Switzerland, 2011; Available online: https://apps.who.int/iris/handle/10665/85839 (accessed on 6 June 2022).
24. International Diabetes Federation. *The IDF Consensus Worldwide Definition of the Metabolic Syndrome*; International Diabetes Federation: Brussels, Belgium, 2006; Available online: https://www.idf.org/e-library/consensus-statements/60-idfconsensus-worldwide-definitionof-the-metabolic-syndrome.html (accessed on 6 June 2022).
25. Zimmet, P.Z.; Alberti, K.G. Introduction: Globalization and the non-communicable disease epidemic. *Obesity* **2006**, *14*, 1–3. [CrossRef]
26. Chiang, C.E.; Wang, T.D.; Ueng, K.C.; Lin, T.H.; Yeh, H.I.; Chen, C.Y.; Wu, Y.J.; Tsai, W.C.; Chao, T.H.; Chen, C.H.; et al. 2015 guidelines of the Taiwan Society of Cardiology and the Taiwan Hypertension Society for the management of hypertension. *J. Chin. Med. Assoc.* **2015**, *78*, 1–47. [CrossRef]
27. Williams, B.; Mancia, G.; Spiering, W.; Agabiti Rosei, E.; Azizi, M.; Burnier, M.; Clement, D.L.; Coca, A.; de Simone, G.; ESC Scientific Document Group; et al. 2018 ESC/ESH Guidelines for the management of arterial hypertension. *Eur. Heart J.* **2018**, *39*, 3021–3104. [CrossRef] [PubMed]
28. World Health Organization & International Diabetes Federation. *Definition and Diagnosis of Diabetes Mellitus and Intermediate Hyperglycaemia: Report of a WHO/IDF Consultation*; World Health Organization: Geneva, Switzerland, 2006; Available online: https://apps.who.int/iris/handle/10665/43588 (accessed on 6 June 2022).
29. American Diabetes Association. Classification and Diagnosis of Diabetes. *Diabetes Care* **2017**, *40*, S11–S24. [CrossRef]
30. Riemsma, R.; Corro Ramos, I.; Birnie, R.; Büyükkaramikli, N.; Armstrong, N.; Ryder, S.; Duffy, S.; Worthy, G.; Al, M.; Severens, J.; et al. Integrated sensor-augmented pump therapy systems [the MiniMed®Paradigm™ Veo system and the Vibe™ and G4®PLATINUM CGM (continuous glucose monitoring) system] for managing blood glucose levels in type 1 diabetes: A systematic review and economic evaluation. *Health Technol. Assess.* **2016**, *20*, 1–251. [CrossRef] [PubMed]
31. Hoelzel, W.; Weykamp, C.; Jeppsson, J.O.; Miedema, K.; Barr, J.R.; Goodall, I.; Hoshino, T.; John, W.G.; Kobold, U.; Little, R.; et al. IFCC Working Group on HbA1c Standardization. IFCC reference system for measurement of hemoglobin A1c in human blood and the national standardization schemes in the United States, Japan, and Sweden: A method-comparison study. *Clin. Chem.* **2004**, *50*, 166–174. [CrossRef] [PubMed]
32. Eastwood, S.V.; Mathur, R.; Atkinson, M.; Brophy, S.; Sudlow, C.; Flaig, R.; de Lusignan, S.; Allen, N.; Chaturvedi, N. Algorithms for the Capture and Adjudication of Prevalent and Incident Diabetes in UK Biobank. *PLoS ONE* **2016**, *11*, e0162388. [CrossRef]
33. Expert Panel on Detection, Evaluation, and Treatment of High Blood Cholesterol in Adults. Executive Summary of The Third Report of The National Cholesterol Education Program (NCEP) Expert Panel on Detection, Evaluation, And Treatment of High Blood Cholesterol In Adults (Adult Treatment Panel III). *JAMA* **2001**, *285*, 2486–2497. [CrossRef] [PubMed]
34. Li, Y.H.; Ueng, K.C.; Jeng, J.S.; Charng, M.J.; Lin, T.H.; Chien, K.L.; Wang, C.Y.; Chao, T.H.; Liu, P.Y.; Writing Group of 2017 Taiwan Lipid Guidelines for High Risk Patients; et al. 2017 Taiwan lipid guidelines for high risk patients. *J. Formos. Med. Assoc.* **2017**, *116*, 217–248. [CrossRef]
35. US Preventive Services Task Force; Bibbins-Domingo, K.; Grossman, D.C.; Curry, S.J.; Davidson, K.W.; Epling, J.W., Jr.; García, F.A.; Gillman, M.W.; Kemper, A.R.; Krist, A.H.; et al. Screening for Lipid Disorders in Children and Adolescents: US Preventive Services Task Force Recommendation Statement. *JAMA* **2016**, *316*, 625–633. [CrossRef]
36. Chang, H.Y.; Pan, W.H.; Yeh, W.T.; Tsai, K.S. Hyperuricemia and gout in Taiwan: Results from the Nutritional and Health Survey in Taiwan (1993–1996). *J. Rheumatol.* **2001**, *28*, 1640–1646.

37. Desideri, G.; Castaldo, G.; Lombardi, A.; Mussap, M.; Testa, A.; Pontremoli, R.; Punzi, L.; Borghi, C. Is it time to revise the normal range of serum uric acid levels? *Eur. Rev. Med. Pharmacol. Sci.* **2014**, *18*, 1295–1306.

38. Kim, S.Y.; Guevara, J.P.; Kim, K.M.; Choi, H.K.; Heitjan, D.F.; Albert, D.A. Hyperuricemia and coronary heart disease: A systematic review and meta-analysis. *Arthritis Care Res.* **2010**, *62*, 170–180. [CrossRef]

39. Grundy, S.M.; Brewer, H.B., Jr.; Cleeman, J.I.; Smith, S.C., Jr.; Lenfant, C.; American Heart Association, & National Heart, Lung, and Blood Institute. Definition of metabolic syndrome: Report of the National Heart, Lung, and Blood Institute/American Heart Association conference on scientific issues related to definition. *Circulation* **2004**, *109*, 433–438. [CrossRef]

40. Pan, W.H.; Lee, M.S.; Chuang, S.Y.; Lin, Y.C.; Fu, M.L. Obesity pandemic, correlated factors and guidelines to define, screen and manage obesity in Taiwan. *Obes. Rev.* **2008**, *9*, 22–31. [CrossRef]

41. WHO Consultation on Obesity (1997: Geneva, Switzerland), World Health Organization. Division of Noncommunicable Diseases & World Health Organization. *Programme of Nutrition, Family and Reproductive Health. Obesity: Preventing and Managing the Global Epidemic: Report of a WHO Consultation on Obesity, Geneva, 3–5 June 1997.* Available online: https://apps.who.int/iris/handle/10 665/63854 (accessed on 6 June 2022).

42. Kawamoto, R.; Tabara, Y.; Kohara, K.; Miki, T.; Kusunoki, T.; Abe, M.; Katoh, T. Hematological parameters are associated with metabolic syndrome in Japanese community-dwelling persons. *Endocrine* **2013**, *43*, 334–341. [CrossRef]

43. Nebeck, K.; Gelaye, B.; Lemma, S.; Berhane, Y.; Bekele, T.; Khali, A.; Haddis, Y.; Williams, M.A. Hematological parameters and metabolic syndrome: Findings from an occupational cohort in Ethiopia. *Diabetes Metab. Syndr.* **2012**, *6*, 22–27. [CrossRef]

44. Lohsoonthorn, V.; Jiamjarasrungsi, W.; Williams, M.A. Association of Hematological Parameters with Clustered Components of Metabolic Syndrome among Professional and Office Workers in Bangkok, Thailand. *Diabetes Metab. Syndr.* **2007**, *1*, 143–149. [CrossRef]

45. Tapio, J.; Vähänikkilä, H.; Kesäniemi, Y.A.; Ukkola, O.; Koivunen, P. Higher hemoglobin levels are an independent risk factor for adverse metabolism and higher mortality in a 20-year follow-up. *Sci. Rep.* **2021**, *11*, 19936. [CrossRef]

46. He, S.; Gu, H.; Yang, J.; Su, Q.; Li, X.; Qin, L. Hemoglobin concentration is associated with the incidence of metabolic syndrome. *BMC Endocr. Disord.* **2021**, *21*, 53. [CrossRef]

47. Betteridge, D.J. What is oxidative stress? *Metabolism* **2000**, *49*, 3–8. [CrossRef]

48. Fernández-Sánchez, A.; Madrigal-Santillán, E.; Bautista, M.; Esquivel-Soto, J.; Morales-González, A.; Esquivel-Chirino, C.; Durante-Montiel, I.; Sánchez-Rivera, G.; Valadez-Vega, C.; Morales-González, J.A. Inflammation, oxidative stress, and obesity. *Int. J. Mol. Sci.* **2011**, *12*, 3117–3132. [CrossRef]

49. Furukawa, S.; Fujita, T.; Shimabukuro, M.; Iwaki, M.; Yamada, Y.; Nakajima, Y.; Nakayama, O.; Makishima, M.; Matsuda, M.; Shimomura, I. Increased oxidative stress in obesity and its impact on metabolic syndrome. *J. Clin. Investig.* **2004**, *114*, 1752–1761. [CrossRef]

50. Dinh, Q.N.; Drummond, G.R.; Sobey, C.G.; Chrissobolis, S. Roles of inflammation, oxidative stress, and vascular dysfunction in hypertension. *Biomed. Res. Int.* **2014**, *2014*, 406960. [CrossRef]

51. Son, M.; Park, J.; Park, K.; Yang, S. Association between hemoglobin variability and incidence of hypertension over 40 years: A Korean national cohort study. *Sci. Rep.* **2020**, *10*, 12061. [CrossRef]

52. Rosenfeld, M.E. Inflammation, lipids, and free radicals: Lessons learned from the atherogenic process. *Semin. Reprod. Endocrinol.* **1998**, *16*, 249–261. [CrossRef]

53. Tamariz, L.J.; Young, J.H.; Pankow, J.S.; Yeh, H.C.; Schmidt, M.I.; Astor, B.; Brancati, F.L. Blood viscosity and hematocrit as risk factors for type 2 diabetes mellitus: The atherosclerosis risk in communities (ARIC) study. *Am. J. Epidemiol.* **2008**, *168*, 1153–1160. [CrossRef]

54. Atsma, F.; Veldhuizen, I.; de Kort, W.; van Kraaij, M.; Pasker-de Jong, P.; Deinum, J. Hemoglobin level is positively associated with blood pressure in a large cohort of healthy individuals. *Hypertension* **2012**, *60*, 936–941. [CrossRef]

55. Billiet, L.; Doaty, S.; Katz, J.D.; Velasquez, M.T. Review of hyperuricemia as new marker for metabolic syndrome. *ISRN Rheumatol.* **2014**, *2014*, 852954. [CrossRef]

56. Chen, L.Y.; Zhu, W.H.; Chen, Z.W.; Dai, H.L.; Ren, J.J.; Chen, J.H.; Chen, L.Q.; Fang, L.Z. Relationship between hyperuricemia and metabolic syndrome. *J. Zhejiang Univ. Sci. B* **2007**, *8*, 593–598. [CrossRef]

57. Ghio, A.J.; Ford, E.S.; Kennedy, T.P.; Hoidal, J.R. The association between serum ferritin and uric acid in humans. *Free Radic. Res.* **2005**, *39*, 337–342. [CrossRef]

58. Mainous, A.G., 3rd; Knoll, M.E.; Everett, C.J.; Matheson, E.M.; Hulihan, M.M.; Grant, A.M. Uric acid as a potential cue to screen for iron overload. *J. Am. Board Fam. Med.* **2011**, *24*, 415–421. [CrossRef]

59. Smith, E.; March, L. Global Prevalence of Hyperuricemia: A Systematic Review of Population-Based Epidemiological Studies [abstract]. *Arthritis Rheumatol.* **2015**, *67*. Available online: https://acrabstracts.org/abstract/global-prevalence-of-hyperuricemia-a-systematic-review-of-population-based-epidemiological-studies/ (accessed on 15 July 2022).

60. Wu, S.H.; Liu, Z.; Ho, S.C. Metabolic syndrome and all-cause mortality: A meta-analysis of prospective cohort studies. *Eur. J. Epidemiol.* **2010**, *25*, 375–384. [CrossRef] [PubMed]

61. Bao, W.; Rong, Y.; Rong, S.; Liu, L. Dietary iron intake, body iron stores, and the risk of type 2 diabetes: A systematic review and meta-analysis. *BMC Med.* **2012**, *10*, 119. [CrossRef] [PubMed]

62. Liu, J.; Li, Q.; Yang, Y.; Ma, L. Iron metabolism and type 2 diabetes mellitus: A meta-analysis and systematic review. *J. Diabetes Investig.* **2020**, *11*, 946–955. [CrossRef] [PubMed]

63. Kabat, G.C.; Kim, M.Y.; Verma, A.K.; Manson, J.E.; Lessin, L.S.; Kamensky, V.; Lin, J.; Wassertheil-Smoller, S.; Rohan, T.E. Association of Hemoglobin Concentration With Total and Cause-Specific Mortality in a Cohort of Postmenopausal Women. *Am. J. Epidemiol.* **2016**, *183*, 911–919. [CrossRef]

64. Lee, G.; Choi, S.; Kim, K.; Yun, J.M.; Son, J.S.; Jeong, S.M.; Kim, S.M.; Park, S.M. Association of Hemoglobin Concentration and Its Change With Cardiovascular and All-Cause Mortality. *J. Am. Heart Assoc.* **2018**, *7*, e007723. [CrossRef]

65. Patel, K.V. Variability and heritability of hemoglobin concentration: An opportunity to improve understanding of anemia in older adults. *Haematologica* **2008**, *93*, 1281–1283. [CrossRef]

66. Chen, C.; Cohrs, C.M.; Stertmann, J.; Bozsak, R.; Speier, S. Human beta cell mass and function in diabetes: Recent advances in knowledge and technologies to understand disease pathogenesis. *Mol. Metab.* **2017**, *6*, 943–957. [CrossRef]

67. Raju, K.; Venkataramappa, S.M. Primary Hemochromatosis Presenting as Type 2 Diabetes Mellitus: A Case Report with Review of Literature. *Int. J. Appl. Basic Med. Res.* **2018**, *8*, 57–60. [CrossRef]

68. Siegelman, E.S.; Mitchell, D.G.; Rubin, R.; Hann, H.W.; Kaplan, K.R.; Steiner, R.M.; Rao, V.M.; Schuster, S.J.; Burk, D.L., Jr.; Rifkin, M.D. Parenchymal versus reticuloendothelial iron overload in the liver: Distinction with MR imaging. *Radiology* **1991**, *179*, 361–366. [CrossRef]

69. McLaren, G.D.; Muir, W.A.; Kellermeyer, R.W. Iron overload disorders: Natural history, pathogenesis, diagnosis, and therapy. *Crit. Rev. Clin. Lab. Sci.* **1983**, *19*, 205–266. [CrossRef]

70. Laakso, M. Cardiovascular disease in type 2 diabetes from population to man to mechanisms: The Kelly West Award Lecture. *Diabetes Care* **2010**, *33*, 442–449. [CrossRef]

71. American Diabetes Association. Glycemic Targets: Standards of Medical Care in Diabetes—2021. *Diabetes Care* **2021**, *44*, S73–S84. [CrossRef]

72. Bae, J.C.; Suh, S.; Jin, S.M.; Kim, S.W.; Hur, K.Y.; Kim, J.H.; Min, Y.K.; Lee, M.S.; Lee, M.K.; Jeon, W.S.; et al. Hemoglobin A1c values are affected by hemoglobin level and gender in non-anemic Koreans. *J. Diabetes Investig.* **2014**, *5*, 60–65. [CrossRef]

73. Dagogo-Jack, S. Pitfalls in the use of HbA$_1$(c) as a diagnostic test: The ethnic conundrum. *Nat. Rev. Endocrinol.* **2010**, *6*, 589–593. [CrossRef]

74. Soliman, A.T.; De Sanctis, V.; Yassin, M.; Soliman, N. Iron deficiency anemia and glucose metabolism. *Acta Biomed.* **2017**, *88*, 112–118. [CrossRef]

75. Mehdi, U.; Toto, R.D. Anemia, diabetes, and chronic kidney disease. *Diabetes Care* **2009**, *32*, 1320–1326. [CrossRef]

76. Ganz, T. Hepcidin, a key regulator of iron metabolism and mediator of anemia of inflammation. *Blood* **2003**, *102*, 783–788. [CrossRef]

77. Pagani, A.; Nai, A.; Silvestri, L.; Camaschella, C. Hepcidin and Anemia: A Tight Relationship. *Front. Physiol.* **2019**, *10*, 1294. [CrossRef]

78. Ganz, T.; Nemeth, E. Hepcidin and iron homeostasis. *Biochim. Biophys. Acta* **2012**, *1823*, 1434–1443. [CrossRef]

79. Martinelli, N.; Traglia, M.; Campostrini, N.; Biino, G.; Corbella, M.; Sala, C.; Busti, F.; Masciullo, C.; Manna, D.; Previtali, S.; et al. Increased serum hepcidin levels in subjects with the metabolic syndrome: A population study. *PLoS ONE* **2012**, *7*, e48450. [CrossRef]

80. Girelli, D.; Nemeth, E.; Swinkels, D.W. Hepcidin in the diagnosis of iron disorders. *Blood* **2016**, *127*, 2809–2813. [CrossRef]

81. Beutler, E.; Waalen, J. The definition of anemia: What is the lower limit of normal of the blood hemoglobin concentration? *Blood* **2006**, *107*, 1747–1750. [CrossRef]

82. Kang, W.; Barad, A.; Clark, A.G.; Wang, Y.; Lin, X.; Gu, Z.; O'Brien, K.O. Ethnic Differences in Iron Status. *Adv. Nutr.* **2021**, *12*, 1838–1853. [CrossRef]

83. Timoteo, V.J.; Chiang, K.M.; Yang, H.C.; Pan, W.H. Common and Ethnic-Specific Genetic Determinants of Hemoglobin Concentration between Taiwanese Han Chinese and European Whites: Findings from Comparative Two-Stage Genome-Wide Association Studies. *J. Nutr. Biochem.* **2022**, in press. [CrossRef]

84. Lok, C.Y.; Merryweather-Clarke, A.T.; Viprakasit, V.; Chinthammitr, Y.; Srichairatanakool, S.; Limwongse, C.; Oleesky, D.; Robins, A.J.; Hudson, J.; Wai, P.; et al. Iron overload in the Asian community. *Blood* **2009**, *114*, 20–25. [CrossRef]

85. Pan, W.H.; Habicht, J.P. The non-iron-deficiency-related difference in hemoglobin concentration distribution between blacks and whites and between men and women. *Am. J. Epidemiol.* **1991**, *134*, 1410–1416. [CrossRef]

Review

Benefits and Risks of Early Life Iron Supplementation

Shasta A. McMillen , Richard Dean, Eileen Dihardja, Peng Ji and Bo Lönnerdal *

Department of Nutrition, University of California, Davis, CA 95616, USA
* Correspondence: bllonnerdal@ucdavis.edu

Abstract: Infants are frequently supplemented with iron to prevent iron deficiency, but iron supplements may have adverse effects on infant health. Although iron supplements can be highly effective at improving iron status and preventing iron deficiency anemia, iron may adversely affect growth and development, and may increase risk for certain infections. Several reviews exist in this area; however, none has fully summarized all reported outcomes of iron supplementation during infancy. In this review, we summarize the risks and benefits of iron supplementation as they have been reported in controlled studies and in relevant animal models. Additionally, we discuss the mechanisms that may underly beneficial and adverse effects.

Keywords: infant nutrition; iron supplement; iron deficiency anemia; growth; neurodevelopment; oxidative stress; trace mineral interactions; gut microbiome

Citation: McMillen, S.A.; Dean, R.; Dihardja, E.; Ji, P.; Lönnerdal, B. Benefits and Risks of Early Life Iron Supplementation. *Nutrients* **2022**, *14*, 4380. https://doi.org/10.3390/nu14204380

Academic Editor: Dietmar Enko

Received: 26 September 2022
Accepted: 17 October 2022
Published: 19 October 2022

Publisher's Note: MDPI stays neutral with regard to jurisdictional claims in published maps and institutional affiliations.

1. Introduction

Iron is an essential trace element for human life: basic cellular reactions like energy production and DNA replication require iron, and in mammals, iron transports oxygen in the blood as hemoglobin. Insufficient iron intake to meet basic metabolic requirements leads to deficiency. Iron deficiency (ID) affects 10–40% of infants and causes approximately 50% of anemia cases worldwide [1–3].

Infants are especially susceptible to ID and iron deficiency anemia (IDA), both of which disrupt health and development [2–5], including adverse effects on long-term cognition and behavior. Once an infant becomes iron deficient, correcting iron status through dietary intervention prevents anemia, but may not correct disruptions to neurodevelopment and long-term cognitive development because critical phases of brain development occur during infancy [6,7].

Concern about the harms of ID has led to routine use of iron supplements to prevent ID [4,5]. Iron supplements, whether iron drops, multi-nutrient packets (MNPs), fortified formula, or fortified complementary foods, are effective at preventing or treating ID in most infants. Based on the success of iron supplements for preventing ID and IDA, the World Health Organization (WHO) recommends that iron supplements are provided to infants in populations where anemia prevalence exceeds 40% [4]. The same rationale backs the American Academy of Pediatrics' (AAP) recommendations that exclusively breast-fed infants receive iron supplements beginning at 4 months, and that formula-fed infants receive iron-fortified formula [5].

The vast difference in iron intake—between the iron supplemented infant and the un-supplemented infant receiving only breast milk—is important but generally under-recognized. The Dietary Reference Intake (DRI) for iron for infants 0–6 months is 0.3 mg per day, based on the amount provided in breast milk [8]. For healthy infants born at term, liver iron stores in combination with the small amount provided in breast milk supports healthy growth and development up to 6 mo [9–11], but most fortified formulas in the USA contain 40× more iron than breast milk [12]. Even after accounting for differences in bioavailability between breast milk iron and formula iron, formula still provides around 7× more absorbable iron than breast milk. Furthermore, the AAP recommends 1 mg iron/kg daily supplementation for all exclusively or primarily breast-fed infants [5]. Following these recommendations, an iron-supplemented 5 kg infant would receive $17\times$ more iron than what is provided by breast milk.

The WHO and AAP recommendations may lower the risk of ID, but infants with low risk of ID who receive iron supplements may be at risk of adverse effects, including disrupted growth and neurodevelopment, adverse nutrient interactions, and increased morbidity and mortality [13–17]. The mechanisms underlying these effects remain unclear and must be investigated for the risks of iron to be characterized; to predict which infants are the most vulnerable to which outcomes; and to improve efficacy and safety of iron provision [18,19].

Existing reviews in this area are of high quality [13–15,17,20], but they do not provide a full overview of outcomes that have been observed in controlled human and animal studies. The purpose of this review will be to summarize outcomes of iron supplementation between birth and 12 months in infants, as well as the corresponding developmental stage in animal models. With the goal of identifying the most promising directions for future iron intervention research, we discuss likely biological mechanisms underlying risks and benefits of iron supplementation during infancy.

2. Deficiency & Toxicity

Currently, the WHO recommends that infants age 6–23 months of age receive additional iron wherever anemia prevalence is estimated to be >40% [21]. Their recommendation is based on evidence from a meta-analysis of anemia outcomes [20]—part of a large systematic review by Pasricha et al.—which showed that iron was effective at increasing hemoglobin ($p < 0.00001$ for overall effect) and reducing risk of anemia for infants (0.61 relative risk; $p < 0.00001$ for overall effect) (see Appendix Figure A1) [20]. Reductions in anemia prevalence following iron provision are attributed to improvements in iron status, because risk of ID is typically also reduced [20]. Nevertheless, it is necessary to re-evaluate iron prophylaxis and its dose in infant populations—especially those living in areas of low risk of IDA—despite effective anemia prevention, because studies have reported adverse developmental outcomes of iron provision based on the current recommendations [22–26].

2.1. Defining Anemia

It is also necessary to re-evaluate the clinical definition of anemia for infants. The global cutoff for infants and children under 5 years (<110 g Hb/L) has been unchanged since it was defined in a 1968 WHO technical report on nutritional anemias [27,28]. The technical report cites infant data from two studies published in 1954 and 1959 [29,30], both of which included relatively small samples of infants ($n = 237$ and $n = 129$, respectively). In their current guide for assessing anemia [27] the WHO states their cutoff was "validated" by survey data collected in 1976–1980 during the National Health and Nutrition Examination Survey II (NHANES II), published by the United States Center for Disease Control & Prevention (CDC) in 1989 [27,31]. The CDC's anemia cutoffs were derived from the 5th percentile Hb values, calculated from a "nationally representative sample" ($n = 979$) of "healthy" children 1–2 years old [31]. Notably, no Hb values were assessed in infants under 12 months old during this survey [31,32]. In summary, the current definition of anemia for infants is based on very limited evidence; therefore it is necessary to reconsider population-level iron recommendations designed around preventing anemia.

2.2. Defining Iron Deficiency

Hb defines anemia but is not a specific biomarker for iron status [2]. Specific iron biomarkers are serum ferritin (SF), serum iron, total iron binding capacity (TIBC), transferrin saturation, zinc protoporphyrin (ZPP) and soluble transferrin receptor (sTfR) [2]. Table 1 lists commonly used biomarkers as well as their cutoffs and their response to iron supplementation. Based on NHANES II data, the CDC recommends three biomarkers are used (ZPP, sTfR & SF), where at least two abnormal values would indicate ID [33]; however, often it is not feasible for clinicians to measure three biomarkers of iron in infants and young children. Instead, SF is used in combination with Hb to detect IDA. For anemic infants and children under 5 years of age, IDA is diagnosed when ferritin is <12 µg/L [34]. SF is a good biomarker for iron but also an acute phase protein that is elevated during systemic inflammation and thus may mask the presence of ID. Therefore, the WHO recommends a higher SF cutoff (<30 µg/L) to diagnose ID in infants in the presence of infection [34]. Researchers may assess inflammatory status alongside SF (e.g., C-reactive protein) to determine the validity of SF values; however, this method has not yet been standardized and the overall validity of SF as a biomarker for iron status during infancy remains uncertain [35].

Table 1. Common biomarkers for defining anemia and iron deficiency in infants.

Biomarker	Anemia Cutoff	Iron Deficiency Cutoff	Response to Iron Supplementation	References
Hemoglobin	<110 g/L	-	↑ or no change	[27–32]
Serum Ferritin [1]	-	<12 µg/L	↑	[33–35]
Transferrin Saturation [1]	-	<10%	↑	[32,33]
Zinc Protoporphyrin	-	80 µmol/mol heme	↓	[33]
Soluble Transferrin Receptor	-	8.3 mg/L	↓	[33]

[1] Also elevated by inflammation [33]. ↑, Increased with iron supplementation; ↓, decreased with iron supplementation.

Concrete evidence remains limited surrounding iron assessment and anemia diagnosis in infants, as well as the prevalence of infant ID or IDA. Thus, current practices for assessing iron status of infants and diagnosing ID or IDA rely on biomarker cutoffs that were defined by outdated, poor-quality evidence. Nevertheless, if one assumes these cutoffs are reliable, then there is evidence that: (1) there is high prevalence of IDA among toddlers 1–3 years of age [2]; (2) ID or IDA during infancy is associated with poorer developmental outcomes, particularly outcomes related to the nervous system and cognitive development [7]; (3) as stated above, controlled iron supplementation trials show reduced risk for ID and IDA [20]. However, whether iron supplementation improves and prevents poorer development outcomes is still unclear [15,20,36].

2.3. Iron Supplementation, Iron Status, & Hematology

There are inherent limitations to diagnosing ID and IDA during infancy, but there is good evidence that providing additional dietary iron will improve iron status. This is further supported by animal models: increased Hb, SF, transferrin saturation and serum iron, as well as increased liver iron concentration (a direct measure of iron stores) have been observed in swine, rats and mice. Studies in animal models support that Hb and iron biomarkers are elevated by iron supplementation but depend on baseline iron status as well as the dose, duration and form of iron supplementation.

Domesticated pigs are born without sufficient iron stores and must receive exogenous iron to prevent anemia (defined as <90 g Hb/L). Typically, 100–200 mg iron is administered to piglets during the first week of life as a single intramuscular or subcutaneous injection of iron dextran [37]. In one neonatal piglet study, non-supplemented piglets were severely anemic (mean Hb 72 g/L) by postnatal day (PD) 8, but piglets that had received an iron dextran injection or 5 days of oral iron had normal Hb levels (99 g/L and 100 g/L, respectively). SF, TIBC and serum iron, as well as spleen, liver, heart and kidney iron levels were also significantly elevated compared to non-supplemented piglets. Notably, greater iron loading in spleen, liver and kidney was observed in the iron dextran group compared to the oral iron group [38]. A small study from our group also found that non-supplemented piglets became severely anemic by PD 14, but iron dextran injections (100 mg iron) or oral iron (10 mg iron/kg BW as ferrous sulfate drops) prevented anemia at this age [39]. Similar results were reported in other pig studies [40–43]. Recently, in a larger and more robustly designed study where the control group received vehicle supplementation without additional iron, Hb was similar at PD 14 between control and iron-supplemented groups; anemia was observed at PD 35 only in the control group (82 g Hb/L), suggesting a long-term effect of iron supplementation [44]. The smaller effect of iron on Hb in the latter study [44] compared to the previous studies [38,39] may be explained by the differences in iron dose (1 vs. 10 or 15–20 mg iron/kg BW, respectively).

Rodent studies show that iron supplementation prior to weaning increases body iron levels, but effects on hematology are inconsistent [45–54]. Our group observed that in rat pups with ID, induced by a maternal low iron diet, iron supplementation corrected Hb and tissue iron levels [45]. Other studies show that hematopoiesis in rodents that were not ID at birth was either increased or unchanged by iron supplementation. Varying the iron dose produces variation in hematology and iron status outcomes, suggesting that iron intake levels affect hematopoiesis [39,41,44,46]. In summary, the extent to which iron supplementation improves iron stores and Hb in the pre-weanling animal depends on baseline iron status as well as the dose and type of iron administered.

2.4. Developmental Regulation of Iron

Homeostatic mechanisms control iron availability in early life, and postnatal growth necessitates rapid expansion of blood volume which increases demand for iron (Figure 1). Erythropoietin (EPO) is synthesized by the kidney in response to low oxygen or ID. Through endocrine signaling, EPO drives erythropoiesis and increases erythroferrone (ERFE) production in the bone marrow. ERFE signaling ensures that there is sufficient iron available for heme synthesis and RBC production by promoting iron absorption and increases circulating levels of iron, which is accomplished through suppression of hepcidin transcription in hepatocytes. Conversely, the iron regulatory hormone hepcidin is upregulated by bone morphogenic protein (BMP6) signaling in response to iron sensing by hepatic endothelial cells— in the absence of suppression by ERFE [55,56]. Hepcidin blocks ferroportin-mediated iron export in enterocytes, iron-storing hepatocytes, and spleen reticuloendothelial macrophages resulting in reduced iron circulation in the blood. This also leads to reduced iron absorption in the small intestine. The hepcidin-ferroportin axis serves as the systemic regulatory mechanism that prevents iron toxicity from dietary overexposure [57,58]. However, recent evidence suggests that this mechanism is not functionally mature in infants: Iron absorption is not well regulated in response to iron over-supplementation during the first year of life [9,10]. The same appears to be true for pre-weanling mice [59], rats [45,60] and piglets [39]. These animal studies show that intestinal ferroportin is hypo-responsive to hepcidin-induced degradation and permits elevated iron absorption during early development, despite substantial hepatic iron deposition [39,45,48,59,60]. This suggests infants are more vulnerable to iron overload.

2.5. Oxidative Stress Results from Iron Overload

Iron is a pro-oxidative element and iron overload in cells disrupts the oxidative balance by generating reactive oxygen species (ROS). Iron catalyzes the conversion of hydrogen peroxide into the highly oxidizing species hydroxyl radical. Iron overload thereby causes lipid, protein and DNA oxidation, which can ultimately result in cell death. This type of cell death caused by iron-induced lipid peroxidation and ROS accumulation is termed ferroptosis [61,62]. Mutations in the hepcidin-ferroportin pathway cause hereditary hemochromatosis (HH), an iron overload disease that demonstrates the pathological effects of iron toxicity. During HH, iron accumulates in the liver, where extreme iron overload initiates fibrosis, then cirrhosis and loss of liver function [63,64], eventually leading to complications and death if untreated [63,64]. In HH patients, liver fibrosis is believed to result from iron-induced oxidative stress [65]. Iron overload leads to extrahepatic iron loading, negatively affecting functions of other tissues. Thus, iron overload resulting from blunted regulation of iron absorption in early life may explain how excess iron can be harmful to development, but whether excess iron in early life causes tissue iron overload and oxidative stress remains to be investigated. Iron-toxicity injuries to developing organs like the liver would explain delays in growth and other adverse effects of iron supplementation in young children [10,13].

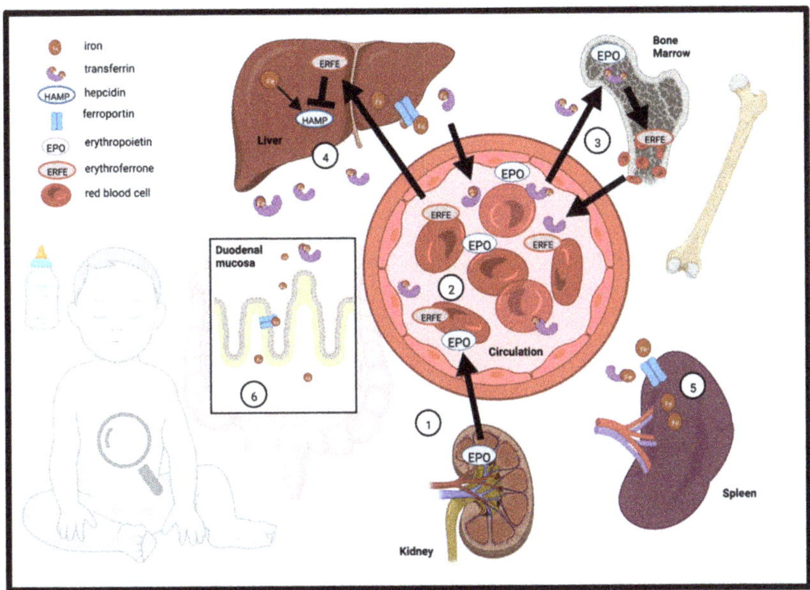

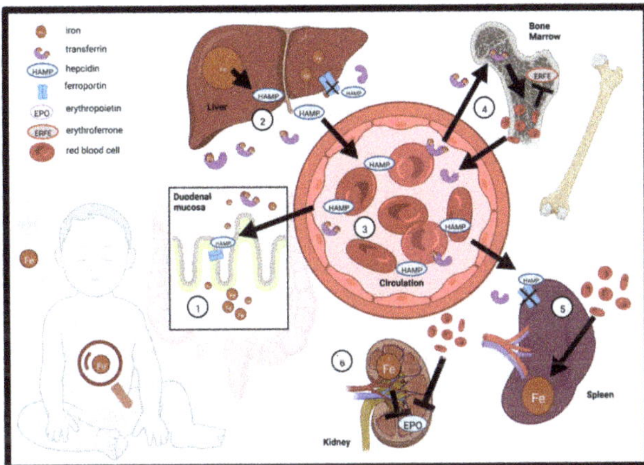

Figure 1. Iron regulation in the infant in response to iron supplementation. TOP—Iron regulation in the breast-fed infant in the absence of iron supplementation (**1**) The kidney secretes erythropoietin (EPO) to support the expansion of blood volume in response to low iron and low oxygen sensing; (**2**) EPO enters circulation and travels to the bone marrow, (**3**) where it drives erythropoiesis and secretion of erythroferrone (ERFE); (**4**) ERFE travels to the liver and suppresses hepcidin (HAMP) production, which allows for increased transferrin-bound iron in circulation; suppression of HAMP allows export of iron from the (**5**) spleen via ferroportin, as well as increased intestinal absorption of iron through ferroportin, both of which further increases iron in circulation. BOTTOM—Iron regulation in response to iron supplementation: (**1**) iron from supplementation is absorbed through the duodenal mucosa, is picked up by transferrin, and travels to the liver (**2**), increasing iron stores and up-regulating HAMP; (**3**) HAMP enters the circulation; (**4**) transferrin-bound iron supports increased erythropoiesis; in the spleen (**5**) erythrocytes are recycled but iron is sequestered because HAMP prevents export through ferroportin; (**6**) elevated iron and blood volume suppresses EPO production in the kidney.

One double-blinded RCT investigated whether the amount of iron in formula alters blood markers of oxidative stress in infants [66]. Infants consuming 4 mg iron/L (as lactoferrin and FS) had greater plasma glutathione peroxidase activity (a marker of antioxidant activity) than those receiving more iron (6.9 mg iron/L as FS). The higher activity may have been due to higher levels of selenium in the 4 mg iron/L formulas, because selenium is a required component of glutathione peroxidase. When controlling for copper and selenium, there was no difference in glutathione peroxidase activity due to iron levels. Another RCT in Sweden and Honduras found that daily iron supplementation from 4–9 mo of age (at 1 mg/kg body weight, the current recommended dose) reduced plasma copper-zinc superoxide dismutase (SOD) activity, which is an antioxidant marker as well as an indicator of copper status [26].

Few other studies in human infants have reported effects on oxidative stress markers, but animal studies provide some insight. Nearly all iron supplementation studies that have measured oxidative stress in pre-weanling animal models have focused on oxidative stress in the CNS [39,48,49,53,67–74]. Our group observed no significant effect on hippocampal oxidative stress in pre-weanling rats or weanling piglets [39,48]. Dong et al. found that in piglets—which are born with ID—iron supplementation decreased expression of pro-inflammatory cytokines in the liver and spleen while increasing expression of genes involved in anti-oxidative activity [38]. Intriguingly, this effect was unique to the piglet group orally receiving ferrous glycine chelate iron, while the iron dextran injection group actually had increased expression of interleukin-1β and had no effect on antioxidant gene expression in the liver. Although both forms of iron had similar effects on Hb and SF, injection of iron dextran further increased hepatic and extrahepatic iron loading, and this likely contributed to the elevated inflammatory and oxidative stress markers [38]. Inconsistent oxidative stress effects have been observed in various brain regions in aging rodents following excess neonatal iron supplementation [49,53,67,69–74]. Few studies assessed tissue iron content or iron status when determining long-term oxidative stress effects of neonatal iron exposure. Kaur et al. observed increased oxidative stress in the substantia nigra (SN) of aged mice (12 mo) but not young adult mice (2 mo) following neonatal iron exposure, and this was associated with increased SN iron levels and reduced CNS motor circuit (nigrostriatal) activity [49]. Additional studies in human infants and animal models are necessary to understand how the pro-oxidative effects of iron might play a role in the growth and development outcomes of iron supplementation.

3. Growth & Development

3.1. Growth Effects of Dietary Iron Excess

Controlled studies have shown that iron supplementation of iron-replete infants negatively impacts their growth [22,24,75], but this effect has not been consistent in all studies [20,76]. A randomized placebo-controlled trial (RCT) reported iron supplementation from 4–9 mo reduced length-gain and head circumference-gain to 9 mo in Swedish infants who had low risk of ID [22]. A separate RCT in Indonesia found that iron provision reduced weight-for-age and length-for-age z-scores of iron-replete infants [24]. Another RCT in South East Asia found that iron supplementation from 6 to 12 months reduced length-for-age, but only in infants who had a healthy birth weight at baseline [75]. However, a more recent RCT from our group did not find any effects on growth metrics for healthy, full term Swedish infants from 6 weeks to 6 months [77]. It should be noted, though, that the previous three studies [22,24,75] all provided iron as drops, whereas the latter one provided iron in infant formula. A systematic review and meta-analysis of randomized controlled studies in children age 4–23 months reported negative effects of iron on weight and length gain [20], while another systematic review and meta-analysis of studies in children age 6–23 months did not find an effect on growth [76]. The difference in age of introduction of iron supplementation may explain this discrepancy–considerably more iron may be absorbed during 4–6 months of age when regulation of iron absorption is immature [9,10]. Yet, another comparable review and meta-analysis will investigate growth effects in iron-replete infants [36]. To date, there are insufficient studies to conclude the effect of iron supplementation on the growth of healthy, iron-replete infants. Nevertheless, the finding that iron is disruptive to growth in some cases demands further investigation into these effects.

3.2. Neurodevelopmental Outcomes of Iron Supplementation

The cognitive and behavioral effects of iron administration are also inconsistent [15]. Iron provision may prevent ID-related disruptions to nervous system development, but may be harmful to iron-replete infants, leading to long-term cognitive and behavioral deficits [78,79].

A well-powered, double-blind RCT conducted in Chile observed improved iron status and metrics of behavioral and social development in infants fed high-iron formula levels (12 vs. 2.3 mg iron/L as FS) from 6–12 mo of age. However, the pooling of breast-fed and formula-fed groups and the poor control of iron intake in this study muddles the interpretation of these results [80,81]. Moreover, despite exclusion of infants with IDA, ID may have been common at baseline. A follow-up study found increased response to reward, language abilities, and motor function in 10-year-olds who had been pooled into the high-iron group as infants. The authors did not report whether baseline iron status or estimated daily iron intake influenced behavioral outcomes of iron provision [82]. An additional follow-up study of this trial reported adverse cognitive and behavior effects in 16-year-olds who had received high-iron formula during infancy [79]. A small RCT in Canada found a positive effect of iron on Bayley's scores of cognitive development [83], but iron intake from formula was poorly controlled and drop-out rates relatively high in this study. Another small RCT in Spain found that adding iron to cow's milk improved the iron status of infants who were already iron-replete at baseline, but did not affect mental and psychomotor development metrics [84]. Thus, the impact of iron supplementation on long-term cognitive function is still unclear.

Ideally, supplement dose would be determined by an infant's baseline iron status and optimized for healthy brain development, but this requires robust, well-powered studies. Unfortunately, few well-powered studies have measured baseline iron status or stratified results according to baseline iron status. One follow-up [78] of the same RCT above [80,81] found that after exclusion of anemic infants, baseline Hb predicted the effects of formula iron (12 vs. 2.3 mg/L) on cognitive development scores: infants with higher hemoglobin levels at baseline had poorer development scores at 10 years of age if they received high-iron formula, while infants with lower hemoglobin at baseline had improved development scores [78].

In a meta-analysis of RCTs, Pasricha et al. found that iron supplementation of all children aged 4–23 months did not affect Bayley's mental or psychomotor development scores. Indeed, they observed a positive effect on Bayley scores when iron was provided to iron-deficient children, but stated there were insufficient well-powered studies to conclude whether iron provision is beneficial or harmful to iron-replete infants [20]. An upcoming systematic review from Hare et al. and meta-analysis may provide further insight on this matter [36]. Animal studies provide some compelling evidence that excess iron is harmful to brain development and leads to long-term cognitive and psychomotor deficits (discussed below); however, more human studies are needed to confirm these effects [15].

3.3. Mechanisms Underlying Neurodevelopmental Effects of Iron Supplementation

Iron is required not only for postnatal proliferation and differentiation of the central nervous system (CNS)—which begins prenatally and continues postnatally—but also for CNS-specific pathways, including neurotransmitter synthesis and myelination [85]. Brain regions with greater metabolic need for iron are programmed to import iron more rapidly than other regions. By this reasoning, such regions may permit excess iron loading and influence susceptibility to iron toxicity-induced oxidative stress. Oxidative stress damages CNS cells by triggering apoptosis, ferroptosis and necrosis.

The adult hippocampus is heavily myelinated, and the infant hippocampus requires relatively large amounts of iron because myelin synthesis is iron-demanding and peaks at this age. Myelin sheaths in the CNS are formed by oligodendrocytes, which wrap their myelin around neuronal axons, surrounding and insulating them to reduce axon resistance and accelerate signaling speed. Oligodendrocytes and their precursors must import and store sufficient iron for myelination, which is why ID leads to insufficient myelination. This may explain how ID during infancy leads to long-term cognitive and behavioral deficits; however, myelination is only one of many iron-demanding processes that take place in the CNS during the first year of life [86].

One study in pre-weanling rats observed that excess iron increased total iron content in the cortex, hippocampus, substantia nigra, thalamus, deep cerebellum and pons, but not in the striatum at PD 21. In contrast, supplying iron after weaning increased iron in the hippocampus and pons at PD 35, but not in other regions. Moreover, pre-weanling rats supplemented through PD 35 had elevated iron levels in the cortex, hippocampus, pons and superficial cerebellum [54]. These findings provide evidence that brain regions are differentially affected by iron supplementation.

Another study investigated how the timing of excess iron exposure affected oxidative stress in various brain regions. A gastric gavage of iron (10 mg iron/kg BW as ferrous succinate) was administered daily to rats PD 5–7, PD 10–12, PD 19–21 (pre-weaning), or PD 30–32 (post-weaning), and brain regions were assessed for oxidative stress at 3–5 mo of age (adulthood) [73]. They observed that pre-weanling iron exposure caused oxidative stress in the hippocampus, cortex and substantia nigra, suggesting a lasting effect of early life brain iron accumulation. Furthermore, CNS oxidative stress in this study was associated with impaired recognition memory. The hippocampus is part of the brain circuitry that encodes learning and memory—including spatial mapping and social cognition—and was also the region most consistently affected by oxidative stress in this study. The results from this study [73] are in agreement with recent studies from our group in piglets [39,87] and suggest that excess iron provision causes iron loading and oxidative stress in the hippocampus, associated with adverse effects on long-term cognitive function.

In a long-term animal study, oxidative stress was measured in brain regions of aging rats that were exposed to excess iron as neonates (oral gavage of 120 mg iron/kg BW as carbonyl iron). They found that pre-weanling iron overexposure elevated substantia nigra malondialdehyde (MDA) content (a marker for lipid peroxidation) and reduced glutathione content (a marker for antioxidant activity) at PD 400. These changes were associated with reduced dopamine neurotransmitter content in the striatum, as well as alterations in motor behavior, suggesting that excess iron in early life may lead to long-term dysfunction of the nigrostriatal pathway, a brain circuit involved in controlling movement, memory and response to reward [68]. Additional animal studies are needed to confirm these effects; however, these findings are congruent to cognitive and behavior effects in human infants [78].

4. Trace Mineral Interactions

4.1. Iron Deficiency May Mask Copper or Zinc Deficiency

Prolonged copper or zinc deficiency leads to iron deficiency. Copper is a required component of hephaestin and ceruloplasmin, which operate as co-transporters of iron [88]. Copper deficiency progressively diminishes the activity of these co-transporters, thereby reducing intestinal iron absorption, which causes iron deficiency. Similarly, zinc deficiency reduces iron absorption by suppressing the expression of the iron importer DMT1 and the iron exporter ferroportin [89–91].

In cases when ID derives from copper or zinc deficiency, iron supplementation would not be an effective treatment, as iron intake may be sufficient. Conversely, excess dietary trace mineral intake, such as excess iron, can affect absorption and metabolism of other trace minerals [92].

4.2. Iron Competes with Other Trace Minerals for Absorption & Metabolism

Excess iron may disrupt absorption and metabolism of other trace minerals. In a secondary analysis of a randomized, placebo-controlled trial, serum zinc decreased in infants after 6 months of iron supplementation, but only in infants that were iron-replete at baseline (6 months). However, a study in non-anemic Kenyan infants did not find an effect on serum zinc or zinc absorption with the addition of iron in micronutrient powder [93].

Copper-zinc superoxide (CuZnSOD) dismutase activity, a marker of copper status, was reduced in iron vs. placebo-supplemented infants at 9 months; however, no effect on serum copper was observed [26]. Insufficient research exists to ascertain that excess iron influences infant zinc and copper status, but similarities in biochemistry and pathways of absorption among iron, zinc, copper and manganese may explain how excess iron intake would disrupt trace mineral metabolism.

A pre-weanling rat supplementation study from our group demonstrated that tissue levels of zinc, copper, and manganese were altered by excess iron supplementation [46]. Pre-weanling rats with high iron intake had reduced liver copper levels and elevated levels of zinc in the liver, kidney, brain, and intestine compared to a vehicle control group. Prolonged supplementation with excess iron reduced zinc and copper levels in rat brains, reduced zinc and manganese in spleen tissue and caused elevated zinc in the liver. Lower levels of iron supplementation affected trace mineral levels to a lesser extent in the pre-weanling rats, suggesting that excess early life iron supplementation stimulates dysregulation of trace mineral metabolism. Nevertheless, additional studies are needed to determine how and when excess iron influences or disrupts trace mineral status in infants.

4.3. Trace Minerals and Oxidative Stress

The transporters DMT1, ZIP8, and ZIP14 import divalent metals including iron, copper, zinc, and manganese [45,94–96], therefore it is possible that high levels of iron may outcompete other divalent metals for import, and this may explain alterations in availability of these minerals in response to excess iron supplementation. Likely a secondary effect of cellular iron loading is the upregulation of trace metal binding and storage proteins such as copper-zinc SOD, manganese SOD, metallothioneins (MT) and ceruloplasmin (CP). MTs, CP, and copper-zinc or manganese SODs require these metals to function as ROS scavengers and their upregulation is induced by oxidative stress. By this reasoning iron loading may upregulate antioxidant, metal-binding proteins including MTs, SODs and CP by inducing oxidative stress. Since zinc and copper are needed for basic metabolism, growth and resistance to infection, disrupting their availability to growing organs and tissues would disrupt development and health [8]. However, it remains to be investigated if mineral interactions in the context of excess iron are linked to the adverse growth and development effects of excess iron.

5. Morbidity & Mortality

5.1. Iron Affects Morbidity & Mortality of Infants & Children

Approximately 90% of iron from FS, a common iron supplement, remains in the gastrointestinal (GI) tract until it is excreted [9]. The GI side effects of FS iron supplements are well-established: after pooling data from 43 studies, a meta-analysis of GI side effects of FS for adults estimated an 11% incidence rate for nausea, 12% for constipation and 8% for diarrhea [97]. Another systematic review and meta-analysis estimated that 1 in 3 adults who received FS supplementation experiences some adverse effects [98]. It seems likely that infants would be affected similarly, but this has not been fully investigated.

Iron provision has been associated with increased risk of diarrhea and respiratory infection in some studies [17,20]. If excess iron increases infection risk, this may explain how growth is negatively affected by iron supplementation in some cases [22,24,75]. A previous study from our group found that diarrhea frequency increased and growth was reduced for infants in Sweden and Honduras who had normal Hb levels at baseline, while the opposite was true for infants who were anemic at baseline [22]. For many studies reporting no effect of iron supplementation on diarrhea frequency, results are not stratified according to baseline iron status (provided that baseline iron status was measured in the study) [99–104]. One recent study found that infants who were treated with antibiotics experienced greater frequency of diarrhea if they were also receiving MNP with iron, as compared to infants who were treated with antibiotics while receiving MNP without iron, as part of a larger double-blinded RCT [25]. Increased iron availability in the gut may have increased the proliferation of diarrhea-causing

Clostridium difficile in infants receiving high-iron formula and iron drops [105]. The bioavailability of iron (i.e., the extent to which iron is absorbed or passed through the gut) may be influenced significantly by intervention methodology: supplementation vs. fortification with iron, form of iron used, and timing of iron administration [9,11,17]. In their review of diarrhea outcomes [17], Ghanchi et al. suggested that supplementation may increase risk for diarrhea when compared to fortification; conversely, more expensive forms of iron (such as NaFeEDTA) may lower the risk for diarrhea compared to iron salts. Furthermore, common foods introduced as part of the complementary diet after 6 months of age influence iron absorption: grains, beans, and legumes contain indigestible phytates that reduce iron bioavailability, and citric or ascorbic acids in foods can augment iron absorption [106]. However, there are still an insufficient number of comparative studies to define the safest iron intervention methods for infants. In summary, current evidence suggests that baseline microbiota, iron status and iron intervention methodology are essential for predicting whether iron may increase morbidity in infants.

5.2. Gut Development & the Gut Microbiota

Alterations to the gut microbiota may contribute to GI side effects of iron supplements, as well as growth and development outcomes. Infancy is a critical period for symbiotic gut microbiota colonization and recent studies show that iron supplements alter the gut microbiota in ways that may be unfavorable to infant GI health [23,105,107,108]. Enteropathogens invade more easily during this age due to immature barrier function of the intestinal mucosa [109], leading to diarrhea or other infections [17,110]. Bacteria translocating across the mucosa trigger pro-inflammatory signaling, perhaps leading to diarrhea, both of which are likely to impair the nutrient absorption capacity of the GI tract [111]. Prolonged GI inflammation or diarrhea might therefore reduce an infant's growth rate, suggesting GI effects are mechanistically related to adverse growth and development effects of excess iron.

An important aspect of development involves healthy colonization of the gut with commensal microbes, because the gut microbiota provides essential roles to their host's health and development [112–114]. Besides maternal microbiota and birth method, the infant diet is the major determinant of gut colonization [114–117]. Breastfeeding and breast milk support healthy gut microbiota development by providing prebiotic oligosaccharides that preferentially craft the infant gut so that it is dominated by commensal *Bifidobacterium infantis*, which serves multiple health and development roles [118–121]. *Bifidobacterium infantis* has been shown to suppress the proliferation of pathogens and improve the integrity of the mucosal barrier, preventing inflammation and diarrhea [121]. Multiple studies have shown that iron reduces the abundance of commensal bacteria (including *Bifidobacterium infantis*) and elevates pathogen-associated bacteria [16,25,107]. Since commensal gut bacteria are so important for health and development, disrupting healthy colonization with commensals might explain some adverse outcomes of iron [108,112,122]. The gut microbiome of iron-replete infants may be more adversely affected by iron, but only one study has investigated gut microbiota outcomes in healthy, iron-replete infants [105].

A double-blind RCT of iron in micronutrient powder (MNP) given to Kenyan infants found increased abundance of *Clostridium* and *Escherichia/Shigella*–including increased pathogenic strains of *E. coli*–as well as elevated calprotectin, a measure of GI inflammation [23]. An additional robustly designed, double-blinded placebo controlled trial tested the effects of iron in MNP which was provided to 6-months old Kenyan infants for 3 months. In contrast to infants who received MNP without iron, infants who consumed MNP with iron had reduced abundance of commensal bacteria *Bifidobacterium* over time, while maintaining the abundance of *Escherichia* [107]. Another study from this group that was part of a large double-blind RCT in Kenya concluded that the addition of galacto-oligosaccharides (GOS) to the MNP with iron prevented its adverse effects on the microbiome. Despite the small sample size in this study, the results provide compelling evidence that iron adversely alters microbiome development by disruption of colonization by commensal bacteria [108]. A separate analysis that was part of this RCT followed gut microbiota changes and diarrhea outcomes in infants participating in the trial that had to be treated with antibiotics.

Antibiotics were not as effective at suppressing the growth of enteropathogens or reducing diarrhea incidence in infants who were receiving MNP with iron, as compared to antibiotic-treated infants receiving MNP without iron [25]. These findings suggest that infants receiving iron supplements would be more susceptible to enteropathogens and have more diarrhea despite antibiotic treatment [35].

Disruptions to gut microbiota development lead to adverse effects on infant health, including alterations to GI development, metabolic signaling, brain development and immune system development [112,122]. Therefore, further studies are necessary to define how excess iron-induced alterations to gut microbiota development during infancy impact infant health and growth [13,16]. Considering that infant gut microbiota development is so important for overall development and that increased iron levels in the gut may cause adverse GI side effects and gut microbiota dysbiosis, it seems likely that the gut microbiota is involved in the adverse development effects of excess iron. Additional studies in animal models should characterize effects of excess iron on gut microbiota development and generate hypotheses about iron-induced alterations to the microbiota that may be causing adverse health and development outcomes.

6. Conclusions

Our conclusions are summarized graphically in Figure 2. Iron supplementation during infancy improves iron status, thereby reducing the risk of developing ID or IDA. However, the capacity of exogenous iron provision to disrupt health and development of otherwise healthy infants who are iron-replete is unclear because few existing studies have specifically measured iron status at baseline. Further, translationally optimized animal models are needed to investigate the mechanisms behind the adverse effects to infant health. Excess iron provision may delay growth and neurodevelopment and increase susceptibility to disease and infection, and it is likely that iron toxicity, mineral interactions and alterations in the gut microbiota are behind these outcomes.

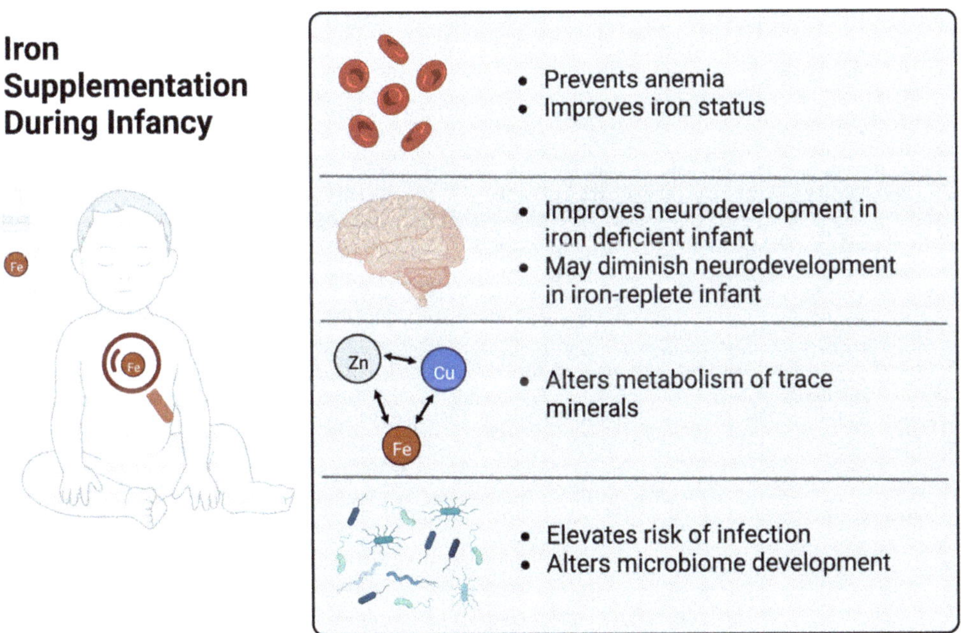

Figure 2. Iron Supplementation During Infancy. A graphical summary of the beneficial and adverse effects of iron supplementation of infants.

Author Contributions: Conceptualization, S.A.M. and B.L.; methodology, S.A.M., P.J. and B.L.; investigation, S.A.M., R.D. and E.D.; writing—original draft preparation, S.A.M.; writing—review and editing, P.J. and B.L.; visualization, S.A.M.; supervision, B.L. All authors have read and agreed to the published version of the manuscript.

Funding: This research received no external funding.

Data Availability Statement: The data presented in this study are not publicly available. The data are available on request from the corresponding author.

Acknowledgments: We thank Patricia Oteiza for providing helpful feedback on this review.

Conflicts of Interest: The authors declare no conflict of interest.

References

1. Miller, J.L. Iron Deficiency Anemia: A Common and Curable Disease. *Cold Spring Harb. Perspect. Med.* **2013**, *3*, a011866. [CrossRef] [PubMed]
2. Burke, R.; Leon, J.; Suchdev, P. Identification, Prevention and Treatment of Iron Deficiency during the First 1000 Days. *Nutrients* **2014**, *6*, 4093–4114. [CrossRef] [PubMed]
3. Yang, Z.; Lönnerdal, B.; Adu-Afarwuah, S.; Brown, K.H.; Chaparro, C.M.; Cohen, R.J.; Domellöf, M.; Hernell, O.; Lartey, A.; Dewey, K.G. Prevalence and Predictors of Iron Deficiency in Fully Breastfed Infants at 6 Mo of Age: Comparison of Data from 6 Studies. *Am. J. Clin. Nutr.* **2009**, *89*, 1433–1440. [CrossRef]
4. World Health Organization. *Iron Deficiency Anaemia: Assessment, Prevention, and Control. A Guide for Programme Managers*; World Health Organization: Geneva, Switzerland, 2001.
5. Shelov, S.P. *American Academy of Pediatrics. Caring for Your Baby and Young Child: Birth to Age Five*; Bantam: New York, NY, USA, 2009; ISBN 978-0-553-38630-1.
6. Lozoff, B.; Beard, J.; Connor, J.; Barbara, F.; Georgieff, M.; Schallert, T. Long-Lasting Neural and Behavioral Effects of Iron Deficiency in Infancy. *Nutr. Rev.* **2006**, *64*, S34–S43; discussion S72–S91. [CrossRef] [PubMed]
7. East, P.; Doom, J.R.; Blanco, E.; Burrows, R.; Lozoff, B.; Gahagan, S. Iron Deficiency in Infancy and Neurocognitive and Educational Outcomes in Young Adulthood. *Dev. Psychol.* **2021**, *57*, 962–975. [CrossRef] [PubMed]
8. DRI. *Dietary. Reference Intakes for Vitamin A, Vitamin K, Arsenic, Boron, Chromium, Copper, Iodine, Iron, Manganese, Molybdenum, Nickel, Silicon, Vanadium, and Zinc: A Report of the Panel on Micronutrients ... and the Standing Committee on the Scientific Evaluation of Dietary Reference Intakes, Food and Nutrition Board, Institute of Medicine*; Institute of Medicine (U.S.), Ed.; National Academy Press: Washington, DC, USA, 2001; ISBN 978-0-309-07279-3.
9. Domellöf, M.; Lönnerdal, B.; Abrams, S.A.; Hernell, O. Iron Absorption in Breast-Fed Infants: Effects of Age, Iron Status, Iron Supplements, and Complementary Foods. *Am. J. Clin. Nutr.* **2002**, *76*, 198–204. [CrossRef]
10. Lönnerdal, B. Development of Iron Homeostasis in Infants and Young Children. *Am. J. Clin. Nutr.* **2017**, *106*, 1575S–1580S. [CrossRef]
11. Lönnerdal, B.; Georgieff, M.K.; Hernell, O. Developmental Physiology of Iron Absorption, Homeostasis, and Metabolism in the Healthy Term Infant. *J. Pediatr.* **2015**, *167*, S8–S14. [CrossRef]
12. Lönnerdal, B. Iron, Zinc, Copper, and Manganese in Infant Formulas. *Arch. Pediatr. Adolesc. Med.* **1983**, *137*, 433. [CrossRef]
13. Lonnerdal, B. Excess Iron Intake as a Factor in Growth, Infections, and Development of Infants and Young Children. *Am. J. Clin. Nutr.* **2017**, *106*, 1681S–1687S. [CrossRef]
14. Wessling-Resnick, M. Excess Iron: Considerations Related to Development and Early Growth. *Am. J. Clin. Nutr.* **2017**, *106*, 1600S–1605S. [CrossRef]
15. Agrawal, S.; Berggren, K.L.; Marks, E.; Fox, J.H. Impact of High Iron Intake on Cognition and Neurodegeneration in Humans and in Animal Models: A Systematic Review. *Nutr. Rev.* **2017**, *75*, 456–470. [CrossRef] [PubMed]
16. Paganini, D.; Zimmermann, M.B. The Effects of Iron Fortification and Supplementation on the Gut Microbiome and Diarrhea in Infants and Children: A Review. *Am. J. Clin. Nutr.* **2017**, *106*, 1688S–1693S. [CrossRef] [PubMed]
17. Ghanchi, A.; James, P.T.; Cerami, C. Guts, Germs, and Iron: A Systematic Review on Iron Supplementation, Iron Fortification, and Diarrhea in Children Aged 4–59 Months. *Curr. Dev. Nutr.* **2019**, *3*, nzz005. [CrossRef]
18. Allen, L.H. Iron Supplements: Scientific Issues Concerning Efficacy and Implications for Research and Programs. *J. Nutr.* **2002**, *132*, 813S–819S. [CrossRef] [PubMed]
19. Dietary Guidelines Advisory Committee. *Scientific Report of the 2020 Dietary Guidelines Advisory Committee: Advisory Report to the Secretary of Agriculture and the Secretary of Health and Human Services*; U.S. Department of Agriculture, Agricultural Research Service: Washington, DC, USA, 2020; p. 786.
20. Pasricha, S.-R.; Hayes, E.; Kalumba, K.; Biggs, B.-A. Effect of Daily Iron Supplementation on Health in Children Aged 4–23 Months: A Systematic Review and Meta-Analysis of Randomised Controlled Trials. *Lancet Glob. Health* **2013**, *1*, e77–e86. [CrossRef]
21. *Guideline: Daily Iron Supplementation in Infants and Children*; World Health Organization: Geneva, Switzerland, 2016; ISBN 978-92-4-154952-3.

22. Dewey, K.G.; Domellöf, M.; Cohen, R.J.; Landa Rivera, L.; Hernell, O.; Lönnerdal, B. Iron Supplementation Affects Growth and Morbidity of Breast-Fed Infants: Results of a Randomized Trial in Sweden and Honduras. *J. Nutr.* **2002**, *132*, 3249–3255. [CrossRef]
23. Jaeggi, T.; Kortman, G.A.M.; Moretti, D.; Chassard, C.; Holding, P.; Dostal, A.; Boekhorst, J.; Timmerman, H.M.; Swinkels, D.W.; Tjalsma, H.; et al. Iron Fortification Adversely Affects the Gut Microbiome, Increases Pathogen Abundance and Induces Intestinal Inflammation in Kenyan Infants. *Gut* **2015**, *64*, 731–742. [CrossRef]
24. Lind, T.; Seswandhana, R.; Persson, L.-A.; Lönnerdal, B. Iron Supplementation of Iron-Replete Indonesian Infants Is Associated with Reduced Weight-for-Age. *Acta Paediatr.* **2008**, *97*, 770–775. [CrossRef]
25. Paganini, D.; Uyoga, M.A.; Kortman, G.A.M.; Cercamondi, C.I.; Winkler, H.C.; Boekhorst, J.; Moretti, D.; Lacroix, C.; Karanja, S.; Zimmermann, M.B. Iron-Containing Micronutrient Powders Modify the Effect of Oral Antibiotics on the Infant Gut Microbiome and Increase Post-Antibiotic Diarrhoea Risk: A Controlled Study in Kenya. *Gut* **2019**, *68*, 645–653. [CrossRef]
26. Domellöf, M.; Dewey, K.G.; Cohen, R.J.; Lönnerdal, B.; Hernell, O. Iron Supplements Reduce Erythrocyte Copper-Zinc Superoxide Dismutase Activity in Term, Breastfed Infants. *Acta Paediatr.* **2005**, *94*, 1578–1582. [CrossRef]
27. World Health Organization. *Hemoglobin Concentrations for the Diagnosis of Anemia and Assessment of Severity*; World Health Organization: Geneva, Switzerland, 2011.
28. Nutritional Anemias. Report of a WHO Scientific Group. 1968. Available online: https://apps.who.int/iris/handle/10665/40707 (accessed on 26 July 2022).
29. Sturgeon, P. Studies of Iron Requirements in Infants and Children. *Pediatrics* **1954**, *13*, 107–125. [CrossRef]
30. Sturgeon, P. Studies of Iron Requirements in Infants III. Influence of Supplemental Iron during Normal Pregnancy on Mother and Infant B. The Infant. *Br. J. Haematol* **1959**, *5*, 45–55. [CrossRef]
31. Center for Disease Control & Prevention. CDC Criteria for Anemia in Children and Childbearing-Aged Women. *Morb. Mortal. Wkly. Rep.* **1989**, *38*, 400–404.
32. Pilch, S.M.; Senti, F.R.; Assessment of the Iron Nutritional Status of the U.S. Population Based on Data Collected in the Second National Health and Nutrition Examination Survey. 1984. Available online: https://agris.fao.org/agris-search/search.do?recordID=US201300391414 (accessed on 19 August 2022).
33. Baker, R.D.; Greer, F.R. Diagnosis and Prevention of Iron Deficiency and Iron-Deficiency Anemia in Infants and Young Children (0–3 Years of Age). *Pediatrics* **2010**, *126*, 1040–1050. [CrossRef]
34. World Health Organization. *Serum Ferritin Concentrations for the Assessment of Iron Status and Iron Deficiency in Populations*; World Health Organization: Geneva, Switzerland, 2011.
35. Garcia-Casal, M.N.; Pasricha, S.-R.; Martinez, R.X.; Lopez-Perez, L.; Peña-Rosas, J.P. Serum or Plasma Ferritin Concentration as an Index of Iron Deficiency and Overload. *Cochrane Database Syst. Rev.* **2021**, *2021*, CD011817. [CrossRef]
36. Hare, D.J.; Braat, S.; Cardoso, B.R.; Morgan, C.; Szymlek-Gay, E.A.; Biggs, B.-A. Health Outcomes of Iron Supplementation and/or Food Fortification in Iron-Replete Children Aged 4–24 Months: Protocol for a Systematic Review and Meta-Analysis. *Syst. Rev.* **2019**, *8*, 253. [CrossRef]
37. *Nutrient Requirements of Swine*, 11th ed.; National Research Council, (U.S.) (Ed.) National Academies Press: Washington, DC, USA, 2012; ISBN 978-0-309-22423-9.
38. Dong, Z.; Wan, D.; Li, G.; Zhang, Y.; Yang, H.; Wu, X.; Yin, Y. Comparison of Oral and Parenteral Iron Administration on Iron Homeostasis, Oxidative and Immune Status in Anemic Neonatal Pigs. *Biol. Trace Elem. Res.* **2020**, *195*, 117–124. [CrossRef]
39. Ji, P.; Lonnerdal, B.; Kim, K.; Jinno, C.N. Iron Oversupplementation Causes Hippocampal Iron Overloading and Impairs Social Novelty Recognition in Nursing Piglets. *J. Nutr.* **2019**, *149*, 398–405. [CrossRef]
40. Egeli, A.K.; Framstad, T. Effect of an Oral Starter Dose of Iron on Haematology and Weight Gain in Piglets Having Voluntary Access to Glutamic Acid-Chelated Iron Solution. *Acta Vet. Scand.* **1998**, *39*, 359–365. [CrossRef]
41. Furugouri, K.; Kawabata, A. Iron Absorption in Nursing Piglets. *J. Anim. Sci.* **1975**, *41*, 1348–1354. [CrossRef]
42. Loh, T.J.; Leong, K.; Too, H.; Mah, C.; Choo, P. The Effects of Iron Supplementation in Preweaning Piglets. *Malays. J. Nutr.* **2001**, *7*, 41–49.
43. Webster, W.R.; Dimmock, C.K.; O'Rourke, P.K.; Lynch, P.J. Evaluation of Oral Iron Galactan as a Method of Iron Supplementation for Intensively Housed Sucking Piglets. *Aust. Vet. J.* **1978**, *54*, 345–348. [CrossRef]
44. Perng, V.; Li, C.; Klocke, C.R.; Navazesh, S.E.; Pinneles, D.K.; Lein, P.J.; Ji, P. Iron Deficiency and Iron Excess Differently Affect Dendritic Architecture of Pyramidal Neurons in the Hippocampus of Piglets. *J. Nutr.* **2021**, *151*, 235–244. [CrossRef]
45. Leong, W.-I.; Bowlus, C.L.; Tallkvist, J.; Lönnerdal, B. DMT1 and FPN1 Expression during Infancy: Developmental Regulation of Iron Absorption. *Am. J. Physiol.-Gastrointest. Liver Physiol.* **2003**, *285*, G1153–G1161. [CrossRef]
46. Leong, W.-I.; Bowlus, C.L.; Tallkvist, J.; Lönnerdal, B. Iron Supplementation during Infancy—Effects on Expression of Iron Transporters, Iron Absorption, and Iron Utilization in Rat Pups. *Am. J. Clin. Nutr.* **2003**, *78*, 1203–1211. [CrossRef]
47. Alexeev, E.E.; He, X.; Slupsky, C.M.; Lönnerdal, B. Effects of Iron Supplementation on Growth, Gut Microbiota, Metabolomics and Cognitive Development of Rat Pups. *PLoS ONE* **2017**, *12*, e0179713. [CrossRef]
48. McMillen, S.; Lönnerdal, B. Postnatal Iron Supplementation with Ferrous Sulfate vs. Ferrous Bis-Glycinate Chelate: Effects on Iron Metabolism, Growth, and Central Nervous System Development in Sprague Dawley Rat Pups. *Nutrients* **2021**, *13*, 1406. [CrossRef]

49. Kaur, D.; Peng, J.; Chinta, S.J.; Rajagopalan, S.; Di Monte, D.A.; Cherny, R.A.; Andersen, J.K. Increased Murine Neonatal Iron Intake Results in Parkinson-like Neurodegeneration with Age. *Neurobiol. Aging* **2007**, *28*, 907–913. [CrossRef]
50. Fredriksson, A.; Archer, T. Subchronic Administration of Haloperidol Influences the Functional Deficits of Postnatal Iron Administration in Mice. *Neurotox. Res.* **2006**, *10*, 123–129. [CrossRef]
51. Fredriksson, A.; Archer, T. Effect of Postnatal Iron Administration on MPTP-Induced Behavioral Deficits and Neurotoxicity: Behavioral Enhancement by L-Dopa-MK-801 Co-Administration. *Behav. Brain Res.* **2003**, *139*, 31–46. [CrossRef]
52. Fredriksson, A.; Archer, T. Postnatal Iron Overload Destroys NA-DA Functional Interactions. *J. Neural Transm.* **2007**, *114*, 195–203. [CrossRef]
53. Berggren, K.L.; Lu, Z.; Fox, J.A.; Dudenhoeffer, M.; Agrawal, S.; Fox, J.H. Neonatal Iron Supplementation Induces Striatal Atrophy in Female YAC128 Huntington's Disease Mice. *J. Huntingt. Dis.* **2016**, *5*, 53–63. [CrossRef]
54. Piñero, D.J.; Li, N.-Q.; Connor, J.R.; Beard, J.L. Variations in Dietary Iron Alter Brain Iron Metabolism in Developing Rats. *J. Nutr.* **2000**, *130*, 254–263. [CrossRef]
55. Muckenthaler, M.U.; Rivella, S.; Hentze, M.W.; Galy, B. A Red Carpet for Iron Metabolism. *Cell* **2017**, *168*, 344–361. [CrossRef]
56. Wang, C.-Y.; Babitt, J.L. Liver Iron Sensing and Body Iron Homeostasis. *Blood* **2019**, *133*, 18–29. [CrossRef]
57. Billesbølle, C.B.; Azumaya, C.M.; Kretsch, R.C.; Powers, A.S.; Gonen, S.; Schneider, S.; Arvedson, T.; Dror, R.O.; Cheng, Y.; Manglik, A. Structure of Hepcidin-Bound Ferroportin Reveals Iron Homeostatic Mechanisms. *Nature* **2020**, *586*, 807–811. [CrossRef]
58. Nemeth, E. Hepcidin Regulates Cellular Iron Efflux by Binding to Ferroportin and Inducing Its Internalization. *Science* **2004**, *306*, 2090–2093. [CrossRef]
59. Frazer, D.M.; Wilkins, S.J.; Darshan, D.; Mirciov, C.S.G.; Dunn, L.A.; Anderson, G.J. Ferroportin Is Essential for Iron Absorption During Suckling, But Is Hyporesponsive to the Regulatory Hormone Hepcidin. *Cell. Mol. Gastroenterol. Hepatol.* **2017**, *3*, 410–421. [CrossRef]
60. Darshan, D.; Wilkins, S.J.; Frazer, D.M.; Anderson, G.J. Reduced Expression of Ferroportin-1 Mediates Hyporesponsiveness of Suckling Rats to Stimuli That Reduce Iron Absorption. *Gastroenterology* **2011**, *141*, 300–309. [CrossRef]
61. Chen, X.; Yu, C.; Kang, R.; Tang, D. Iron Metabolism in Ferroptosis. *Front. Cell Dev. Biol.* **2020**, *8*, 590226. [CrossRef]
62. Li, J.; Cao, F.; Yin, H.; Huang, Z.; Lin, Z.; Mao, N.; Sun, B.; Wang, G. Ferroptosis: Past, Present and Future. *Cell Death Dis.* **2020**, *11*, 88. [CrossRef]
63. Niederau, C.; Fischer, R.; Sonnenberg, A.; Stremmel, W.; Trampisch, H.J.; Strohmeyer, G. Survival and Causes of Death in Cirrhotic and in Noncirrhotic Patients with Primary Hemochromatosis. *N. Engl. J. Med.* **1985**, *313*, 1256–1262. [CrossRef]
64. Deugnier, Y.M.; Loréal, O.; Turlin, B.; Guyader, D.; Jouanolle, H.; Moirand, R.; Jacquelinet, C.; Brissot, P. Liver Pathology in Genetic Hemochromatosis: A Review of 135 Homozygous Cases and Their Bioclinical Correlations. *Gastroenterology* **1992**, *102*, 2050–2059. [CrossRef]
65. Houglum, K.; Ramm, G.A.; Crawford, D.H.; Witztum, J.L.; Powell, L.W.; Chojkier, M. Excess Iron Induces Hepatic Oxidative Stress and Transforming Growth Factor? 1 in Genetic Hemochromatosis. *Hepatology* **1997**, *26*, 605–610. [CrossRef]
66. Lönnerdal, B.; Hernell, O. Iron, Zinc, Copper and Selenium Status of Breast-Fed Infants and Infants Fed Trace Element Fortified Milk-Based Infant Formula. *Acta Paediatr.* **1994**, *83*, 367–373. [CrossRef]
67. Budni, P.; de Lima, M.N.M.; Polydoro, M.; Moreira, J.C.F.; Schroder, N.; Dal-Pizzol, F. Antioxidant Effects of Selegiline in Oxidative Stress Induced by Iron Neonatal Treatment in Rats. *Neurochem. Res.* **2007**, *32*, 965–972. [CrossRef]
68. Chen, H.; Wang, X.; Wang, M.; Yang, L.; Yan, Z.; Zhang, Y.; Liu, Z. Behavioral and Neurochemical Deficits in Aging Rats with Increased Neonatal Iron Intake: Silibinin's Neuroprotection by Maintaining Redox Balance. *Front. Aging Neurosci.* **2015**, *7*, 206. [CrossRef]
69. Agrawal, S.; Fox, J.; Thyagarajan, B.; Fox, J.H. Brain Mitochondrial Iron Accumulates in Huntington's Disease, Mediates Mitochondrial Dysfunction, and Can Be Removed Pharmacologically. *Free Radic. Biol. Med.* **2018**, *120*, 317–329. [CrossRef]
70. Fernandez, L.L.; Carmona, M.; Portero-Otin, M.; Naudi, A.; Pamplona, R.; Schröder, N.; Ferrer, I. Effects of Increased Iron Intake during the Neonatal Period on the Brain of Adult AbetaPP/PS1 Transgenic Mice. *J. Alzheimers Dis.* **2010**, *19*, 1069–1080. [CrossRef]
71. Dal-Pizzol, F.; Klamt, F.; Frota, M.L.J.; Andrades, M.E.; Caregnato, F.F.; Vianna, M.M.; Schröder, N.; Quevedo, J.; Izquierdo, I.; Archer, T.; et al. Neonatal Iron Exposure Induces Oxidative Stress in Adult Wistar Rat. *Brain Res. Dev. Brain Res.* **2001**, *130*, 109–114. [CrossRef]
72. Yu, L.; Wang, X.; Chen, H.; Yan, Z.; Wang, M.; Li, Y. Neurochemical and Behavior Deficits in Rats with Iron and Rotenone Co-Treatment: Role of Redox Imbalance and Neuroprotection by Biochanin A. *Front. Neurosci.* **2017**, *11*, 657. [CrossRef] [PubMed]
73. De Lima, M.N.M.; Polydoro, M.; Laranja, D.C.; Bonatto, F.; Bromberg, E.; Moreira, J.C.F.; Dal-Pizzol, F.; Schröder, N. Recognition Memory Impairment and Brain Oxidative Stress Induced by Postnatal Iron Administration. *Eur. J. Neurosci.* **2005**, *21*, 2521–2528. [CrossRef]
74. Lavich, I.C.; de Freitas, B.S.; Kist, L.W.; Falavigna, L.; Dargél, V.A.; Köbe, L.M.; Aguzzoli, C.; Piffero, B.; Florian, P.Z.; Bogo, M.R.; et al. Sulforaphane Rescues Memory Dysfunction and Synaptic and Mitochondrial Alterations Induced by Brain Iron Accumulation. *Neuroscience* **2015**, *301*, 542–552. [CrossRef]
75. Dijkhuizen, M.A.; Winichagoon, P.; Wieringa, F.T.; Wasantwisut, E.; Utomo, B.; Ninh, N.X.; Hidayat, A.; Berger, J. Zinc Supplementation Improved Length Growth Only in Anemic Infants in a Multi-Country Trial of Iron and Zinc Supplementation in South-East Asia. *J. Nutr.* **2008**, *138*, 1969–1975. [CrossRef] [PubMed]

76. Petry, N.; Olofin, I.; Boy, E.; Donahue Angel, M.; Rohner, F. The Effect of Low Dose Iron and Zinc Intake on Child Micronutrient Status and Development during the First 1000 Days of Life: A Systematic Review and Meta-Analysis. *Nutrients* **2016**, *8*, 773. [CrossRef]
77. Björmsjö, M.; Hernell, O.; Lönnerdal, B.; Berglund, S.K. Reducing Iron Content in Infant Formula from 8 to 2 Mg/L Does Not Increase the Risk of Iron Deficiency at 4 or 6 Months of Age: A Randomized Controlled Trial. *Nutrients* **2020**, *13*, 3. [CrossRef]
78. Lozoff, B. Iron-Fortified vs Low-Iron Infant Formula: Developmental Outcome at 10 Years. *Arch. Pediatr. Adolesc. Med.* **2012**, *166*, 208. [CrossRef]
79. Gahagan, S.; Delker, E.; Blanco, E.; Burrows, R.; Lozoff, B. Randomized Controlled Trial of Iron-Fortified versus Low-Iron Infant Formula: Developmental Outcomes at 16 Years. *J. Pediatr.* **2019**, *212*, 124–130. [CrossRef]
80. Lozoff, B.; De Andraca, I.; Castillo, M.; Smith, J.B.; Walter, T.; Pino, P. Behavioral and Developmental Effects of Preventing Iron-Deficiency Anemia in Healthy Full-Term Infants. *Pediatrics* **2003**, *112*, 846–854. [CrossRef]
81. Walter, T.; Pino, P.; Pizarro, F.; Lozoff, B. Prevention of Iron-Deficiency Anemia: Comparison of High- and Low-Iron Formulas in Term Healthy Infants after Six Months of Life. *J. Pediatr.* **1998**, *132*, 635–640. [CrossRef]
82. Lozoff, B.; Castillo, M.; Clark, K.M.; Smith, J.B.; Sturza, J. Iron Supplementation in Infancy Contributes to More Adaptive Behavior at 10 Years of Age. *J. Nutr.* **2014**, *144*, 838–845. [CrossRef]
83. Friel, J.K.; Aziz, K.; Andrews, W.L.; Harding, S.V.; Courage, M.L.; Adams, R.J. A Double-Masked, Randomized Control Trial of Iron Supplementation in Early Infancy in Healthy Term Breast-Fed Infants. *J. Pediatr.* **2003**, *143*, 582–586. [CrossRef]
84. Iglesias Vázquez, L.; Canals, J.; Voltas, N.; Jardí, C.; Hernández, C.; Bedmar, C.; Escribano, J.; Aranda, N.; Jiménez, R.; Barroso, J.M.; et al. Does the Fortified Milk with High Iron Dose Improve the Neurodevelopment of Healthy Infants? Randomized Controlled Trial. *BMC Pediatr.* **2019**, *19*, 315. [CrossRef]
85. McCann, S.; Perapoch Amadó, M.; Moore, S.E. The Role of Iron in Brain Development: A Systematic Review. *Nutrients* **2020**, *12*, 2001. [CrossRef]
86. Todorich, B.; Pasquini, J.M.; Garcia, C.I.; Paez, P.M.; Connor, J.R. Oligodendrocytes and Myelination: The Role of Iron. *Glia* **2009**, *57*, 467–478. [CrossRef]
87. Ji, P.; B Nonnecke, E.; Doan, N.; Lönnerdal, B.; Tan, B. Excess Iron Enhances Purine Catabolism Through Activation of Xanthine Oxidase and Impairs Myelination in the Hippocampus of Nursing Piglets. *J. Nutr.* **2019**, *149*, 1911–1919. [CrossRef]
88. Chen, H.; Huang, G.; Su, T.; Gao, H.; Attieh, Z.K.; McKie, A.T.; Anderson, G.J.; Vulpe, C.D. Decreased Hephaestin Activity in the Intestine of Copper-Deficient Mice Causes Systemic Iron Deficiency. *J. Nutr.* **2006**, *136*, 1236–1241. [CrossRef]
89. Sreedhar, B.; Nair, K.M. Modulation of Aconitase, Metallothionein, and Oxidative Stress in Zinc-Deficient Rat Intestine during Zinc and Iron Repletion. *Free Radic. Biol. Med.* **2005**, *39*, 999–1008. [CrossRef]
90. El Hendy, H. Effect of Dietary Zinc Deficiency on Hematological and Biochemical Parameters and Concentrations of Zinc, Copper, and Iron in Growing Rats. *Toxicology* **2001**, *167*, 163–170. [CrossRef]
91. Kondaiah, P.; Yaduvanshi, P.S.; Sharp, P.A.; Pullakhandam, R. Iron and Zinc Homeostasis and Interactions: Does Enteric Zinc Excretion Cross-Talk with Intestinal Iron Absorption? *Nutrients* **2019**, *11*, 1885. [CrossRef] [PubMed]
92. Ha, J.-H.; Doguer, C.; Collins, J.F. Consumption of a High-Iron Diet Disrupts Homeostatic Regulation of Intestinal Copper Absorption in Adolescent Mice. *Am. J. Physiol.-Gastrointest. Liver Physiol.* **2017**, *313*, G353–G360. [CrossRef] [PubMed]
93. Esamai, F.; Liechty, E.; Ikemeri, J.; Westcott, J.; Kemp, J.; Culbertson, D.; Miller, L.V.; Hambidge, K.M.; Krebs, N.F. Zinc Absorption from Micronutrient Powder Is Low but Is Not Affected by Iron in Kenyan Infants. *Nutrients* **2014**, *6*, 5636–5651. [CrossRef] [PubMed]
94. Liu, Q.; Barker, S.; Knutson, M.D. Iron and Manganese Transport in Mammalian Systems. *Biochim. Et Biophys. Acta (BBA)-Mol. Cell Res.* **2021**, *1868*, 118890. [CrossRef]
95. Aydemir, T.B.; Cousins, R.J. The Multiple Faces of the Metal Transporter ZIP14 (SLC39A14). *J. Nutr.* **2018**, *148*, 174–184. [CrossRef]
96. Wang, C.-Y.; Jenkitkasemwong, S.; Duarte, S.; Sparkman, B.K.; Shawki, A.; Mackenzie, B.; Knutson, M.D. ZIP8 Is an Iron and Zinc Transporter Whose Cell-Surface Expression Is Up-Regulated by Cellular Iron Loading. *J. Biol. Chem.* **2012**, *287*, 34032–34043. [CrossRef]
97. Tolkien, Z.; Stecher, L.; Mander, A.P.; Pereira, D.I.A.; Powell, J.J. Ferrous Sulfate Supplementation Causes Significant Gastrointestinal Side-Effects in Adults: A Systematic Review and Meta-Analysis. *PLoS ONE* **2015**, *10*, e0117383. [CrossRef]
98. Cancelo-Hidalgo, M.J.; Castelo-Branco, C.; Palacios, S.; Haya-Palazuelos, J.; Ciria-Recasens, M.; Manasanch, J.; Pérez-Edo, L. Tolerability of Different Oral Iron Supplements: A Systematic Review. *Curr. Med. Res. Opin.* **2013**, *29*, 291–303. [CrossRef]
99. Lind, T.; Lönnerdal, B.; Stenlund, H.; Gamayanti, I.L.; Ismail, D.; Seswandhana, R.; Persson, L.-A. A Community-Based Randomized Controlled Trial of Iron and Zinc Supplementation in Indonesian Infants: Effects on Growth and Development. *Am. J. Clin. Nutr.* **2004**, *80*, 729–736. [CrossRef]
100. Nagpal, J.; Sachdev, H.P.S.; Singh, T.; Mallika, V. A Randomized Placebo-Controlled Trial of Iron Supplementation in Breastfed Young Infants Initiated on Complementary Feeding: Effect on Haematological Status. *J. Health Popul. Nutr.* **2004**, *22*, 203–211.
101. Bora, R.; Ramasamy, S.; Brown, B.; Wolfson, J.; Rao, R. Effect of Iron Supplementation from Neonatal Period on the Iron Status of 6-Month-Old Infants at-Risk for Early Iron Deficiency: A Randomized Interventional Trial. *J. Matern. Fetal Neonatal Med.* **2019**, *34*, 1421–1429. [CrossRef] [PubMed]
102. Ermis, B.; Demirel, F.; Demircan, N.; Gurel, A. Effects of Three Different Iron Supplementations in Term Healthy Infants after 5 Months of Life. *J. Trop. Pediatr.* **2002**, *48*, 280–284. [CrossRef] [PubMed]

103. Smuts, C.M.; Dhansay, M.A.; Faber, M.; van Stuijvenberg, M.E.; Swanevelder, S.; Gross, R.; Benadé, A.J.S. Efficacy of Multiple Micronutrient Supplementation for Improving Anemia, Micronutrient Status, and Growth in South African Infants. *J. Nutr.* **2005**, *135*, 653S–659S. [CrossRef]
104. Silva, D.G.; Franceschini, S.d.C.C.; Sigulem, D.M. Growth in Non-Anemic Infants Supplemented with Different Prophylactic Iron Doses. *J. Pediatr.* **2008**, *84*, 365–372. [CrossRef] [PubMed]
105. Simonyté Sjödin, K.; Domellöf, M.; Lagerqvist, C.; Hernell, O.; Lönnerdal, B.; Szymlek-Gay, E.A.; Sjödin, A.; West, C.E.; Lind, T. Administration of Ferrous Sulfate Drops Has Significant Effects on the Gut Microbiota of Iron-Sufficient Infants: A Randomised Controlled Study. *Gut* **2019**, *68*, 2095–2097. [CrossRef]
106. Lynch, S.; Pfeiffer, C.M.; Georgieff, M.K.; Brittenham, G.; Fairweather-Tait, S.; Hurrell, R.F.; McArdle, H.J.; Raiten, D.J. Biomarkers of Nutrition for Development (BOND)-Iron Review. *J. Nutr.* **2018**, *148*, 1001S–1067S. [CrossRef]
107. Tang, M.; Frank, D.N.; Hendricks, A.E.; Ir, D.; Esamai, F.; Liechty, E.; Hambidge, K.M.; Krebs, N.F. Iron in Micronutrient Powder Promotes an Unfavorable Gut Microbiota in Kenyan Infants. *Nutrients* **2017**, *9*, 776. [CrossRef]
108. Paganini, D.; Uyoga, M.A.; Kortman, G.A.M.; Cercamondi, C.I.; Moretti, D.; Barth-Jaeggi, T.; Schwab, C.; Boekhorst, J.; Timmerman, H.M.; Lacroix, C.; et al. Prebiotic Galacto-Oligosaccharides Mitigate the Adverse Effects of Iron Fortification on the Gut Microbiome: A Randomised Controlled Study in Kenyan Infants. *Gut* **2017**, *66*, 1956–1967. [CrossRef]
109. Chin, A.M.; Hill, D.R.; Aurora, M.; Spence, J.R. Morphogenesis and Maturation of the Embryonic and Postnatal Intestine. *Semin. Cell Dev. Biol.* **2017**, *66*, 81–93. [CrossRef]
110. Black, R.E.; Heidkamp, R. Causes of Stunting and Preventive Dietary Interventions in Pregnancy and Early Childhood. In *Nestlé Nutrition Institute Workshop Series*; Colombo, J., Koletzko, B., Lampl, M., Eds.; S. Karger AG: Basel, Switzerland, 2018; Volume 89, pp. 105–113, ISBN 978-3-318-06351-6.
111. Brown, K.H. Diarrhea and Malnutrition. *J. Nutr.* **2003**, *133*, 328S–332S. [CrossRef]
112. Dominguez-Bello, M.G.; Godoy-Vitorino, F.; Knight, R.; Blaser, M.J. Role of the Microbiome in Human Development. *Gut* **2019**, *68*, 1108–1114. [CrossRef] [PubMed]
113. Koenig, J.E.; Spor, A.; Scalfone, N.; Fricker, A.D.; Stombaugh, J.; Knight, R.; Angenent, L.T.; Ley, R.E. Succession of Microbial Consortia in the Developing Infant Gut Microbiome. *Proc. Natl. Acad. Sci. USA* **2011**, *108*, 4578–4585. [CrossRef] [PubMed]
114. Moore, R.E.; Townsend, S.D. Temporal Development of the Infant Gut Microbiome. *Open Biol.* **2019**, *9*, 190128. [CrossRef] [PubMed]
115. Mueller, N.T.; Bakacs, E.; Combellick, J.; Grigoryan, Z.; Dominguez-Bello, M.G. The Infant Microbiome Development: Mom Matters. *Trends Mol. Med.* **2015**, *21*, 109–117. [CrossRef]
116. Nielsen, S.; Nielsen, D.S.; Lauritzen, L.; Jakobsen, M.; Michaelsen, K.F. Impact of Diet on the Intestinal Microbiota in 10-Month-Old Infants. *J. Pediatr. Gastroenterol. Nutr.* **2007**, *44*, 613–618. [CrossRef]
117. O'Sullivan, A.; He, X.; McNiven, E.M.S.; Haggarty, N.W.; Lönnerdal, B.; Slupsky, C.M. Early Diet Impacts Infant Rhesus Gut Microbiome, Immunity, and Metabolism. *J. Proteome Res.* **2013**, *12*, 2833–2845. [CrossRef]
118. Zivkovic, A.M.; German, J.B.; Lebrilla, C.B.; Mills, D.A. Human Milk Glycobiome and Its Impact on the Infant Gastrointestinal Microbiota. *Proc. Natl. Acad. Sci. USA* **2011**, *108*, 4653–4658. [CrossRef]
119. Bode, L. Human Milk Oligosaccharides: Every Baby Needs a Sugar Mama. *Glycobiology* **2012**, *22*, 1147–1162. [CrossRef]
120. Pacheco, A.R.; Barile, D.; Underwood, M.A.; Mills, D.A. The Impact of the Milk Glycobiome on the Neonate Gut Microbiota. *Annu. Rev. Anim. Biosci.* **2015**, *3*, 419–445. [CrossRef]
121. Fukuda, S.; Toh, H.; Hase, K.; Oshima, K.; Nakanishi, Y.; Yoshimura, K.; Tobe, T.; Clarke, J.M.; Topping, D.L.; Suzuki, T.; et al. Bifidobacteria Can Protect from Enteropathogenic Infection through Production of Acetate. *Nature* **2011**, *469*, 543–547. [CrossRef]
122. Xu, J.; Gordon, J.I. Honor Thy Symbionts. *Proc. Natl. Acad. Sci. USA* **2003**, *100*, 10452–10459. [CrossRef] [PubMed]

 *nutrients* 

Article

Effect of Ferric Carboxymaltose Supplementation in Patients with Heart Failure with Preserved Ejection Fraction: Role of Attenuated Oxidative Stress and Improved Endothelial Function

Annachiara Mollace [1,†], Roberta Macrì [1,†], Rocco Mollace [1,2,†], Annamaria Tavernese [1], Micaela Gliozzi [1], Vincenzo Musolino [3], Cristina Carresi [1], Jessica Maiuolo [3], Martina Nicita [1], Rosamaria Caminiti [1], Sara Paone [1], Francesco Barillà [4], Maurizio Volterrani [2] and Vincenzo Mollace [1,2,*]

[1] Institute of Research for Food Safety & Health (IRC-FSH), Department of Health Science, University Magna Graecia, 88100 Catanzaro, Italy
[2] IRCCS San Raffaele Pisana, 88163 Rome, Italy
[3] Laboratoy of Pharmaceutical Biology, IRC-FSH Department of Health Sciences, University "Magna Græcia" of Catanzaro, Campus Universitario di Germaneto, 88100 Catanzaro, Italy
[4] Department of Systems Medicine, Tor Vergata University, 00133 Rome, Italy
* Correspondence: mollace@libero.it
† These authors contributed equally to this work.

Citation: Mollace, A.; Macrì, R.; Mollace, R.; Tavernese, A.; Gliozzi, M.; Musolino, V.; Carresi, C.; Maiuolo, J.; Nicita, M.; Caminiti, R.; et al. Effect of Ferric Carboxymaltose Supplementation in Patients with Heart Failure with Preserved Ejection Fraction: Role of Attenuated Oxidative Stress and Improved Endothelial Function. *Nutrients* 2022, 14, 5057. https://doi.org/10.3390/nu14235057

Academic Editor: Dietmar Enko

Received: 22 October 2022
Accepted: 25 November 2022
Published: 28 November 2022

Publisher's Note: MDPI stays neutral with regard to jurisdictional claims in published maps and institutional affiliations.

Abstract: Both clinical and experimental evidence shows that iron deficiency (ID) correlates with an increased incidence of heart failure (HF). Moreover, data on iron supplementation demonstrating a beneficial effect in subjects with HF have mostly been collected in patients undergoing HF with reduced ejection fraction (HFrEF). Relatively poor data, however, exist on the potential of iron supplementation in patients with HF with preserved ejection fraction (HFpEF). Here, we report on data emerging from a multicentric, double-blind, randomized, placebo-controlled study investigating the effect of IV supplementation with a placebo or ferric carboxymaltose (FCM) on 64 subjects with HFpEF. ID was detected by the measurement of ferritin levels. These data were correlated with cardiac performance measurements derived from a 6 min walking test (6MWT) and with echocardiographic determinations of diastolic function. Moreover, an EndoPAT analysis was performed to correlate cardiac functionality with endothelial dysfunction. Finally, the determination of serum malondialdehyde (MDA) was performed to study oxidative stress biomarkers. These measurements were carried out before and 8 weeks after starting treatment with a placebo (100 mL of saline given i.v. in 10 min; $n = 32$) or FCM at a dose of 500 mg IV infusion ($n = 32$), which was given at time 0 and repeated after 4 weeks. Our data showed that a condition of ID was more frequently associated with impaired diastolic function, worse 6MWT and endothelial dysfunction, an effect that was accompanied by elevated MDA serum levels. Treatment with FCM, compared to the placebo, improved ferritin levels being associated with an improved 6MWT, enhanced cardiac diastolic function and endothelial reactivity associated with a significant reduction in MDA levels. In conclusion, this study confirmed that ID is a frequent comorbidity in patients with HFpEF and is associated with reduced exercise capacity and oxidative stress-related endothelial dysfunction. Supplementation with FCM determines a significant improvement in diastolic function and the exercise capacity of patients with HFpEF and is associated with an enhanced endothelial function and a reduced production of oxygen radical species.

Keywords: iron deficiency; heart failure; HF with preserved ejection fraction; ferric carboxymaltose; ferritin; transferrin saturation; 6 min walking test; EndoPAT analysis; serum malondialdehyde

1. Introduction

Iron deficiency (ID) represents an emerging issue for public health and is correlated with many disease states, including myocardial dysfunction [1,2]. To date, epidemiological data have confirmed that 50% of anemia cases may be clearly correlated with ID [3,4], representing the most relevant basis for this nutritional deficiency state.

The development of ID with or without anemia occurs as a consequence of different pathophysiological conditions [5,6], as well as malabsorption that correlates with chronic degenerative diseases [7,8]. Moreover, a condition of inflammation and oxidative stress in HF associated with inappropriate iron absorption has been found to contribute to reduced iron stores [1,2,9].

While the correlation between ID and myocardial dysfunction has clearly been shown, the potential for iron supplementation in counteracting myocardial damage and its consequences is still controversial.

Evidence exists that IV iron supplementation produces benefits in HF [10–12]. In particular, clear benefits of IV administration of ferric carboxymaltose (FCM) in patients suffering from HF and ID, with and without anemia, have been found with the study FAIR-HF [13]. In particular, the study showed that IV administration of a single bolus of FCM in patients with ID and HF with an ejection fraction < 40–45% produces an improvement in exercise capacity, independent of the presence of anemia [13].

These results have also been confirmed by results emerging from the CONFIRM-HF study [14], thus showing that iron supplementation improves the quality of life (QoL) and reduces the rate of hospitalization of patients with HF with reduced ejection fraction (HFrEF), with similar adverse effect and mortality compared to placebo. Limited data, however, exist on iron supplementation and HF with preserved ejection fraction (HFpEF), which represents an emerging issue in patients with metabolic disorders [15,16].

Several studies have shown that iron-deficient HFpEF patients suffered a similar impact as the health-related QoL of HFrEF patients [17,18]. Furthermore, in both HFrEF and HFpEF, concomitant ID indicates the same prognosis in terms of increased mortality from all causes, regardless of the presence or absence of anemia [19,20]. Against this epidemiological finding, there is a lack of evidence of better results after intravenous iron therapy. In a single-blind, randomized clinical trial, the anemia in HFpEF patients was treated with epoetin alfa or oral iron and was reported to have no effect on the left-diastolic ventricular volume and left ventricular mass; furthermore, no improvement in submaximal exercise capacity or QoL was observed [21]. In addition, in confirmation of these results, Kasner et al. showed no association between functional iron deficiency and exercise capacity in patients with HFpEF [22]. Finally, a condition of endothelial dysfunction seems to potentially contribute to the pathophysiological mechanisms underlying HFpEF, though its role in producing altered heart function needs to be better clarified [23,24].

In this context, this study aimed to verify the possible impact of IV administration of FCM on cardiac performance and on reliable indicators of diastolic function in patients with HFpEF. This was correlated with oxidative stress and endothelial function, which represent reliable biomarkers of vascular impairment occurring at the early stages of myocardial dysfunction in subjects in which HFpEF and ID occur simultaneously.

2. Materials and Methods

2.1. Study Design

A multicentric, double-blind, randomized, placebo-controlled study was conducted on a population of 64 subjects suffering from HFpEF classified according to the ESC Guidelines 2021 [25] and, in particular, echocardiographic parameters associated with normal ejection fraction, including left atrium (LA) size (as expressed by LA volume index, LAVI, >32 mL/m^2), mitral E velocity > 90 cm/s, septal velocity < 9 cm/s, E/e′ ratio > 9, which have been shown to better correlate with an increased mortality rate (Figure 1).

Study design

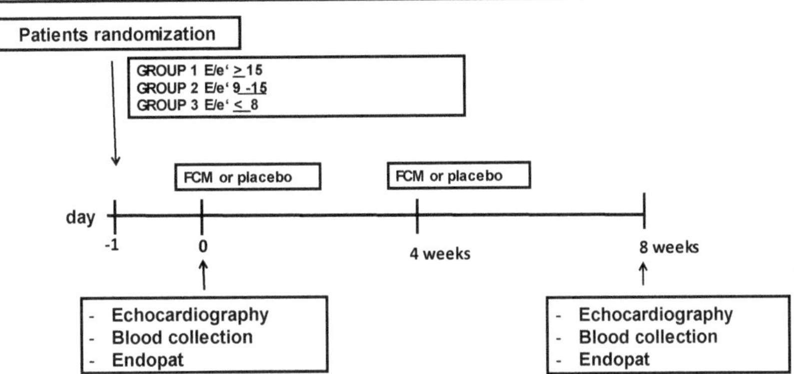

Figure 1. Study design. The multicentric, double-blind, randomized, placebo-controlled study was conducted on 64 subjects suffering from HFpEF classified according to the ESC Guidelines 2021.

In this cohort of patients, the prevalence, incidence, and impact of ID were compared with the degree of diastolic function impairment, reactive vasodilatation (as detected via EndoPAT measurements), and oxidative stress serum biomarkers, respectively. In particular, patients were divided into three groups according to their diastolic function as assessed by the evaluation of the E/e' ratio (by means of echocardiography at time 0; see Section 2.2 for the methodology): Group 1: E/e' ratio $\geq$ 15 [n = 24]; Group 2: E/e' ratio between 9 and 14 [n = 24]; Group 3: E/e' ratio $\leq$ 8 [n = 16].

The groups of patients studied were fairly homogeneous in age (69.7 $\pm$ 9.4 years) and sex (m/f ratio in the 3 groups 34/30), with the presence of metabolic syndrome in 70% of the subjects enrolled and moderately overweight. A proportion of 44% had moderate arterial hypertension on treatment with ACE-i/ARB and/or beta-blockers. Patients with valvular pathology or atrial fibrillation were excluded from the study. None of the patients were smokers at the time of the study (Table 1).

Table 1. Distribution of 64 patients entering the study among the different study groups.

Study Groups	Without ID n = 26		With ID n = 38	
	Placebo	FCM	Placebo	FCM
Group 1 E/e' $\geq$ 15	3	3	9	9
Group 2 E/e' 9–16	5	5	7	7
Group 3 E/e' $\leq$ 8	5	5	3	3

In addition, the effect of IV administration of the placebo (100 mL of saline given i.v. in 10 min) or ferric carboxymaltose (10 mL of Ferinject corresponding to 500 mg iron diluted into 100 mL of saline given i.v. in 10 min) infused at time 0 and after four weeks was evaluated. The outcome of the treatment was verified 8 weeks after starting the administration of FCM. All the patients completed the study. Moreover, the use of an iron-containing supplement was not allowed either before or during the study.

The patients were enrolled at the Metabolic Diseases Outpatient Clinic of the IRC-FSH Center of the "Magna Graecia" University of Catanzaro, in collaboration with the UTIC B of the Umberto I Polyclinic in Rome and the Unit of Cardiology at IRCCS San Raffaele, Rome. All subjects provided written informed consent at the time of enrollment, and the protocol was approved by the local ethics committee and met all the principles of the Declaration of Helsinki. The protocol is registered on the ISRCTN registry (study ID. ISRCTN12833814).

A condition of ID was defined on the basis of serum ferritin levels < 100 nanograms/mL or with levels between 100 and 299 µg/L in the presence of transferrin saturation (TSAT) < 20% [26]. The degree of anemia was defined in the presence of Hb levels < 13 g/dL in men and <12 g/dL in women. In addition, a 6 min walking test (6MWT) was performed as described by Guyatt et al. [26]. In particular, in a long unobstructed corridor, using standardized patient instructions, patients covered as much distance as possible during the allotted 6 min. The total distance covered was measured and recorded. Staff overseeing the 6MWT were unaware of the hemodynamic results.

These determinations were performed at time 0 and at 8 weeks after starting the study. Care was taken that patients maintained the same dietary regimen during the observation period. In analogy with the sample, on the day of the visit at time 0 and after 8 weeks, for the definition of the degree of diastolic function, the patients had an echocardiographic examination, took a vascular reactivity test using the EndoPAT method, and had an additional sample taken for the determination of the levels of serum malonyldialdehyde (MDA).

2.2. Echocardiographic Evaluation of Diastolic Function

The diastolic function parameters used for detecting HFpEF were represented by the mitral E velocity, E/e' ratio, and left atrium volume. In particular, E velocity is detected transmitrally. The transmitral PW spectrum includes two velocity peaks generated by differences in the atrioventricular pressure during the diastolic phase of the cardiac cycle. The first peak (wave E) corresponds to rapid ventricular filling, the one observed when, following isovolumetric relaxation, the atrial pressure far exceeds the ventricular pressure, generating an acceleration in the flow in this sense. The extent of the E peak depends on ventricular relaxation, left atrial pressure, and ventricular protodiastolic elastic return; these parameters determine the faster or slower speed of rapid filling.

The speed with which the atrio-ventricular pressure gradient tends to cancel itself following the first phase of diastole influences the speed and deceleration time (DT) of the E wave. The faster the ventricular pressure increase in protostole, the slower the deceleration time of the E wave.

Tissue Doppler (TD) is a method that has been used to assess the speed of movement of the myocardial wall. The study of diastolic function using this technique is based on the assessment of the displacement speed of the mitral annulus, with image acquisitions through the apical projection in four chambers.

The standard pattern of pulsed TD at the level of the mitral annulus includes three components: a positive (S) systolic wave, that is, directed toward the apex, and two diastolic waves, one early protodiastolic (e') and the other late (a') during atrial systole. Taken together, these rates and time intervals provide quantitative information on systolic longitudinal ventricular function and diastolic function.

Compared to the transmitral flow Doppler, the pulsed annular TD is less dependent on variations in ventricular preload. For this reason, the calculation of the E/e' ratio is a reliable parameter for the non-invasive estimation of filling pressures.

An E/e' ratio of less than 8 is often associated with a normal filling pressure, while a ratio above 15 is observed in the case of increased filling pressures. Values between 8 and 15 need further investigation to be interpreted correctly.

Left Atrial (LA) volume was measured from standard apical 2- and 4-chamber views at end-systole. LA borders were traced using planimetry. The borders consisted of the walls of the left atrium and a line drawn across the mitral annulus. Attention was given to bridge the ostia of pulmonary veins (when visualized) to not include the veins in the measurement. If seen, the LA appendage was excluded from measurement. The biplane method of disks was used to calculate LA volume, and measurements were rounded to the nearest integer. Left Atrial Volume Index (LAVI) was calculated by dividing LA volume by body surface area and expressed as mL/m^2.

2.3. Determinations of Endothelial Function

Endothelial function was assessed using the EndoPAT 2000 technique, which measures PAT using the reactive hyperemia index (RHI, arbitrary units). Briefly, the examination was performed after 20 min of rest in a chair tilted at an angle of approximately 45° at room temperature, with a blood pressure cuff placed on the non-dominant upper arm (study arm), while the other arm served as a control. The hands were placed on the chair stands with the palms facing down so that the fingers hung freely.

The EndoPAT probes were then placed on the tip of each index finger of both hands. The probes were prevented from touching any other fingers or objects and were then electronically inflated. The PAT signal was continuously recorded on a personal computer during the test. The baseline pulse width was measured from each fingertip for five minutes. After a five-minute baseline recording on each arm, the arterial flow was then stopped in the experimental arm by rapidly inflating the cuff to the occlusion pressure of 200 or 60 mmHg plus systolic blood pressure (whichever was greater). After five minutes of occlusion, the cuff pressure was rapidly reduced, and post-occlusion recording continued for an additional five minutes in the experimental and control arms. The pulse width response to hyperemia was automatically calculated from hyperemia in the finger of the experimental arm as a ratio of post-deflation mean pulse ingestion to obtain the RH–PAT ratio or PAT ratio. EndoPAT 2000 measured not only endothelial function with the RHI but also assessed arterial stiffness by measuring the peripheral augmentation index (PAIx) from the radial pulse wave analysis. PAIx is automatically calculated as the ratio of the difference between the early and late systolic peaks of the waveform to the early and late systolic peak of the waveform related to the systolic peak, expressed as a percentage.

2.4. Measurements of MDA in Patients with HFpEF

Oxidative stress was assessed by quantifying the reactivity of thiobarbituric acid (TBA) as MDA in a spectrophotometer. To 0.5 mL of serum, 0.5 mL of 30% trichloroacetic acid (TCA) (Merck, Rahway, NJ, USA) was added and centrifuged at 3000 rpm for 5 min, and the supernatant was collected. Subsequently, 0.5 mL of supernatant was added to 0.5 mL of 1% TBA (Merck) in a boiling water bath for 30 min, after which the tubes were kept in an ice water bath for 10 min. The resulting absorbance of chromogen was determined at a wavelength of 532 nm at room temperature with respect to the blank reference. The concentration of MDA was read from the standard calibration curve plotted using 1, 1, 3, 3 "tetra-ethoxy propane (TEP)". The amount of lipid peroxidation was expressed as MDA (μg/dL), using a molar extinction coefficient for MDA of $1.56 \times 105\ \text{M}^{-1}\ \text{cm}^{-1}$ [27].

2.5. Statistics

Data are presented as mean $\pm$ standard deviation (SD). StatView 5.0 (SAS Institute, inc., Cary, NC, USA) and the Statistical Package for the Social Sciences (SPSS version 21) were used for the statistical analyses. Analysis of variance (ANOVA), followed by post hoc comparisons, Student's unpaired *t*-test, Fisher's exact test, Pearson's simple regression and logistic regression were used as appropriate. A two-tailed *p*-value < 0.05 indicated statistical significance.

3. Results

The study, conducted on 64 patients with HfpEF, documented, at baseline, a condition of ID found in 38/64 patients corresponding to 53.1% of the total subjects entering the study (Table 2). The population of patients with ID showed, under basal conditions, a moderate prevalence of male subjects (21 M vs. 17 W). On the other hand, no significant differences in BMI, Fasting Serum Glucose, serum triglycerides, LDL cholesterol, HDL cholesterol, creatinine, and Left Ventricle Ejection Fraction were found among patients with and without ID (Table 2), and treatment with Placebo or FCM did not produce any change in these parameters. However, treatment with FCM, compared to placebo, improved diastolic function indexes in patients with ID. In fact, the mean values of E/e' ratio and

LAVI were improved (from 14.1 ± 2.1 to 10.2 ± 2.5 and from 38.1 ± 12.8 to 31.2 ± 10.7, respectively; Table 2).

Table 2. Basal levels of principal biomarkers of metabolic balance and the demographics of patients with HFpEF (either with or without ID) enrolled in the study. Table also shows data on the effect of treatment with placebo and FCM.

Parameter	Without ID $n = 26$	With ID $n = 38$
Sex (M/W)	13/13	21/17
Age (years)	61.9 ± 8.9	63.2 ± 9.6
BMI (Kg/m^2)		
Basal	27.5 ± 5.6	26.5 ± 5.2
Placebo	$27.2 + 4.6$	$26.8 + 5.0$
FCM	$26.8 + 5.1$	$26.6 + 4.5$
FSG (mg/dL)		
Basal	114 ± 11.8	115.8 ± 11.2
Placebo	110 ± 8.1	112 ± 9.4
FCM	110 ± 8.9	112 ± 10.1
Triglycerides (mg/dL)		
Basal	195 ± 21.5	196 ± 21.8
Placebo	190 ± 18.4	192 ± 15.6
FCM	191 ± 16.9	190 ± 17.8
Creatinine (mg/dL)		
Basal	1.0 ± 0.9	1.1 ± 0.8
Placebo	1.1 ± 0.7	1.0 ± 0.7
FCM	1.0 ± 0.8	1.0 ± 0.9
LDL-Cholesterol (mg/dL)		
Basal	94.2 ± 23.8	96.1 ± 24.3
Placebo	95.1 ± 21.2	95.4 ± 24.2
FCM	94.2 ± 20.6	95.2 ± 21.9
HDL-Cholesterol (mg/dL)		
Basal	38.0 ± 6.8	39.1 ± 6.1
Placebo	40.2 ± 6.5	41.5 ± 6.4
FCM	39.7 ± 7.1	42.4 ± 6.3
LVEF (%)		
Basal	54.6 ± 6.0	53.5 ± 5.8
Placebo	54.9 ± 5.4	54.3 ± 5.5
FCM	55.6 ± 5.2	56.4 ± 5.7
E/e′ ratio		
Basal	9.2 ± 2.4	14.4 ± 2.7
Placebo	9.1 ± 2.6	14.1 ± 2.1
FCM	8.8 ± 2.2	11.8 ± 2.5 *
LAVI (mL/m^2)		
Basal	37.8 ± 12.7	39.2 ± 14.1
Placebo	36.6 ± 10.3	38.1 ± 12.8
FCM	37.4 ± 11.7	34.4 ± 13.7

BMI: body mass index; FSG: fasting serum glucose; LVEF: left ventricular ejection fraction; LAVI: Left Atrial Volume Index. * $p < 0.05$ Placebo vs. FCM.

Moreover, the distribution of patients into three groups made according to the degree of diastolic dysfunction (as detected on the basis of E/e′ ratio) was, overall, consistent with the data of the most recent literature on this area of interest [28,29]. In fact, the patients with severe diastolic dysfunction (Group 1; E/e′ < 8) were 24/64, equal to 31%, while the subjects enrolled in Groups 2 and 3, with a medium or low degree of dysfunction, were 24/64 and 16/64, equal to 31 and 25%, respectively (Tables 1 and 3). On the other hand, patients in Group 1 showed, at baseline, lower ferritin levels as compared with Groups 2 and 3, thereby confirming that a condition of ID is associated, in patients with HFpEF, with an impairment in cardiac diastolic functionality (Table 3). This effect was associated with reduced exercise capacity as expressed by reduced distance covered during 6MWT, an effect accompanied by the highest serum concentration of MDA, and by altered parameters of endothelial functionality obtained via EndoPAT analysis (Tables 3 and 4). Thus, patients with HFpEF and marked diastolic impairment showed elevated ID-associated oxidative stress and endothelial dysfunction.

Table 3. The effect of placebo or FCM on ferritin, TSAT, 6MWT, and MDA levels in patients with HFpEF according to their diastolic function as expressed by E/e' ratio. Data are expressed as mean ± S.E.M.

Parameter	Group 1 E/e' ≥ 15 $n = 24$	Group 2 E/e' 9–14 $n = 24$	Group 3 E/e' ≤ 8 $n = 16$
Ferritin (ng/mL)			
Basal	60.3 ± 13.3	110.2 ± 14.6	98.2 ± 14.1
Placebo	62.3 ± 12.4	112.2 ± 13.6	120.4 ± 14.6
FCM	129.1 ± 13.9 *	132.2 ± 14.3	145.1 ± 15.3
6 mWT (mt/6 min)			
Basal	291.1 ± 22.6	335.4 ± 21.3	395.4 ± 28.6
Placebo	294.7 + 20.7	340.8 + 20.5	401.5 + 24.4
FCM	354.4 + 23.7 *	365.9 + 21.1	421.6 + 24.3
MDA (µg/dL)			
Basal	1.1 ± 0.1	1.3 ± 0.2	1.6 ± 0.1
Placebo	1.0 ± 0.2	1.4 ± 0.2	1.5 ± 0.2
FCM	0.7 ± 0.1 *	0.8 ± 0.1 *	0.9 ± 0.2 *

6 mWT—6 min walking test; MDA–malondialdehyde; * $p < 0.05$ Placebo vs. FCM.

Table 4. Parameters of endothelial functionality measured via EndoPAT methodology (see methods) at time 0 (basal) and after 8 weeks from starting the treatment with FCM or the placebo in patients with HFpEF classified according to the E/e' ratio (Groups 1, 2, and 3). Data are expressed as mean ± S.E.M.

EndoPAT Index	Group 1 E/e' ≥ 15 $n = 24$	Group 2 E/e' 9–14 $n = 24$	Group 3 E/e' ≤ 8 $n = 16$
RHI			
Basal	1.50 ± 0.3	1.90 ± 0.4	1.95 ± 0.2
After Placebo	1.54 ± 0.4	1.93 ± 0.3	1.98 ± 0.3
After FCM	2.15 ± 0.11 *	2.12 + 0.4	2.21 ± 0.06
fRHI			
Basal	0.20 ± 0.05	0.24 ± 0.03	0.28 ± 0.02
After Placebo	0.21 ± 0.03	0.23 ± 0.02	0.25 ± 0.03
After FCM	0.28 + 0.04 *	0.24 ± 0.04	0.29 ± 0.03
AI			
Basal	7.6 ± 3.2	8.3 ± 2.6	9.6 ± 2.5
After Placebo	8.1 ± 2.6	8.4 ± 2.3	9.5 ± 2.2
After FCM	9.1 ± 4.2	8.4 ± 3.8	9.6 ± 3.5

RHI—reactive hyperemia index; fRHI—Framingham reactive hyperemia index; AI—augmentation index; * $p < 0.05$ FCM vs. placebo.

According to the degree of diastolic impairment, administration of FCB i.v. at time 0 and after 4 weeks from entering the study was able to improve ID levels, oxidative stress, and impaired endothelial function, compared to the placebo. Indeed, FCM restored the ferritin levels in patients with HFpEF associated with high degree diastolic dysfunction, an effect associated with improved cardiac performance as showed by the better response at the 6MWT (Table 3). In addition, this effect was associated with a reduction in MDA levels

(Table 3) and with an enhanced endothelial-dependent response, as studied by means of the EndoPAT procedure (Table 4).

4. Discussion

Our data show, for the first time, that supplementation with FCM administered via i.v. infusion in such a way as to bring ferritin within physiological limits in a cohort of patients affected by HFpEF leads to an improvement in cardiac performance evaluated by means of a 6MWT. This improvement was mainly found to occur in patients in which diastolic function was particularly compromised (Group 1 of patients enrolled in the study), as assessed by the Echo cardio color Doppler study and in particular through reliable indicators such as the E/e' ratio and LAVI.

Moreover, the study showed that the improvement in cardiac performance, seen after supplementation with IV FCM, was associated with a marked improvement in endothelial function and in the degree of oxidative stress that characterized the basal conditions in patients at the time of their enrollment in the study.

In particular, at 8 weeks after the administration of FCM (500 mg/i.v. as a single bolus at time 0, repeated after 4 weeks), there was a marked improvement in vascular reactivity through the EndoPAT study, an effect accompanied by a significant reduction in MDA levels, a reliable index of the level of lipid peroxidation and, finally, of oxidative stress.

Diastolic dysfunction is known to represent the pathophysiological mechanism characterizing the more severe forms of HFpEF [27,30]. In particular, the reduced ability to release myocardial fibers in diastole, even in the presence of a preserved contractile capacity, determines a reduction in cardiac performance as a whole, which is associated with the basal symptoms of HF (moderate exertional dyspnea, signs of pulmonary congestion, perimalleolar edema, etc.), clearly evident at the time of the execution of the 6MWT. This condition has pathophysiological bases very different from the contractile compromise that is found in conditions of HF with low or moderately reduced EF. In fact, the condition of HF consequent to ischemic or toxic events that reduce the mass of contractile myocardium, useful for maintaining a good hemodynamic balance, is to be ascribed to a loss (often due to apoptosis) of the mass of efficient contractile myocardium or to its dysfunction, due to an impairment in the functionality of myofibrils (primary or secondary).

Conversely, diastolic dysfunction seems to have its pathophysiological origin in the progressive loss of the "lusitropic" properties of the myocardium, associated with progressive fibrosis rather than with the absolute or relative loss of the contractile myocardium [30].

The pathophysiological bases of this condition are, to date, not fully clarified. In fact, the "phenotype" of the early stages of the pathology associated with HFpEF still remains to be defined, during which the development of myocardial compromise is resolved toward an impairment of the lusitropic properties of the myocardium rather than toward the cellular apoptosis of cardiomyocytes.

However, recent studies make it possible to identify the *"primum movens"* of this pathophysiological evolution in mitochondrial oxidative stress associated with impaired iron metabolism and endothelial dysfunction, leading, in the late stages, to myocardial impairment. In particular, Zhang et al. showed that in the heart tissue of patients with HF, myocardial iron concentration was lower compared to that detected in non-failing hearts [28]. This effect was associated with the suppression of the respiratory chain and Krebs cycle enzymatic activities, which had a significant correlation with depleted iron stores, an effect that was accompanied by an increased concentration of MDA, thus showing that a condition of oxidative stress is associated with ID. This was also combined with decreased antioxidant enzymes in the myocardial tissue of ID hearts, as shown by reduced superoxide dismutase and glutathione peroxidase. Finally, they also found that iron uptake is impaired in cardiomyocytes, as expressed by decreased translocation to the sarcolemma, while the transmembrane fraction of ferroportin positively correlated with cardiac impairment. Thus, ID associated with HF seems to correlate with mitochondrial dysfunction and oxidative damage [29,31]. This fits very well with our data. In particular, the consistent reduction in ferritin levels. The reduction found in patients with a consistent reduction in diastolic function and reduced exercise capacity is associated with elevated levels of MDA, thus confirming that oxidative stress is associated with HFpEF. This effect was counteracted by restoring ferritin levels via FCM administration, thereby leading to improved exercise capacity, as found with improved performance by means of the 6mWT found after FCM treatment.

Our data also show that endothelial functionality contributes to the pathophysiology of ID-associated disorders and HFpEF. Indeed, FCM administration was able to restore altered ferritin levels and exercise capacity in patients with HFpEF and produced an improvement in endothelial-dependent vasodilatation as investigated by EndoPAT analysis in our patients at time 0 and after FCM treatment. In fact, a lower capacity to respond to flow restriction with a vasodilatory response was found in Group 1, in which a significant correlation between ID and diastolic dysfunction was found. In this group of patients, restoring ferritin levels via FCM was accompanied by improvement in all the EndoPAT parameters of endothelial-mediated vasodilatation, showing that endothelial dysfunction has a crucial role in ID-related HFpEF. This correlates with evidence suggesting that endothelial impairment accompanies HFpEF [29,31–33]. In particular, evidence was provided that a reduction in the number of endothelial progenitor cells (EPCs) may be found in patients with HFpEF, thereby showing that the repair of endothelial functionality may represent a potential benefit in such a class of patients. On the other hand, recent evidence in women with severe ID status suggests that there is a possible impairment in endothelial function and oxidative stress mediated by altered HDL lipoproteins, which is counteracted by i.v. iron administration; however, the mechanism of action needs to be better assessed.

In conclusion, this study confirmed that ID is a frequent comorbidity in patients with HFpEF and is associated with reduced exercise capacity and oxidative stress-related endothelial dysfunction. Supplementation with FCM determines a significant improvement in the cardiac performance of patients with HFpEF enrolled in this study, an effect that was associated with enhanced endothelial function and reduced production of oxygen radical species. Further studies, however, based on a larger number of patients, are required in order to better clarify the potential for iron supplementation in patients with HFpEF.

Author Contributions: Conceptualization, A.M., R.M. (Roberta Macrì) and V.M. (Vincenzo Mollace); Data curation, S.P.; Formal analysis, M.N.; Funding acquisition, V.M. (Vincenzo Mollace); Investigation, A.T., V.M. (Vincenzo Musolino), C.C. and R.C.; Methodology, F.B. and M.V.; Supervision, M.G.; Validation, F.B. and M.V. (Maurizio Volterrani); Writing—original draft, R.M. (Rocco Mollace) and J.M. All authors have read and agreed to the published version of the manuscript.

Funding: The work was supported by public resources from the Italian Ministry of Research. (PON-MIUR 03PE000_78_1 and PONMIUR 03PE000_78_2; POR Calabria FESR FSE 2014–2020 Asse 12-Azioni 10.5.6 e 10.5.12.)

Institutional Review Board Statement: The study was conducted in accordance with the Declaration of Helsinki and approved by the local ethics committee. The protocol is registered on the ISRCTN registry (Study ID. ISRCTN12833814).

Informed Consent Statement: Informed consent was obtained from all the subjects involved in the study.

Data Availability Statement: The data presented in this study are available upon request from the corresponding author.

Conflicts of Interest: The authors declare no conflict of interest.

Communication

Hepcidin Status in Cord Blood: Observational Data from a Tertiary Institution in Belgium

Michael Ceulemans [1,2,3,*], Joline Van de Vel [4], Dorine W. Swinkels [5,6,7], Coby M. M. Laarakkers [5,6], Jaak Billen [8], Kristel Van Calsteren [9,10] and Karel Allegaert [1,3,10,11]

[1] Clinical Pharmacology and Pharmacotherapy, Department of Pharmaceutical and Pharmacological Sciences, KU Leuven, 3000 Leuven, Belgium
[2] Teratology Information Service, Netherlands Pharmacovigilance Centre Lareb, 5237 MH Hertogenbosch, The Netherlands
[3] L-C&Y, Child and Youth Institute, KU Leuven, 3000 Leuven, Belgium
[4] Faculty of Pharmaceutical Sciences, KU Leuven, 3000 Leuven, Belgium
[5] Translational Metabolic Laboratory (TML 830), Department of Laboratory Medicine, Radboud University Medical Center, 6500 HB Nijmegen, The Netherlands
[6] Hepcidinanalysis.com, 6525 GA Nijmegen, The Netherlands
[7] Sanquin Blood Bank, Plesmanlaan 125, 1066 CX Amsterdam, The Netherlands
[8] Department of Laboratory Medicine, University Hospitals Leuven Gasthuisberg, 3000 Leuven, Belgium
[9] Department of Obstetrics & Gynecology, University Hospitals Leuven Gasthuisberg, 3000 Leuven, Belgium
[10] Woman and Child, Department of Development and Regeneration, KU Leuven, 3000 Leuven, Belgium
[11] Department of Hospital Pharmacy, Erasmus MC, 3000 CA Rotterdam, The Netherlands
* Correspondence: michael.ceulemans@kuleuven.be; Tel.: +32-16-37-72-27

Abstract: The hormone hepcidin plays an important role in intestinal iron absorption and cellular release. Cord blood hepcidin values reflect fetal hepcidin status, at least at the time of delivery, but are not available for the Belgian population. Therefore, we aimed (1) to provide the first data on cord blood hepcidin levels in a Belgian cohort and (2) to determine variables associated with cord blood hepcidin concentrations. A cross-sectional, observational study was performed at the University Hospital Leuven, Belgium. Cord blood samples were analyzed using a combination of weak cation exchange chromatography and time-of-flight mass spectrometry. Descriptive statistics, Spearman correlation tests, and Mann–Whitney U tests were performed. In total, 61 nonhemolyzed cord blood samples were analyzed. The median hepcidin level was 17.6 µg/L (IQR: 18.1; min-max: 3.9–54.7). A moderate correlation was observed between cord blood hepcidin and cord blood ferritin (r = 0.493) and hemoglobin (r = −0.342). Cord blood hepcidin was also associated with mode of delivery ($p = 0.01$), with higher hepcidin levels for vaginal deliveries. Nonetheless, larger studies are needed to provide more evidence on the actual clinical value and benefit of cord blood hepcidin measurements.

Keywords: pregnancy; cord blood; iron status; iron regulation; hepcidins

Citation: Ceulemans, M.; Van de Vel, J.; Swinkels, D.W.; Laarakkers, C.M.M.; Billen, J.; Van Calsteren, K.; Allegaert, K. Hepcidin Status in Cord Blood: Observational Data from a Tertiary Institution in Belgium. *Nutrients* 2023, 15, 546. https://doi.org/10.3390/nu15030546

Academic Editor: Dietmar Enko

Received: 12 December 2022
Revised: 11 January 2023
Accepted: 16 January 2023
Published: 20 January 2023

1. Introduction

Iron requirements are increased during pregnancy as a result of the increased number of red blood cells and circulating blood volume [1]. Iron plays a crucial role in oxygen transport to the fetus. The fetus also stores iron during pregnancy that will be used during its first months of life [2]. Hence, pregnant women are at risk of iron deficiency anemia [2]. This should be avoided at all times, as iron deficiency anemia during pregnancy and early childhood is associated with maternal, perinatal, and neonatal morbidity, including altered cognitive and neurobehavioral outcomes in offspring [2–4].

To meet the high iron requirements during pregnancy [5], both iron absorption from the maternal diet as well as cellular iron release within the body should be sufficient. The hepatic 25-amino acid peptide hormone hepcidin hereby plays an important role. Hepcidin inhibits dietary iron absorption in the duodenum and iron release from macrophages and hepatocytes [6]. So, increased hepcidin concentrations result in lower levels of circulating iron. Hepcidin concentrations below or above a specific threshold may be indicative of an increased risk of iron deficiency in women [7]. Although the application of hepcidin is currently mainly limited to research settings, as a regulator of iron homeostasis, hepcidin may be, for example, a useful biomarker of the (oral) bioavailability of iron supplementation in pregnancy, guiding clinicians to appropriately prescribe and monitor iron therapy in pregnant women [5].

Previous studies have shown that maternal hepcidin values decrease in the second and third gestational trimester, probably to meet the increased iron needs due to maternal, fetal, and placental weight gain and growth [8–10]. However, in pregnancies complicated with inflammatory conditions such as preeclampsia, maternal hepcidin levels were higher compared to women with uncomplicated pregnancies, potentially limiting the amount of iron available for transplacental transfer [11].

Moreover, cord blood hepcidin values reflect fetal hepcidin status, at least at the time of delivery. Previous studies have found a correlation between cord blood hepcidin levels and cord blood iron status [12–20]. Likewise, cord blood hepcidin has been associated with maternal variables (e.g., body mass index or BMI) [13,21], pregnancy outcomes (e.g., gestational age at birth, mode of delivery) [14–16], and neonatal outcomes (e.g., birth weight) [17], regardless of some conflicting results across studies. Hence, insight into the relationship between cord blood hepcidin and maternal variables/pregnancy–neonatal outcomes could be instructive to contribute to identifying neonates at high(er) risk of low iron status.

To date, observational data on cord blood hepcidin levels are not available for (pregnant) women living in Belgium. Furthermore, the application of different hepcidin assays with varying calibrators used across previous studies impedes the comparison and utility of hepcidin values, limiting its future research and clinical potential [22,23]. Therefore, by using a validated and internationally accepted hepcidin analysis method [23,24], this study aimed (1) to provide observational data on cord blood hepcidin levels in a Belgian cohort and (2) to determine variables associated with cord blood hepcidin concentrations.

2. Materials and Methods

A cross-sectional, observational study was performed at the University Hospital Leuven, Belgium between March and September 2017. Cord blood samples were obtained at delivery from women who participated during pregnancy in the PREVIM study exploring the extent and type of "PREgnancy related use of VItamins and Medication" [25,26]. To be eligible for study participation, women must be at least 18 years old, understand Dutch, French, or English, and have visited the obstetrics department of the university hospital as part of the routine antenatal care during the ongoing pregnancy.

Cord blood samples were collected using serum tubes and subsequently centrifuged and aliquoted at the Department of Laboratory Medicine of the University Hospital. Aliquots of 2 mL were kept between 2–8° degrees for maximum 7 days, followed by storing at −80°. Previously, we have shown that in serum samples, which were kept at 4° for 0–7 days and at −80° for less than 2 years, concentrations of hepcidin-25 remained stable [24].

In December 2017, the aliquots were analyzed in one batch at the "Hepcidinanalysis" lab of the Radboud University Medical Center in the Netherlands. To assess hepcidin-25 concentrations, a validated hepcidin analysis method was used based on a combination of weak cation exchange chromatography and time-of-flight mass spectrometry (WCX-TOF MS) [24]. Peptide spectra were generated on a Microflex LT matrix-enhanced laser desorption/ionization TOF MS platform (Bruker Daltonics). A stable hepcidin-25 + 40 isotope was used as the internal standard for quantification. The lower limit of quantification (LOQ) was 0.5 nM (nmol/L); samples with values below the LOQ were given the value zero. Hepcidin-25 concentrations were provided as nM and converted to µg/L (by multiplying with a factor 2.7894).

In our study, the measurements of cord blood hepcidin (2017) were performed before the assay was further standardized (2019) by using secondary matrix-based reference material that was value-assigned by a primary reference material [27,28]. However, as the standardization only slightly altered serum hepcidin values obtained by the WCX-TOF MS method (i.e., standardized results were a factor 1.054 higher compared to historic results obtained without standardization), the results obtained in our study could be considered as results obtained by a standardized hepcidin assay.

To determine variables associated with cord blood hepcidin, the following variables for which a potential relationship has been shown in earlier studies [12–21] were assessed: (1) 'cord blood iron status' (i.e., hemoglobin, ferritin, serum iron, transferrin and transferrin saturation); (2) 'pre-gestational BMI'; (3) 'pregnancy outcomes' (i.e., onset of labor, mode of delivery); and 4) 'neonatal outcomes' (i.e., prematurity (born < 37 weeks), low birth weight (<2500 g), small for gestational age (SGA; <10th birth percentile) [29], and large for gestational age (LGA; > 90th birth percentile) [29]. Customized birth weight centiles were calculated, accounting for gestational age, fetal sex, parity, and single/multiple pregnancies. Data were retrieved from hospital medical records shortly after childbirth.

All data were analyzed using descriptive statistics (i.e., median, interquartile range (IQR), absolute numbers, and percentages). The relationship between cord blood hepcidin and the continuous variables was tested using the Spearman correlation test. A correlation coefficient $r < 0.3$ was considered a weak relationship, between 0.3 and 0.7 a moderate relationship, and > 0.7 a strong relationship. The relationship between cord blood hepcidin and the noncontinuous variables was assessed using Mann–Whitney U tests. The variable onset of labor was dichotomized into 'spontaneous' and 'not spontaneous' (i.e., induction of labor and none/elective caesarean section). The results were considered significant if $p < 0.05$. Data were analyzed using SPSS Statistics version 28 (IBM Corp, Armonk, NY, USA).

Ethical approval was obtained from the EC Research UZ/KU Leuven (S59516; 25 November 2016). All women provided, while being pregnant, their written informed consent for cord blood sampling at delivery and data collection from hospital medical records.

3. Results

In total, cord blood samples were collected from 72 individual women. Median gestational age at birth was 39 weeks (IQR: 1.7). Median pregestational BMI was 22.78 (IQR: 5.91). Two women had a multiple pregnancy. Table 1 provides an overview of the study participants' baseline maternal and pregnancy-related characteristics as well as pregnancy outcomes, according to cord blood hepcidin levels of the nonhemolyzed samples. With regard to pregnancy complications, the following complications were reported in this cohort: preeclampsia ($N = 3$), gestational diabetes ($N = 3$), placenta previa ($N = 1$), hyperemesis gravidarum ($N = 1$), polyhydramnios ($N = 1$) and cardiac complications ($N = 1$).

In total, 74 neonates were born. Overall, median birth weight was 3290 g (IQR: 478 g) ($N = 73$). The other neonatal outcomes are summarized in Table 2.

In total, 73 cord blood samples were available for hepcidin analysis. One sample did not contain a sufficient amount of blood that was needed for the analysis. In 12 samples, hemolysis had occurred. Only one (hemolyzed) sample had a hepcidin level < LOQ. Overall, the median hepcidin level in the nonhemolyzed ($N = 61$) and hemolyzed ($N = 12$) cord blood samples was 17.6 µg/L (IQR: 18.1; min-max: 3.9–54.7) and 9.6 µg/L (IQR: 11.1; min-max: 0.0–28.2), respectively ($p = 0.012$). Given the difference in hepcidin values depending on the occurrence or absence of hemolysis, hepcidin analyses were only performed with the nonhemolyzed samples.

Table 3 shows the results of the cord blood iron status parameters. No single woman had a C-reactive protein (CRP) value of >5mg/L in their cord blood sample ($N = 73$). Among the three women suffering from preeclampsia as pregnancy complication and delivered at 39w6d, 37w6d, and 38w, the corresponding hepcidin levels in their (nonhemolyzed) cord blood samples were 5.3, 12.0, and 37.1 µg/L, respectively, showing large interindividual variability.

Table 1. Overview of study participants' baseline characteristics and pregnancy outcomes, according to cord blood hepcidin levels of the nonhemolyzed samples.

Variable	% (*n*)	Cord Blood Hepcidin Level
Maternal and pregnancy-related variables		
Gravidity (*N* = 72)		
Primigravida	37.5 (27)	17.9 (19.8)
Multigravida	62.5 (45)	17.6 (19.0)
Parity (*N* = 72)		
Nullipara	61.1 (44)	18.0 (18.0)
Primi- or multipara	38.9 (28)	15.5 (15.8)
Pregestational body mass index (*N* = 70)		
<25 kg/m^2	67.1 (47)	15.6 (19.1)
$\geq$ 25 kg/m^2	32.9 (23)	19.4 (14.3)
Onset of pregnancy (*N* = 72)		
Spontaneous	87.5 (63)	18.0 (15.2)
Assisted	12.5 (9)	9.1 (14.4)
Singleton or multiple pregnancy (*N* = 72)		
Single	97.2 (70)	17.0 (17.9)
Multiple	2.8 (2)	24.1 (/)
Pregnancy outcomes		
Onset of labor (*N* = 71)		
Spontaneous	50.7 (36)	18.4 (18.8)
Induction of labor	33.8 (24)	18.0 (15.1)
None (elective caesarian section)	15.5 (11)	7.3 (9.8)
Mode of delivery (*N* = 72)		
Vaginal	79.2 (57)	18.4 (16.5)
Elective caesarean section	15.3 (11)	7.3 (9.8)
Secondary caesarean section	5.6 (4)	17.9 (/)
Gestational age at delivery (*N* = 72)		
<34 weeks	0.0 (0)	/
34–37 weeks	8.3 (6)	9.2 (19.7)
$\geq$37 weeks	91.7 (66)	17.9 (16.2)

The results are shown as % (*n*). Hepcidin levels are shown as median (interquartile range, IQR) and are expressed in µg/L.

Table 2. Overview of the neonatal outcomes (*N* = 74) *.

Variable	% (*n*)
Prematurity [1]	10.8 (8)
Low birth weight [2]	1.4 (1)
Small for gestational age (SGA) [3]	8.2 (6)
Large for gestational age (LGA) [3]	12.3 (9)

The results are shown as % (*n*). * For the variables low birth weight, small for gestational age and large for gestational age, there was one missing value. [1] Prematurity was defined as being born <37 weeks gestational age. [2] Low birth weight was defined as a birth weight < 2500 g. [3] The results for small for gestational age (SGA; < 10th birth percentile) and large for gestational age (LGA; > 90th birth percentile) were calculated based on the customized criteria of the Study Centre on Perinatal Epidemiology (SPE) in Belgium, thereby accounting for gestational age, fetal sex, parity, and single/multiple pregnancies [29].

Table 3. Overview of the iron status parameters in cord blood [1].

Variable	N	Median (IQR)
Hepcidin (µg/L)	61	17.6 (18.1)
Hemoglobin (g/dL) [2]	47	15.4 (1.8)
Ferritin (µg/L)	61	218 (179)
Iron (µg/dL)	61	152 (45)
Transferrin (g/L)	61	1.77 (0.35)
Transferrin saturation (%)	61	61 (21)

The results are shown as median (and interquartile range, IQR). The units for each variable are shown between brackets. [1] Only nonhemolyzed cord blood samples were considered for the analysis. [2] Some missing data exist for the variable hemoglobin.

Table 4 provides a detailed overview of the results of the correlation tests assessing a potential relationship between cord blood hepcidin and the continuous variables related to cord blood iron status and maternal pregestational BMI. Overall, a moderate positive relationship was found between cord blood hepcidin and ferritin ($r = 0.493$). Furthermore, a moderate negative relationship was observed between cord blood hepcidin and hemoglobin ($r = -0.342$). No other strong/moderate correlations were found with cord blood hepcidin.

Table 4. Relationship between cord blood hepcidin and (1) iron status and (2) pregestational BMI.

Variable	r [1]	*p*-Value
Cord blood iron status parameters		
Hemoglobin	−0.342	0.02
Ferritin	0.493	<0.001
Iron	0.118	0.36
Transferrin	−0.098	0.45
Transferrin saturation	0.118	0.37
Maternal variable		
Pregestational BMI[2]	0.241	0.07

[1] The results were calculated using the Spearman correlation test and are reported using the correlation coefficient r and *p*-value. [2] BMI = body-mass index.

Table 5 provides a detailed overview of the results of the association tests assessing a potential relationship between cord blood hepcidin and pregnancy/neonatal outcomes. The only association with cord blood hepcidin was found for the variable mode of delivery ($p = 0.01$), with higher hepcidin levels for vaginal deliveries compared to elective caesarean sections. Although no other associations were identified, the absolute difference in cord blood hepcidin levels with respect to prematurity is notable (8.7 µg/L), with premature neonates (born < 37 weeks) showing lower hepcidin levels.

Table 5. Relationship between cord blood hepcidin and pregnancy/neonatal outcomes.

Variable	Median (IQR)	*p*-Value [1]
Pregnancy outcomes		
Onset of labor		
Spontaneous	18.4 (18.8)	0.23
Not spontaneous	15.8 (16.9)	
Mode of delivery [2]		
Vaginal	18.4 (16.5)	0.01
Elective caesarean section	7.3 (9.8)	
Neonatal outcomes		
Prematurity [3]		
Yes	9.2 (19.7)	0.21
No	17.9 (16.2)	

Table 5. *Cont.*

Variable	Median (IQR)	*p*-Value [1]
Small for gestational age (SGA) [4]		
Yes	25.9 (21.5)	0.19
No	17.6 (18.4)	
Large for gestational age (LGA) [4]		
Yes	19.8 (18.7)	0.89
No	17.6 (17.9)	

[1] The results were calculated using Mann–Whitney U tests and are reported using median (and interquartile range, IQR) and *p*-values. [2] The four women with a secondary caesarean section were not considered in order to not obscure the findings. [3] Prematurity was defined as being born < 37 weeks gestational age. [4] The results for small for gestational age (SGA; < 10th birth percentile) and large for gestational age (LGA; > 90th birth percentile) were calculated based on the customized criteria of the Study Centre on Perinatal Epidemiology in Belgium, thereby accounting for gestational age, fetal sex, parity, and single/multiple pregnancies [29].

Moreover, only one neonate had a birth weight of < 2500 g. As a result, no bivariate analysis was performed for the variable 'low birth weight'. In the (hemolyzed) cord blood sample of this single neonate, an absolute hepcidin value of 7.25 μg/L was observed, which is much lower than the median hepcidin value in the overall cohort (i.e., 17.6 μg/L).

4. Discussion

This cross-sectional, observational study performed at a tertiary hospital in Belgium and using a validated hepcidin analysis method based on weak cation exchange chromatography and time-of-flight mass spectrometry, providing results similar to a standardized assay [24,27,28], aimed (1) to provide observational data on cord blood hepcidin levels in a Belgian cohort and (2) to determine variables associated with cord blood hepcidin levels.

Overall, 61 nonhemolyzed cord blood samples were collected and batch-analyzed. In total, a median hepcidin value of 17.6 μg/L was found, along with substantial variability in hepcidin levels among women (min-max: 3.9–54.7 μg/L). The observed median value is somewhat in line with the results of a previous study using the same method of analysis measuring hepcidin levels in cord blood of mothers with or without placental malaria infection and/or maternal anemia [30]. However, additional comparisons with hepcidin levels observed in other studies is difficult, given that many other studies used different (immunoassay or mass spectrometry) techniques and/or calibrators [13,15,17,19,31,32].

Moreover, in our cohort, we found a potential relationship between cord blood hepcidin and cord blood ferritin and hemoglobin, and mode of delivery. First, for cord blood ferritin, a moderate positive correlation was observed, in line with previous studies pointing at a relationship between cord blood hepcidin and cord blood iron status [12–20]. During pregnancy, maternal hepcidin concentrations also correlate with indicators of maternal iron status [5], which is also the case for healthy adult individuals [33]. Second, a negative correlation was found between hepcidin and hemoglobin levels in cord blood, similar to previous research [15]. Third, an association with mode of delivery was observed, with higher cord blood hepcidin levels for vaginal deliveries compared to elective caesarean sections, as shown earlier [14]. Finally, the observed trend towards higher cord blood hepcidin levels among term versus preterm neonates has also been previously shown [8,14].

Our study has some strengths. To our knowledge, this was the first study measuring hepcidin values in cord blood samples of pregnant women living in Belgium. Second, we used a validated and internationally accepted hepcidin analysis method [24], providing values similar to those obtained from a standardized assay. This enables the comparison with the findings of validated hepcidin assays used elsewhere, provided that they are standardized by using the same second reference material for calibration [27,28]. This will eventually reduce confusion in this field and ultimately allow for a global comparison of hepcidin values measured in cord blood. To achieve this, we collaborated with lab experts of the Radboud University Medical Center, who have extensive knowledge of and experience in hepcidin analysis, standardization, and interpretation of the results. International collaboration should always be pursued in this area, as it not only enables the sharing of expertise but also facilitates the application of fully validated analysis methods. Such approach may accelerate the acquisition of knowledge and insight into the actual clinical value of cord blood hepcidin. Third, we explored the potential relationship of cord blood hepcidin with maternal variables and pregnancy/neonatal outcomes that have previously been shown to be associated, despite conflicting results. So, we aimed to contribute to the replication of previous findings instead of testing numerous variables and running a (higher) risk of accidental findings (i.e., type I errors).

Some limitations should also be addressed. First, we acknowledge that the total number of 61 samples remains rather limited, which is partially explained/exacerbated by hemolysis of some samples. Due to the limited sample size, it cannot be excluded that for some variables, no significant results could be found (i.e., type II error). The limited sample size further avoided performing regression analyses and adjusting for potential confounders. Hence, we could not draw firm conclusions on potential relationships between cord blood hepcidin and other variables, and therefore consider our findings mainly explorative. A sample size or power calculation was not performed. Second, no maternal serum hepcidin levels in pregnancy and at the time of delivery were measured due to logistic and financial restraints. Future studies should further investigate the relative contribution of maternal/fetal hepcidin to placental iron transport [5,34]. Third, in our cohort, only few women were affected by preeclampsia or other pregnancy complications, hindering us to explore their relationship with cord blood hepcidin. Fourth, no (reliable) data were available on maternal iron status and the use of iron-containing medicines and/or supplements at the time of delivery, nor on ethnicity, although most—if not all—participants were probably Caucasian. Finally, we assumed that hemolysis of the samples occurred completely at random. Still, the finding that hepcidin levels were lower in hemolyzed samples could not be explained based on our relatively small cohort and requires further investigation.

Considering all the strengths and limitations of our work, our observational data and explorative analyses could hopefully contribute, to some extent, to establishing reference values of cord blood hepcidin [14] and to a better understanding of the potential of cord blood hepcidin for clinical and research purposes. Nevertheless, as various international round-robin tests performed by the Radboud group showed substantial differences in absolute levels measured by different assays—which decreased after standardization with their reference material [22,27,28,32,35]—specific attention should be paid to the harmonization and standardization of different hepcidin analysis methods.

5. Conclusions

In this observational study, hepcidin concentrations were measured in 61 nonhemolyzed cord blood samples using a validated weak cation exchange chromatography and time-of-flight mass spectrometry analysis method. Overall, a median hepcidin value of 17.6 µg/L was found, with a substantial variability in hepcidin levels among women. Moreover, a moderate positive and negative correlation was observed for cord blood hepcidin and cord blood ferritin and hemoglobin. A third potential association was identified for mode of delivery, with higher cord blood hepcidin levels for vaginal deliveries compared to elective caesarean sections. Although this exploratory study provided the first Belgian data on cord blood hepcidin levels, given its relatively limited sample size, larger studies collecting sufficient data on potential confounders are needed to provide more evidence on the actual value and benefit of cord blood hepcidin measurements for clinical and research purposes.

Author Contributions: Conceptualization, M.C., K.V.C. and K.A.; methodology, M.C., D.W.S., C.M.M.L., K.V.C. and K.A.; formal analysis, M.C., D.W.S., C.M.M.L. and J.V.d.V.; investigation, M.C.; resources, M.C., K.V.C. and K.A.; data curation, M.C.; writing—original draft preparation, M.C. and J.V.d.V.; writing—review and editing, J.V.d.V., D.W.S., J.B., K.V.C. and K.A.; supervision, K.V.C. and K.A.; project administration, M.C.; funding acquisition, K.A. All authors have read and agreed to the published version of the manuscript.

Funding: This research received no external funding. The research activities of Michael Ceulemans were supported by the Faculty of Pharmaceutical Sciences of the KU Leuven.

Institutional Review Board Statement: The study was conducted in accordance with the Declaration of Helsinki and approved by the Ethics Committee (EC) Research of UZ/KU Leuven (S59516, 25 November 2016).

Informed Consent Statement: Written informed consent was obtained from all subjects involved in the study.

Data Availability Statement: Not applicable.

Acknowledgments: The authors would like to thank Karolien Claes and Ria Goossens for their help and support as part of the PREVIM study, and to the staff of the delivery room and the obstetrics department of the University Hospitals Leuven for their help with recruitment and sampling. The authors are also grateful to Master students Rozelien Peyls, Caroline Houben, and Emma Bauwens for their contribution to data collection as part of the PREVIM study; to Christophe Matthys and Nele Steenackers for their reflections on hepcidin-related research; and to Jan Verbakel for his scientific advice.

Conflicts of Interest: Dorine W. Swinkels and Coby M.M. Laarakkers are employees of the Radboud University Medical Center, that via the www.hepcidinanalysis.com initiative offers high-quality hepcidin measurements to the scientific, medical and pharmaceutical communities at a fee for service basis. The other authors declare no conflict of interest.

References

1. Fisher, A.L.; Nemeth, E. Iron homeostasis during pregnancy. *Am. J. Clin. Nutr.* **2017**, *106*, 1567s–1574s. [CrossRef]
2. Cao, C.; O'Brien, K.O. Pregnancy and iron homeostasis: An update. *Nutr. Rev.* **2013**, *71*, 35–51. [CrossRef] [PubMed]
3. East, P.; Doom, J.R.; Blanco, E.; Burrows, R.; Lozoff, B.; Gahagan, S. Iron deficiency in infancy and neurocognitive and educational outcomes in young adulthood. *Dev. Psychol.* **2021**, *57*, 962–975. [CrossRef] [PubMed]
4. Benson, C.S.; Shah, A.; Frise, M.C.; Frise, C.J. Iron deficiency anaemia in pregnancy: A contemporary review. *Obstet. Med.* **2021**, *14*, 67–76. [CrossRef]
5. Koenig, M.D.; Tussing-Humphreys, L.; Day, J.; Cadwell, B.; Nemeth, E. Hepcidin and iron homeostasis during pregnancy. *Nutrients* **2014**, *6*, 3062–3083. [CrossRef] [PubMed]
6. Ganz, T.; Nemeth, E. Hepcidin and iron homeostasis. *Biochim. Biophys. Acta* **2012**, *1823*, 1434–1443. [CrossRef] [PubMed]
7. Galetti, V.; Stoffel, N.U.; Sieber, C.; Zeder, C.; Moretti, D.; Zimmermann, M.B. Threshold ferritin and hepcidin concentrations indicating early iron deficiency in young women based on upregulation of iron absorption. *eClinicalMedicine* **2021**, *39*, 101052. [CrossRef]
8. van Santen, S.; Kroot, J.J.; Zijderveld, G.; Wiegerinck, E.T.; Spaanderman, M.E.; Swinkels, D.W. The iron regulatory hormone hepcidin is decreased in pregnancy: A prospective longitudinal study. *Clin. Chem. Lab. Med.* **2013**, *51*, 1395–1401. [CrossRef]
9. Guo, Y.; Zhang, N.; Zhang, D.; Ren, Q.; Ganz, T.; Liu, S.; Nemeth, E. Iron homeostasis in pregnancy and spontaneous abortion. *Am. J. Hematol.* **2019**, *94*, 184–188. [CrossRef]
10. Simavli, S.; Derbent, A.U.; Uysal, S.; Turhan, N. Hepcidin, iron status, and inflammation variables among healthy pregnant women in the Turkish population. *J. Matern. Fetal Neonatal Med.* **2014**, *27*, 75–79. [CrossRef]
11. Toldi, G.; Stenczer, B.; Molvarec, A.; Takáts, Z.; Bekő, G.; Rigó, J.; Vásárhelyi, B. Hepcidin concentrations and iron homeostasis in preeclampsia. *Clin. Chem. Lab. Med.* **2010**, *48*, 1423–1426. [CrossRef] [PubMed]
12. Rehu, M.; Punnonen, K.; Ostland, V.; Heinonen, S.; Westerman, M.; Pulkki, K.; Sankilampi, U. Maternal serum hepcidin is low at term and independent of cord blood iron status. *Eur. J. Haematol.* **2010**, *85*, 345–352. [CrossRef] [PubMed]
13. Jones, A.D.; Shi, Z.; Lambrecht, N.J.; Jiang, Y.; Wang, J.; Burmeister, M.; Li, M.; Lozoff, B. Maternal Overweight and Obesity during Pregnancy Are Associated with Neonatal, but Not Maternal, Hepcidin Concentrations. *J. Nutr.* **2021**, *151*, 2296–2304. [CrossRef] [PubMed]
14. Lorenz, L.; Herbst, J.; Engel, C.; Peter, A.; Abele, H.; Poets, C.F.; Westerman, M.; Franz, A.R. Gestational age-specific reference ranges of hepcidin in cord blood. *Neonatology* **2014**, *106*, 133–139. [CrossRef]
15. Ru, Y.; Pressman, E.K.; Guillet, R.; Katzman, P.J.; Bacak, S.J.; O'Brien, K.O. Predictors of anemia and iron status at birth in neonates born to women carrying multiple fetuses. *Pediatr. Res.* **2018**, *84*, 199–204. [CrossRef]
16. Zhang, J.Y.; Wang, J.; Lu, Q.; Tan, M.; Wei, R.; Lash, G.E. Iron stores at birth in a full-term normal birth weight birth cohort with a low level of inflammation. *Biosci. Rep.* **2020**, *40*, BSR20202853. [CrossRef]
17. Ichinomiya, K.; Maruyama, K.; Inoue, T.; Koizumi, A.; Inoue, F.; Fukuda, K.; Yamazaki, Y.; Arakawa, H. Perinatal Factors Affecting Serum Hepcidin Levels in Low-Birth-Weight Infants. *Neonatology* **2017**, *112*, 180–186. [CrossRef]

18. Lee, S.; Guillet, R.; Cooper, E.M.; Westerman, M.; Orlando, M.; Kent, T.; Pressman, E.; O'Brien, K.O. Prevalence of anemia and associations between neonatal iron status, hepcidin, and maternal iron status among neonates born to pregnant adolescents. *Pediatr. Res.* **2016**, *79*, 42–48. [CrossRef]

19. Brickley, E.B.; Spottiswoode, N.; Kabyemela, E.; Morrison, R.; Kurtis, J.D.; Wood, A.M.; Drakesmith, H.; Fried, M.; Duffy, P.E. Cord Blood Hepcidin: Cross-Sectional Correlates and Associations with Anemia, Malaria, and Mortality in a Tanzanian Birth Cohort Study. *Am. J. Trop. Med. Hyg.* **2016**, *95*, 817–826. [CrossRef]

20. Delaney, K.M.; Guillet, R.; Fleming, R.E.; Ru, Y.; Pressman, E.K.; Vermeylen, F.; Nemeth, E.; O'Brien, K.O. Umbilical Cord Serum Ferritin Concentration is Inversely Associated with Umbilical Cord Hemoglobin in Neonates Born to Adolescents Carrying Singletons and Women Carrying Multiples. *J. Nutr.* **2019**, *149*, 406–415. [CrossRef]

21. Cao, C.; Pressman, E.K.; Cooper, E.M.; Guillet, R.; Westerman, M.; O'Brien, K.O. Prepregnancy Body Mass Index and Gestational Weight Gain Have No Negative Impact on Maternal or Neonatal Iron Status. *Reprod. Sci.* **2016**, *23*, 613–622. [CrossRef] [PubMed]

22. van der Vorm, L.N.; Hendriks, J.C.; Laarakkers, C.M.; Klaver, S.; Armitage, A.E.; Bamberg, A.; Geurts-Moespot, A.J.; Girelli, D.; Herkert, M.; Itkonen, O.; et al. Toward Worldwide Hepcidin Assay Harmonization: Identification of a Commutable Secondary Reference Material. *Clin. Chem.* **2016**, *62*, 993–1001. [CrossRef] [PubMed]

23. Girelli, D.; Nemeth, E.; Swinkels, D.W. Hepcidin in the diagnosis of iron disorders. *Blood* **2016**, *127*, 2809–2813. [CrossRef]

24. Laarakkers, C.M.; Wiegerinck, E.T.; Klaver, S.; Kolodziejczyk, M.; Gille, H.; Hohlbaum, A.M.; Tjalsma, H.; Swinkels, D.W. Improved mass spectrometry assay for plasma hepcidin: Detection and characterization of a novel hepcidin isoform. *PLoS ONE* **2013**, *8*, e75518. [CrossRef] [PubMed]

25. Ceulemans, M.; Van Calsteren, K.; Allegaert, K.; Foulon, V. Health products' and substance use among pregnant women visiting a tertiary hospital in Belgium: A cross-sectional study. *Pharmacoepidemiol. Drug Saf.* **2019**, *28*, 1231–1238. [CrossRef]

26. Ceulemans, M.; Van Calsteren, K.; Allegaert, K.; Foulon, V. Beliefs about medicines and information needs among pregnant women visiting a tertiary hospital in Belgium. *Eur. J. Clin. Pharmacol.* **2019**, *75*, 995–1003. [CrossRef]

27. Diepeveen, L.E.; Laarakkers, C.M.M.; Martos, G.; Pawlak, M.E.; Uguz, F.F.; Verberne, K.; van Swelm, R.P.L.; Klaver, S.; de Haan, A.F.J.; Pitts, K.R.; et al. Provisional standardization of hepcidin assays: Creating a traceability chain with a primary reference material, candidate reference method and a commutable secondary reference material. *Clin. Chem. Lab. Med.* **2019**, *57*, 864–872. [CrossRef]

28. Aune, E.T.; Diepeveen, L.E.; Laarakkers, C.M.; Klaver, S.; Armitage, A.E.; Bansal, S.; Chen, M.; Fillet, M.; Han, H.; Herkert, M.; et al. Optimizing hepcidin measurement with a proficiency test framework and standardization improvement. *Clin. Chem. Lab. Med.* **2020**, *59*, 315–323. [CrossRef]

29. Cammu, H.; Martens, G.; Martens, E.; Van Mol, C.; Defoort, P. Perinatale activiteiten in Vlaanderen 2009. 2010. Available online: https://www.zorg-en-gezondheid.be/sites/default/files/2022-04/SPE%20RAPPORT%202009.pdf (accessed on 16 November 2022).

30. Van Santen, S.; de Mast, Q.; Luty, A.J.; Wiegerinck, E.T.; Van der Ven, A.J.; Swinkels, D.W. Iron homeostasis in mother and child during placental malaria infection. *Am. J. Trop. Med. Hyg.* **2011**, *84*, 148–151. [CrossRef]

31. Kulik-Rechberger, B.; Kościesza, A.; Szponar, E.; Domosud, J. Hepcidin and iron status in pregnant women and full-term newborns in first days of life. *Ginekol. Pol.* **2016**, *87*, 288–292. [CrossRef]

32. Kroot, J.J.; van Herwaarden, A.E.; Tjalsma, H.; Jansen, R.T.; Hendriks, J.C.; Swinkels, D.W. Second round robin for plasma hepcidin methods: First steps toward harmonization. *Am. J. Hematol.* **2012**, *87*, 977–983. [CrossRef] [PubMed]

33. Galesloot, T.E.; Vermeulen, S.H.; Geurts-Moespot, A.J.; Klaver, S.M.; Kroot, J.J.; van Tienoven, D.; Wetzels, J.F.; Kiemeney, L.A.; Sweep, F.C.; den Heijer, M.; et al. Serum hepcidin: Reference ranges and biochemical correlates in the general population. *Blood* **2011**, *117*, e218–e225. [CrossRef] [PubMed]

34. Sangkhae, V.; Fisher, A.L.; Wong, S.; Koenig, M.D.; Tussing-Humphreys, L.; Chu, A.; Lelić, M.; Ganz, T.; Nemeth, E. Effects of maternal iron status on placental and fetal iron homeostasis. *J. Clin. Investig.* **2020**, *130*, 625–640. [CrossRef]

35. Kroot, J.J.; Kemna, E.H.; Bansal, S.S.; Busbridge, M.; Campostrini, N.; Girelli, D.; Hider, R.C.; Koliaraki, V.; Mamalaki, A.; Olbina, G.; et al. Results of the first international round robin for the quantification of urinary and plasma hepcidin assays: Need for standardization. *Haematologica* **2009**, *94*, 1748–1752. [CrossRef] [PubMed]

Review

The Link between Iron Turnover and Pharmacotherapy in Transplant Patients

Marcin Delijewski [1,*], Aleksandra Bartoń [2], Beata Maksym [1] and Natalia Pawlas [1]

[1] Department of Pharmacology, Faculty of Medical Sciences in Zabrze, Medical University of Silesia, Jordana 38, 41-808 Zabrze, Poland

[2] Independent Researcher, 40-055 Katowice, Poland

* Correspondence: mdelijewski@sum.edu.pl; Tel.: +48-(32)-2722683

Abstract: Iron is a transition metal that plays a crucial role in several physiological processes. It can also exhibit toxic effects on cells, due to its role in the formation of free radicals. Iron deficiency and anemia, as well as iron overload, are the result of impaired iron metabolism, in which a number of proteins, such as hepcidin, hemojuvelin and transferrin, take part. Iron deficiency is common in individuals with renal and cardiac transplants, while iron overload is more common in patients with hepatic transplantation. The current knowledge about iron metabolism in lung graft recipients and donors is limited. The problem is even more complex when we consider the fact that iron metabolism may be also driven by certain drugs used by graft recipients and donors. In this work, we overview the available literature reports on iron turnover in the human body, with particular emphasis on transplant patients, and we also attempt to assess the drugs' impact on iron metabolism, which may be useful in perioperative treatment in transplantology.

Keywords: iron metabolism; graft; pharmacotherapy; anemia; iron deficiency; iron overload

1. Introduction

Iron Fe^{2+}/Fe^{3+} is the most abundant transition metal in the human body, and it plays an important role in the growth of the organism, energy metabolism and several physiological processes [1]. The iron content of the body normally reaches about 3 to 4 g, and exists mainly in the form of hemoglobin, iron-containing proteins (catalase, cytochromes, myoglobin) and transferrin-bound iron and is stored in the form of hemosiderin and ferritin. Iron is absorbed in the intestine, circulates in blood and cells and is stored mostly in the bone marrow, liver or spleen. The element is lost in shed skin cells, sweat and in the intestine. Menstrual iron loss leads to lower iron stores in women in comparison to men, and therefore women are more prone to become iron deficient [2–4]. Iron can also exhibit toxic effects on cells by catalyzing the reaction of free radical formation [1]. Moreover, iron plays a role in immune modulation, the mechanism of ischemia-reperfusion injury, and in the regulation of organ and graft functions [5].

Iron overload or anemia and iron deficiency are the result of impaired iron metabolism, in which a number of proteins, such as hepcidin and hemojuvelin, take part [1].

Iron deficiency occurs more commonly in individuals with renal, lung and cardiac transplants, while iron overload is more frequent in patients with hepatic transplantation [1,6]. Iron overload is related to poor prognosis in end-stage kidney and liver failure [5]. On the other hand, anemia and iron deficiency are connected with poor prognosis in patients with end-stage heart failure [5]. Moreover, iron deficiency prior to liver transplantation may be even a prognostic factor for the length of intensive care unit stay after operation [7].

Renal disease in the end-stage has also been stated to be associated with hyperferritinemia and the hepatic accumulation of iron [8], and before the introduction of erythropoiesis-stimulating agents in the recipients of renal graft, the iron overload frequency was about 28%. Either phebotomy or iron chelatation may be considered in these cases [5].

Citation: Delijewski, M.; Bartoń, A.; Maksym, B.; Pawlas, N. The Link between Iron Turnover and Pharmacotherapy in Transplant Patients. *Nutrients* **2023**, *15*, 1453. https://doi.org/10.3390/nu15061453

Academic Editor: Dietmar Enko

Received: 16 February 2023
Revised: 11 March 2023
Accepted: 14 March 2023
Published: 17 March 2023

End-stage heart, liver and kidney diseases, and also chronic inflammatory conditions after organ transplantations, are often complicated by the anemia of chronic disease, when high hepcidin levels and proinflammatory cytokines lower iron delivery to the bone marrow cells and inhibit red blood cell production [5]. These effects may be associated with the interaction of hepcidin with ferroportin, taking part in the move of iron from the enterocytes to the blood. Hepcidin causes the internalization and degradation of ferroportin, which can lead, among other things, to the effect of "trapped iron in macrophages".

When we consider the liver as a central store for iron in the body and the place where hepcidin is secreted as a result of increased plasma transferrin saturation and proinflammatory signals, the impact of liver diseases on iron turnover will be more significant. It is known that low transferrin as well as high ferritin levels are related to a worse outcome in patients with acute forms of liver failure [9]. In turn, chronic liver disease with reduced hepatocyte mass leads to the attenuated production of hepcidin and the augmented storage of iron in hepatocytes [5]. Nevertheless, in hereditary hemochromatosis, liver transplantation normalizes the secretion of hepcidin, preventing the regression of iron overload [10].

Perioperative challenges in transplantology, which are critically driven by iron, include such procedures as the selection of patients for transplantation, the management of immunosuppression, and the preservation of the function of graft prior to and after transplantation. Moreover, Schaefer et al. [5] suggest an active management of the status of iron in patients in the periprocedural period.

Disturbed iron metabolism in transplant recipients is often multifactorial. Causes should be found in the blood loss after transplantation, inflammatory processes, viral and microbial infections and drugs taken by those patients, especially immunosuppressive therapy (Table 1). Incorrect iron metabolism may cause a higher risk of anemia complications (Figure 1). Anemia may also increase the risk of cardiovascular events in the case of chronic kidney disease and in renal graft recipients, due to having an estimated similar glomerular filtration rate [11]. Anemia may also lead to death in organ recipients, as cardiovascular disease is the main cause of death in renal transplant individuals [12,13]. Moreover, anemia itself is related to an increased death rate in recipients of renal graft [14,15]. It is also important to underline the impact of inflammation prior to and after transplantation on iron absorption and homeostasis, because in inflammatory conditions serum iron may be low, transferrin saturation may be reduced, and the production of hepcidin is induced by the proinflammatory cytokines. Chronic inflammatory conditions are also related to macrophage iron accumulation [5].

Table 1. The link between pharmacotherapy commonly used in transplantology and anemia.

Drug	Model	Link with Anemia	Data Source
In vitro			
Acetylsalicylic acid	Pharmaceutical interaction	Formation of drug–iron complexes (further study needed to assess the biological significance)	Zhang et al. [16]
	Pharmaceutical interaction	Acetylsalicylic acid may chelate endogenous hepatic iron	Schwarz et al. [17]
Cell cultures			
Acetylsalicylic acid	Microglial cells under inflammatory conditions	Downregulation of hepcidin by inhibition of NF-κB and IL6/JAK2/STAT3 pathways	Li et al. [18]
	Microglial cells under inflammatory conditions	Negative effect on cell iron contents under 'normal' conditions, potential to partly reverse iron imbalance under inflammatory conditions	Xu et al. [19]

Table 1. *Cont.*

Drug	Model	Link with Anemia	Data Source
	Bovine pulmonary artery endothelial cells	Aspirin at low antithrombotic concentrations induced the synthesis of ferritin protein in a time- and concentration-dependent fashion	Oberle et al. [20]
Simvastatin	HepG2 cell line (human liver carcinoma cell line)	Simvastatin significantly suppressed the mRNA expression of hepcidin	Chang et al. [21]
Vitamin C	HepG2 cell line (human liver carcinoma cell line)	Vitamin C directly inhibits hepcidin expression	Chiu et al. [22]
Animal model			
Sirolimus	Healthy rats	Microcytosis and polyglobulia	Diekmann et al. [23]
	Mice	Export-dependent iron-loss on cellular level	Bayeva et al. [24]
Pravastatin	Rat model of cholestasis	Pravastatin raised liver iron content by modulation of heme catabolism and an increase in hepatic iron uptake and storage capacity.	Kolouchova et al. [25]
Acyclovir	Rats	Acyclovir binds endogenous and exogenous iron	Müller [26]
Propranolol	Rats	Reduction of cardiac tissue iron uptake	Kramer et al. [27]
Human model			
Sirolimus	Heart transplant patients	Anemia of chronic disease and functional iron deficiency	McDonald et al. [28]
	Renal transplant patients	Anemia, red blood cell microcytosis	Sofroniadou et al. [29]
	Renal transplant patients	Anemia	Maiorano et al. [30]
	Renal transplant patients	Anemia due to defective IL-10 dependent inflammatory regulation	Thaunat et al. [31]
Everolimus	Renal transplant patients	Anemia, microcytosis	Sánchez Fructuoso et al. [32]
	Liver transplant patients	Anemia	Masetti et al. [33]
Cyclosporine	Liver transplant patients	Anemia	Masetti et al. [33]
	Liver transplant patients	Hemolytic anemia (microangiopathic hemolytic anemia—MAHA) related to injury of microvascular endothelial cells and apoptosis	Kanellopoulou et al. [34]
Cyclosporine/ Sirolimus/ Steroids	Renal transplant patients	Cyclosporine withdrawal followed by sirolimus immunotherapy resulted in significantly less anemia than sirolimus-cyclosporine-steroids therapy	Friend et al. [35]
Tacrolimus	Liver and renal transplant patients	Hemolytic anemia (microangiopathic hemolytic anemia—MAHA) related to injury of microvascular endothelial cells and apoptosis	Kanellopoulou et al. [36]
Prednisone	Renal transplant patients	Pro-erythropoietic, steroid therapy may reduce the severity of anemia in the early post-transplant period	Al-Uzri et al. [37]

Table 1. *Cont.*

Drug	Model	Link with Anemia	Data Source
Mycophenolate mofetil	Renal transplant patients	Megaloblastic anemia	Al-Uzri et al. [37]
	Liver transplant patients	Anemia	Al-Uzri et al. [37]
Enalapril	Renal transplant patients	Anemia	Vlahakos et al. [38]
	Renal transplant patients	Anemia	Graafland et al. [39]
Sulfamethoxazole and trimethoprim	Single patient	Drug-induced immune hemolytic anemia	Frieder et al. [40] Arndt et al. [41]
	Single patient	Drug-induced immune hemolytic anemia with antibodies to both substances	Arndt et al. [42]
	Single patient	Hemolytic anemia	Chisholm-Burns et al. [43]
Ketoconazole, fluconazole	-	-	No data
Azathioprine	Post-kidney transplantation	Serum iron and serum transferrin saturation increased significantly. No evidence of hemolysis	Habas et al. [44]
Azathioprine	Post-liver transplantation	Modulation of purine metabolism	Maheshwari et al. [45]
Azathioprine and prednisolone	Single patient	Successful management of idiopathic pulmonary hemosiderosis	Willms et al. [46]
Basiliximab	-	-	No data
Thymoglobulin	-	-	No data
Rituximab			
Amlodipine	Patients with thalassemia major	Amlodipine decreases iron overload and reduces ferritin levels	Fernandes et al. [47]
Acetylsalicylic acid	A representative cohort of community-dwelling subjects	No association between aspirin use and reduced serum iron or iron saturation	Hammerman-Rozenberg et al. [48]
	Males and females infected or not infected by *H. pylori*	No effect of low-dose aspirin use on ferritin levels (in males); lower ferritin levels in *H. pylori* infected subjects using aspirin, compared with both uninfected and infected non-aspirin users (in females)	Kaffes et al. [49]
	Elderly participants	Aspirin use is associated with lower serum ferritin	Fleming et al. [50]
	Postmenopausal women	19% lower mean serum ferritin in aspirin users than in non-users	Liu et al. [51]
NSAIDs (loxoprofen, diclofenac, ampiroxicam, naproxen, etodolac)	Patients undergoing chronic hemodialysis	The use of non-aspirin NSAIDs may increase the risk of iron deficiency	Wang et al. [52]
Fluvastatin	Dyslipidemic end-stage renal disease patients with renal anemia	Fluvastatin treatment decreased high-sensitive C-reactive protein (hs-CRP) and serum prohepcidin (prohormone of hepcidin) levels	Arabul et al. [53].
Simvastatin	End-stage renal disease patients with renal anemia	Simvastatin did not significantly change the serum prohepcidin, hs-CRP, or IL-6 concentrations	Li et al. [54]
Valgancyclovir	-	-	No data

Table 1. *Cont.*

Drug	Model	Link with Anemia	Data Source
Allopurinol	Patient with chronic kidney disease	Aplastic anemia	Kim et al. [55]
Metformin	Patients treated with metformin	Reduction of vitamin B12	Liu et al. [56]

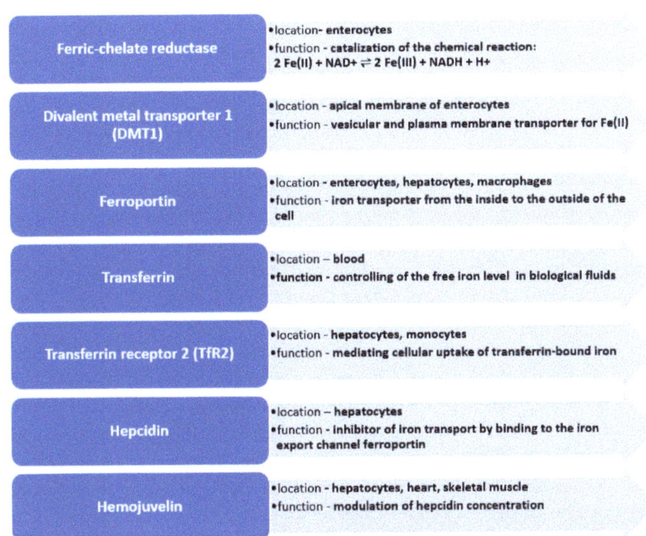

Figure 1. Key enzymes in iron metabolism in human body.

The purpose of this article is to analyze the available data on iron metabolism in transplant patients, with a special emphasis on the influence of pharmacotherapy on iron management in transplantology.

2. Iron Homeostasis

Due to the lack of physiological excretion mechanisms, the iron content in the body depends on the homeostasis of the metal, which is a result of the absorption of iron in the intestines and the release of iron from red blood cells, which is regulated by specific proteins. It is also important to underline that iron absorption can be affected by gut microbiota or the pH of the colon contents, which may also depend on several factors such as age, ethnicity, geography and even lifestyle. Moreover, inflammatory bowel disease may also impair iron absorption [57]. The key proteins involved in iron metabolism are, among others, transferrin, soluble transferrin receptor (sTFR), ferritin, ferroportin, hepcidin and hemojuvelin (Figure 1) [10].

Hepcidin is a peptide hormone and an acute phase protein. It is produced in the liver and plays a role as a key regulator of iron homeostasis. Its secretion is augmented by inflammation and iron loading [11]. It causes the downregulation of the absorption of iron and the release of recycled hemoglobin iron from macrophages. A high hepcidin level is associated with reduced absorption of iron in the intestine. The binding of hepcidin to ferroportin, a cellular iron export pump, leads to its degradation. The ineffective function of hepcidin may cause the manifestation of human hemochromatosis disorders. Normally, increased iron availability increases the expression of hepcidin. Hepcidin interacts with a transmembrane protein ferroportin and is responsible for the process of absorption, storage and the distribution of the element in human body [58]. The secretion of hepcidin is regulated by iron, which in turn regulates the concentration of ferroportin, thus contributing to iron homeostasis. Moreover, during pregnancy, hepcidin regulates the transfer of the iron across the placenta to the fetus [11].

The overexpression of hepcidin leads to severe iron deficiency anemia, causing death within a few hours after birth in transgenic mice [59]. An increased hepcidin level has been found in kidney and heart transplant recipients [60,61]. The augmented level of hepcidin observed by Przybyłowski et al. [62] in patients with heart failure was associated with compromised erythropoiesis contributing to anemia.

High hepcidin concentrations are also present in dialysis patients [63]. This, in turn, may be consistent with findings that they frequently develop iron deficiency [5,64]. Iron storage in the body and inflammation raise the level of hepcidin, while hypoxia and increased erythropoiesis contribute to a reduced expression of hepcidin, leading to an increased absorption of iron. Hepcidin deficiencies can lead to hemochromatosis, which involves the excessive accumulation of iron in tissues [1,58,59].

Additionally, hemojuvelin plays a significant role in the homeostasis of iron. This protein is a hepcidin regulator which signals to promote the expression of hepcidin. Hemojuvelin bound to membrane inhibits iron absorption in the gut through the promotion of hepcidin synthesis [65]. Mutations in several genes encoding hemojuvelin may also cause iron loading syndromes, resembling hepcidin deficiency [66].

Another peptide, transferrin, delivers iron to cells that require it, mainly to red blood cell progenitors. The bone marrow acquires iron for the production of red blood cells, from macrophages that collect iron from recycled red blood cells. The rest of the iron is stored in the cytosolic iron storage protein, ferritin, in hepatocytes. After the secretion of ferritin into the plasma, it constitutes body iron stores [5,11].

Ferroportin is an iron transporter, which is important during the absorption of iron in intestines and its release from cells. Other cells taking part in iron homeostasis, e.g., duodenal enterocytes, hepatocytes, placental cells and macrophages, also contain ferroportin. The internalization and degradation of ferroportin caused by hepcidin leads to iron trapping within the cells [67].

The inactivation of the murine *ferroportin* gene by Donovan et al. [68] caused the accumulation of iron in hepatocytes, enterocytes and macrophages. Moreover, the inactivation of intestinal ferroportin also explained its contribution to the absorption of iron in the intestine. The global inactivation of ferroportin led to early failure in embryonic development, while the inactivation of ferroportin only selectively in the postnatal intestine caused severe iron deficiency, which could be compensated only by the delivery of iron parenterally. The study by Donovan et al. also confirmed that ferroportin is the main iron exporter that is crucial in iron turnover, not only in epithelial cells, but also in iron-recycling macrophages and hepatocytes.

As shown by Nemeth et al. [59], the cellular iron level depends more on the export of iron from cells by ferroportin than the availability of iron undergoing cell influx. The elevated expression of ferroportin in iron deficiency should secure the augmented absorption of iron in the intestine, respecting the fact that iron absorption from the gut can be also influenced by age, gut condition and microbiome, and can also cause effective regaining of the iron that was transferred to the intestinal epithelium from plasma. The study by Donovan et al. [68] also confirmed that ferroportin is the main iron exporter that is crucial in iron turnover, not only in epithelial cells but also in hepatocytes and in iron-recycling macrophages. The trapping of iron in these cells together with impaired iron absorption deepens the anemia.

sTFR is a protein found on the surface of the cell membrane responsible for the binding of transferrin, a process that regulates the delivery of iron to cells. Then, the liberation of iron from the cells (enterocytes or macrophages) takes place with ferroportin. Since sTFR is not an acute phase protein, its study is an alternative to ferritin in cases of the occurrence or suspicion of chronic disease. When assessing iron homeostasis, it is also important to specify parameters such as TIBC, total iron binding capacity (measured or calculated on the basis of transferrin concentration), and TSAT, which means transferrin saturation calculated on the basis of iron concentration and TIBC [1,2,69]. Summing up, iron uptake requires protein transferrin receptor 1 (TfR1), whereas iron liberation needs protein ferroportin 1 (Fpn1) and iron storage requires ferritin [19].

In iron overload, decreased ferroportin expression leads to an increase in iron loss thanks to the elevation of the iron content, which is accumulated by old enterocytes undergoing excretion to the gut lumen. This seems to be an important mechanism of iron regulation as neither the liver nor kidney had mechanisms of iron removal [68]. Iron overload may be a result of mutations in some genes responsible for iron homeostasis (hereditary hemochromatosis), and may be due to refractory anemias, chronic liver diseases, chronic transfusions [70] and eventually some transplantations [1]. Beta thalassemia and sideroblastic anemia classified as "iron-loading" refractory anemias are connected with ineffective erythropoiesis, erythroid hyperplasia and the excessive absorption of iron [70].

3. Redox Activity of Iron

Iron is a transition element which belongs to the d-block and exists in biological systems in oxidation states +2 (ferrous), +1 (ferric) and +4 (ferryl). It binds to several ligands, thanks to its unoccupied d orbitals. Among iron ligands are the atoms of nitrogen, oxygen and sulfur. Iron may change the biological redox potential as well as the electronic spin state, depending on the ligand [71].

The redox status of the graft implicated by iron remains an important factor in the general assessment of the organ prior to the procurement as well as after transplantation. It is known that highly reactive iron may worsen the whole outcome, as it is crucial in the ischemia-reperfusion injury after being released from cells in response to ischemia. Therefore, it has been already well documented that the chelation of iron, e.g., with deferoxamine, is favorable due to the reduction of ischemia-reperfusion injury [5]. Fe^{2+} enters the Fenton reaction and catalyzes the conversion of H_2O_2 into cell-toxic radicals, mostly $OH\cdot$ and OH^-. There is a correlation between the available amount of Fe^{2+} and lipid peroxidation in the liver mitochondrium and microsomes induced by free radicals [72,73]. Moreover, the pulmonary fibrosis is a result of the accumulation of Fe^{2+} on the fiber surface, acting as a Fenton reaction catalyzer [46]. The risk of adverse redox reactions of free iron is increased in individuals suffering from iron overload, and is also connected with an enhanced risk of infection [70].

4. Iron and Immune System

The transplantation of a solid organ is a cause of the activation of the immune system, the intensity of which is a result of iron status. The consequent activation of T-cells is attenuated by iron [5]. It has been stated that individuals with iron deficiency anemia may have decreased IgG levels, interleukin 6 (IL-6) and phagocytic activity, and that there is a significant positive correlation between serum level of iron and IL-6 [74].

The process of T-cells' activation, proliferation and differentiation is mediated through RAS, NF-kappaB and the calcineurin pathway, which are activated by the so called "immunological synapse" formed by CD3, CD71, TfR and the T-cell receptor. The decreased activation of the immune response in iron deficiency is because of attenuated IL-2 levels and CD3 expression. Thus, iron is crucial for the proliferation and differentiation of T-cells. Moreover, a blockade of anti-CD71 antibodies may attenuate T-cell activation due to iron chelation. Similarly, iron deficiency is also connected with reduced CD28 expression, which is also involved in RAS, NF-kappaB and calcineurin pathways [5]. It is also important to underline that the RAS pathway has been found to have the potential to suppress hepcidin expression [5,75].

Iron deficiency diminishes the cellular immune response, whereas iron overload impairs the function of macrophages [5], which remains consistent with the observation of Das et al., according to whom significantly lower levels of CD4+ T-cells as well as reduced

CD4:CD8 ratios have been observed among children with iron deficiency. These effects may be improved after iron supplementation [76].

Inflammation suppresses erythropoiesis due to the sequestration of iron from erythroid precursors. It reduces the flow of iron into the circulation, where it can be used by erythroblasts thanks to the binding of TfR1 with transferrin, which leads to anemia in chronic inflammation [67]. The induction of hepcidin by proinflammatory cytokines during inflammation causes a decrease in blood iron level, which is known as hypoferremia of inflammation [59].

Both iron deficiency as well as iron overload are connected with impaired activation of the immune system. Hyperferritemia is also observed in inflammatory conditions, with accompanied lowered transferrin saturation and low serum iron. High intracellular iron in red blood cells may impact their immune-effector properties, and these cells undergo recycling by macrophages. The accumulation of iron in macrophages occurs in chronic inflammation and correlates with an impaired role as an immune effector [5]. Additionally, IL-6 is necessary in the process of hepcidin synthesis in inflammation [1].

Additionally, ferritin, which is a cellular iron storage protein, may be affected by inflammation processes in the body. It is an acute phase protein, which means that its level increases due to inflammation or in the course of various autoimmune and inflammatory diseases, as well as chronic infections [2].

According to Attia et al. [77], there is a difference in the level of appropriate mature or immature T-lymphocytes between individuals with iron deficiency anemia and control subjects. Children with iron deficiency anemia had significantly higher levels of immature T-cells CD1 a(+) as well as lower levels of mature T-lymphocytes CD4(+) and T-lymphocytes CD8(+) when compared to the control, indicating a defect in T-cell maturation in iron deficiency anemia patients.

5. Iron and Infections

It is also important to underline the involvement of microorganisms in iron homeostasis in the human body, since biochemical pathways utilized by many pathogens may interfere with iron turnover, causing interactions with drugs and becoming an additional factor that may lead to pharmacotherapy failure in transplant patients.

Iron is a key element for the human body and also for microorganisms, being a good redox catalyst for processes such as DNA replication and respiration. On the other hand, its redox potential is also responsible for its toxicity. The pathogens may cause disease after overcoming the so-called nutritional immunity, which is a process of limiting the access to iron by the immune system [70]. Hypoferremia is supposed to increase the resistance of the host organism to microbial infection; on the other hand, it may lead to anemia of inflammation [59].

Iron is an important nutrient for many human bacterial pathogens, which have developed different mechanisms of iron's acquisition such as the production of iron transporters, siderophores (small ferric iron chelators capable of binding iron), heme acquisition systems, and receptors for transferrin and lactoferrin. Siderophores may compete with host transferrin that also binds iron, by binding iron with very high association constants [70]. Siderophores have the ability to take up other iron-chelating substances, so it is important to take into consideration that the treatment of iron overload with such molecules as desferrioxamine may be connected with the increased risk of infection [78]. Siderophores are produced by *Escherichia coli*, *Staphylococcus aureus*, *Bacillus anthracis* and *Legionella pneumophila* [79].

Other mechanisms responsible for iron acquisition by pathogenic microbes are heme uptake systems from erythrocytes. This strategy is used, among others, by *Haemophilus influenzae*, which is incapable of endogenous heme biosynthesis, *Pseudomonas aeruginosa*, *Pseudomonas fluorescens*, *S. aureus*, *Yersinia pestis* and *Yersinia enterocolitica*. Moreover, pathogens also developed iron acquisition mechanisms involving transferrin/lactoferrin receptors. These are *Neisseria meningitides* [70,80].

Predominantly, intracellular bacteria try to acquire iron from macrophages. *Mycobacterium tuberculosis*, after being phagocytosed by alveolar macrophages, obtains host iron. Additionally, pathogenic fungi have developed different mechanisms for the acquisition of iron from host tissues, such as uptake mediated by siderophores, reductive uptake, and heme acquisition [70]. *Saccharomyces cerevisiae, Cryptococcus neoformans* and *Candida albicans* contain ferric reductases. In the process of ferric reduction to ferrous iron, the element shifts from the host molecules that chelate it into the fungal cell. Many pathogenic fungi produce siderophores (*Blastomyces dermatitidis, Histoplasma capsulatum, Aspergillus* spp. and additionally *Rhizopus* spp.) [81]. The heme uptake is also needed for the growing of *C. albicans, C. neoformans,* and *H. capsulatum* [70].

Many viruses, such as human immunodeficiency virus (HIV), hepatitis B virus (HBV), hepatitis C virus (HCV), parvovirus B19 and Epstein–Barr virus, can interfere with myelopoiesis and cause aplastic anemia [82,83]. Parvovirus B19 causes anemia by replicating in erythroid precursors, which proliferate and lead to red blood cell aplasia due to the induction of apoptosis in their precursors [84]. Similarly, Epstein–Barr virus is thought to cause anemia by infecting bone marrow progenitor cells [71].

6. Pharmacotherapy and Iron Turnover in Transplant Patients

The course of iron metabolism can be influenced by the pharmacotherapy. Taking into consideration that pre- and post-transplant patients have to pharmacologically maintain appropriate immune responses, antimicrobial resistance and good graft parameters should avoid the worsening of many other coexisting illnesses; they use various drugs from many different pharmacological groups that often interfere with mechanisms responsible for iron turnover, and may lead to the occurrence of anemia or iron overload.

The use of pharmacotherapy may influence, among other things, the action of hepcidin, transferrin and ferritin and may lead to the chelation of iron or a change in iron's disposition in the human body. The complex evidence for the link between pharmacotherapy used by transplant patients, iron metabolism and anemia has not been studied so far, probably due to a lack of systematic data. Our analysis is presented in Table 1.

It may be possible that certain drugs will directly interact with iron. So-called pharmaceutical interactions between drugs and iron may lead to the formation of complexes that limit the availability of free iron. This is in the case for acetylsalicylic acid [16,17]. Antiviral drugs also have influence on iron turnover, by binding to endogenous or exogenous iron [26].

Drugs may also impact the level of key proteins involved in iron turnover. Reports from studies on cell cultures reveal that acetylsalicylic acid also has the potential to decrease the level of hepcidin and induce the production of ferritin [18,20]. Similarly, simvastatin as well as vitamin C may suppress the expression of hepcidin in cell cultures [21,22].

Cellular iron levels can be affected by immunosuppressive drugs. Data on animal models show that sirolimus may contribute to iron loss on a cellular level [24], which also seems important in cases of human models where the use of sirolimus was correlated with functional iron deficiency and anemia [28,29,85,86]. Similarly, therapy with everolimus and tacrolimus has been found to be associated with the occurrence of anemia [32]. Additionally, cyclosporine may contribute to anemia, as has been stated regarding the liver and lung in transplant patients [34]. Sirolimus, by its mechanism of immunosuppression, blocks the IL-2 post-receptor signals mediating T-cell proliferation, having the ability to bind to the mammalian target of rapamycin and arrest the progression of the cell cycle between the G1 and the S phase [86]. The study of Thaunat et al. indicated that anemia in sirolimus therapy is associated with low serum iron levels, inflammatory states and high serum ferritin levels [31]. Some authors demonstrated that sirolimus-induced anemia occurs due to its direct impact on the homeostasis of iron [30]. On the other hand, introducing prednisone may attenuate the severity of anemia in individuals after renal transplantation. Additionally, treatment with mycophenalate mofetil was associated with the occurrence of anemia in both rena and liver transplant patients [37].

Iron uptake is also a process that can often be influenced by pharmacotherapy. The significant finding by Kramer et al. [27] in an animal model could also be of interest in the case of heart transplantology in patients taking propranolol, as it may reduce iron uptake by cardiac tissue in rats. Regarding statins, pravastatin increases iron uptake in the liver [25] while fluvastatin treatment may decrease serum levels of hepcidin prohormone (prohepcidin) [53]. On the other hand, simvastatin has a minor influence on serum prohepcidin levels, hs-CRP or IL-6, which also take part in iron turnover [54].

The non-specific way in which drugs impact iron turnover can be observed in cardiac and dialysis patients. In renal transplant patients, the use of enalapril was associated with anemia [38]. Another hypertensive drug, amlodipine, was reported to decrease iron overload and to reduce ferritin levels in patients with thalassemia major [47]. Nonsteroidal anti-inflammatory drugs (NSAIDs) (loxoprofen, diclofenac, ampiroxicam, naproxen, etodolac) are supposed to increase the risk of iron deficiency in chronic hemodialysis patients, according to Wang et al. [52]. However, acetylsalicylic acid in a representative cohort did not affect serum iron or iron saturation [48]. Nevertheless, treatment with acetylsalicylic acid was related to lower serum levels of ferritin in a group of elderly subjects [50], in postmenopausal women [51], as well as in *H. pylori* infected subjects [87]. Azathioprine was reported by Habas et al. [44] to cause an increase of serum iron and serum transferring saturation in post-kidney transplantation.

Drug-mediated protein expression may be also involved in iron homeostasis. According to Mleczko-Sanecka et al. [75], the suppression of hepcidin is strongly related to mTOR signaling, since the inhibition of the mTOR kinase with rapamycin may lead to a strong activation of the expression of hepcidin mRNA. Successful treatment of the anemic state by intravenous iron, with simultaneous failure of oral therapy, may suggest that the reason is a hepcidin-mediated reduction of iron absorption from diet [75]. Rapamycin and cycloheximide act by blocking protein synthesis, but cycloheximide may impair the process of translation faster in comparison to rapamycin [49]. Interestingly, in the presence of cycloheximide acting as a translation elongation inhibitor, ferroportin remained on the cell surface while only hepcidin impaired the functioning of ferroportin by causing its internalization [59].

Drugs may also take part in the regulation of iron homeostasis, by modifying the effectiveness of the iron transfer according to the needs of the cell, in the case of restricted iron availability, which is supposed to occur in the mTOR pathway. It is supposed that the mTOR target, tristetraprolin (TTP), may alter the level of TfR1 [88]. Mice treated with rapamycin mediated mTOR inhibitors exhibited augmented levels of TTP and reduced levels of TfR1. This resulted in increased levels of iron and ferritin, with a reduction of cellular iron uptake [24]. It has been also confirmed by Przybyłowski et al. [89] that the treatment of heart transplant recipients with mTOR inhibitors was related to increased concentrations of circulating hepcidin, which may be a reason for the observed lower hemoglobin levels.

7. Pharmacotherapy and Different Types of Anemia in Transplant Patients

Post-transplant anemia is multifactorial. Iron deficiency anemia, on which we focus, is the most common cause of anemia in the world. Some drugs cause anemia in the iron deficiency pathway (Table 1). Other drugs affect red blood cells by causing different types of anemia: megaloblastic anemia, aplastic anemia, hemolytic anemia, thrombocytopenia and agranulocytosis [90] (Figure 2). Finally, it has to be taken into consideration that the activity of several drugs may be also driven by inflammatory conditions that may additionally complicate pharmacotherapy prior to and after organ transplantation.

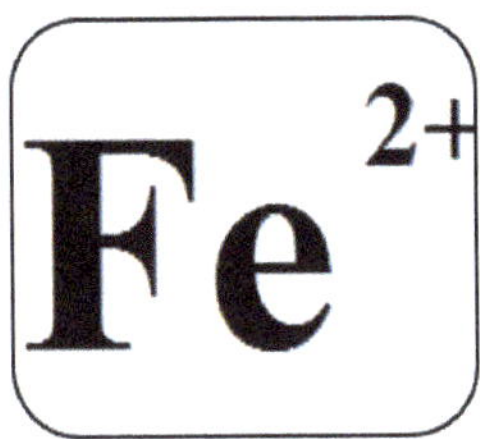
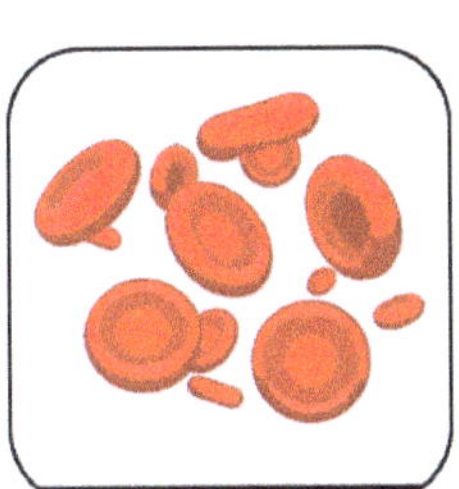

Figure 2. Correlation between pharmacotherapy and anemia.

7.1. Megaloblastic Anemia

Megaloblastic anemia is a disease in which an unproductive hematopoiesis occurs, usually from a deficiency of folic acid and/or vitamin B12 as well as from a metabolic deficiency [91]. This type of anemia can be caused by drugs via decreasing the absorption of folic acid, having folate analogue activity, interfering with pyrimidine synthesis, modulating purine metabolism, decreasing the absorption of vitamin B12 or destroying vitamin B12 [92,93]. Among these drugs are sulfonamides. They are structurally related to *para*-aminobenzoic acid, and are its competitive antagonists. Bacteria need *para*-aminobenzoic acid for the formation of dihydrofolic acid, which is required for the synthesis of folic acid. Folic acid is a substrate for nucleic acid synthesis. Sulfonamides are not toxic to human cells, as they utilize folic acid from the diet, but the long-term use of sulfonamides may cause metabolic deficiency that can result in megaloblastic anemia [94]. Trimetoprim, on the other hand, inhibits the reduction of dihydrofolic acid to tetrahydrofolic acid by binding to dihydrofolate reductase. Tetrahydrofolic acid is then needed for the synthesis of thymidine, and the inhibition of this pathway impairs bacterial DNA synthesis (Figure 3). In this way, co-trimoxazole therapy (sulfamethoxazole and trimethoprim) inhibits two related reactions which are crucial for microorganisms [95].

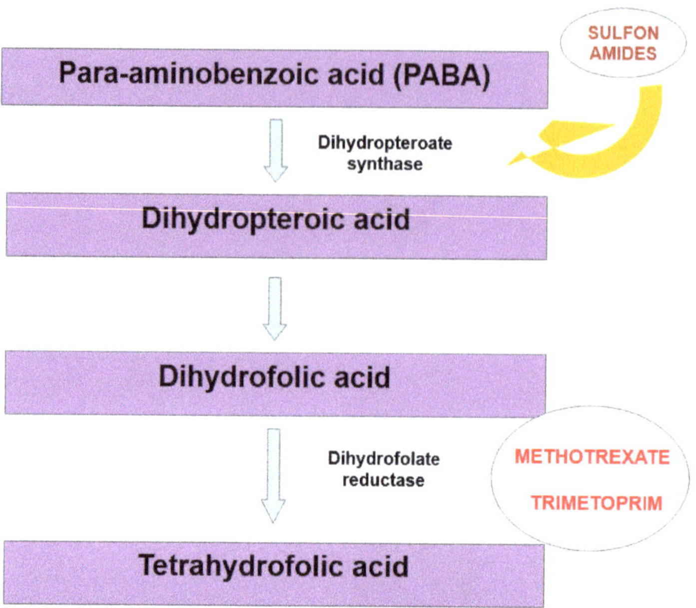

Figure 3. Mechanism that may lead to megaloblastic anemia, due to structural similarity between bacterial and human dihydrofolate reductase.

Methotrexate is used for the treatment of leukemias and lymphomas, for auto-immune diseases such as rheumatoid arthritis, and in transplant patients for immunosuppression and for inhibition of inflammation [96,97]. In the context of anemia, it is important to underline that human and bacterial dihydrofolate reductases are structurally similar, and their impairment in mammalian cells may cause megaloblastic anemia (Figure 3).

Phenytoin belongs to anticonvulsant drugs and is used for the treatment of various types of seizures. It may increase the pH in the small intestine and impair the intestinal conjugate activity, leading to decreased intestinal absorption of folates. It directly competes with folates for their uptake sites. Moreover, phenytoin may reduce the activity of folate inter-converting enzymes and increase the activity of enzymes that catabolize folates. Because of this, and additionally by inhibiting central appetite centers, the drug may cause folate deficiency [98].

Azathioprine, an immunosuppressive drug used for the treatment of rheumatoid arthritis, Crohn's disease, ulcerative colitis and, in transplantology, as a prevention of kidney transplant rejection, inhibits converting reactions among the precursors of purines and suppresses the synthesis of purines [99]. This may result in a deficiency of erythroid precursors from the bone marrow [100]. Similarly, mycophenolate mofetil is another purine synthesis inhibitor. This drug, used in transplantology to prevent graft rejection, impairs the synthesis of purines by blocking the inositol monophosphate dehydrogenase [101]. Nitrous oxide, an inhalatory gas used in anesthesia, may impair the conversion of the reduced form of vitamin B12 to the oxidized form, and thus may also lead to megaloblastic anemia. Since methionine synthase utilizes vitamin B12 in a reduced form (methylcobalamin) in order to transform homocysteine to methionine, and in mitochondria vitamin B12 in the oxidized form (5′-deoxyadenosylcobalamin) is needed for the conversion of methyl-malonylcoenzyme A (CoA) to succinyl CoA, nitrous oxide may lead to the impairment of methylation reactions and DNA synthesis [102]. On the other hand, drugs that have an impact on vitamin B12 absorption are neomycin, metformin, aminosalicylic acid and colchicine, but the risk of inducing megaloblastic anemia by them is rather low [56].

7.2. Aplastic Anemia

There are two classes of aplastic anemia: inherited and acquired aplastic anemia. Acquired aplastic anemia may be induced by chemical exposure, viral infections, exposure to radiation and drugs [103]. Aplastic anemia induced by drugs involves the generation of intermediate metabolites, binding to proteins and DNA, which causes toxic effects on bone marrow and hematopoietic cells (carbamazepine, captopril, furosemide, thiazides, sulfonamides, chlorothiazide and lisinopril) [104]. Additionally, immune-mediated mechanisms as well as direct toxicity may be involved in drug-induced acquired aplastic anemia [105]. Angiotensin converting enzyme inhibitors (ACEIs) such as lisinopril or captopril may have an impact on the production of erythropoietin (EPO) and cause its inhibition, while angiotensin II is known to augment the proliferation of erythroid progenitors [105]. Moreover, carbamazepine, an anticonvulsant used for the treatment of epilepsy and pain associated with neuralgia, exhibits many side effects which include hematopoietic disorders, such as aplastic anemia. Since the mechanism of these side effects has not been explained so far, the involvement of some toxic effects or allergic reactions has been postulated [106,107].

7.3. Hemolytic Anemia

Hemolytic anemia in the drug-induced form may be mediated by metabolic abnormalities in red blood cells, as well as by the production of antibodies. The specific form of anemia called oxidative hemolytic anemia may be caused by glucose-6-phosphate dehydrogenase deficiency, reduced activity of methemoglobin reductase and glutathione peroxidase. Drugs that may induce oxidative hemolytic anemia are nitrofurantoin, sulfacetamide, metformin or ascorbic acid [93], tacrolimus, ciprofloxacin, ceftriaxone and omeprazole. The mechanism may be due to drugs attaching to the surface of red blood cells, resulting in hemolysis. Moreover, drugs may induce changes in the red blood cells' membrane. Another one is the mechanism in which antibodies are developed against drug complexes. The complexes bind to the surface of red blood cells, and impair their functionality [108]. Cephalosporins belonging to β-lactam antibiotics are able to interact with the red blood cells' membrane. First generation cephalosporins may induce hemolysis by causing IgG adsorption to the red blood cells' membrane, which in turn may result in hemolysis. Ceftriaxone is supposed to induce an immune-complex reaction, where IgM antibodies against ceftriaxone are involved, leading to erythrocyte destruction [42,109]. Additionally, chloroquine, an antimalarial drug and a second line treatment for rheumatoid arthritis, as well as nitrofurantoin, an antibiotic used mainly for the treatment of urinary tract infections, may cause the development of hemolytic anemia when the functionality of glucose-6-phosphate dehydrogenase (G6 PD) is impaired, thus becoming unable to protect red blood cells from certain oxidative metabolites and stressors [110,111].

8. Conclusions and Perspectives

Summarizing, anemia caused by drugs directly affecting red blood cells is described widely in the literature, while the problem of iron metabolism and transplantation has been mentioned in the literature by several authors to date. Although the metabolic pathways of iron in the human body are well described, the influence of the transplantation of organs, as well as perioperative pharmacotherapy, which both seem to affect the homeostasis of iron in a very diverse way, are not studied enough.

After organ transplantation, iron deficiency occurs often in patients with renal and cardiac grafts. However, in patients after hepatic transplantation, iron overload is more common [1]. In most cases, authors reviewed the metabolism of iron in kidney, liver and heart transplantations, but the current knowledge about iron metabolism in lung graft recipients and donors is limited. Nevertheless, the overall significance of iron turnover among patients before and after transplantation seems to have hardly been discussed so far and surely needs more attention.

Since iron turnover may be impaired by so many factors, more clinical data on drugs safety related to iron status prior to and after transplantation is needed. This knowledge would be helpful in predicting the risk of iron deficiency and overload related to pharmacotherapy in transplant patients. Moreover, the course of management of patients with iron overload seems to remain a challenge, as there are no evidence-based recommendations for post-transplant patients.

Author Contributions: Conceptualization, M.D. and A.B.; methodology, M.D. and A.B.; validation, M.D. and N.P.; formal analysis, M.D., A.B and B.M.; investigation, M.D., A.B. and B.M; resources, M.D. and A.B; data curation, M.D., A.B., B.M. and N.P.; writing—original draft preparation, M.D. and A.B.; writing—review and editing, M.D.; visualization, M.D. and A.B.; supervision, N.P.; project administration, M.D. and N.P.; funding acquisition, M.D. and N.P. All authors have read and agreed to the published version of the manuscript.

Funding: This research received no external funding.

Institutional Review Board Statement: Not applicable.

Informed Consent Statement: Not applicable.

Data Availability Statement: Not applicable.

Conflicts of Interest: The authors declare no conflict of interest.

References

1. Małyszko, J.; Levin-Laina, N.; Myśliwiec, M.; Przybyłowski, P.; Durlik, M. Iron metabolism in solid-organ transplantation: How far are we from solving the mystery? *Pol. Arch Med. Wewn.* **2012**, *122*, 504–511. [CrossRef] [PubMed]
2. Finch, C.A.; Bellotti, V.; Stray, S. Plasma ferritin determination as a diagnostics tool. *West J. Med.* **1986**, *145*, 657–663.
3. Ganz, T. Iron Metabolism. In *Williams Hematology, 9e*; Kaushansky, K., Lichtman, M.A., Prchal, J.T., Levi, M.M., Press, O.W., Burns, L.J., Caligiuri, M., Eds.; McGraw Hill: New York, NY, USA, 2015.
4. Fact Sheet by the National Institutes of Health (NIH) Office of Dietary Supplements (ODS). Available online: https://ods.od.nih.gov/factsheets/Iron-HealthProfessional/ (accessed on 5 April 2022).
5. Schaefer, B.; Effenberger, M.; Zoller, H. Iron metabolism in transplantation. *Transpl. Int.* **2014**, *27*, 1109–1117. [CrossRef]
6. Modrykamien, A. Anemia post-lung transplantation: Mechanisms and approach to diagnosis. *Chron. Respir. Dis.* **2010**, *7*, 29–34. [CrossRef] [PubMed]
7. Aydogan, M.S.; Ozgül, U.; Erdogan, M.A.; Yucel, A.; Toprak, H.I.; Durmus, M.; Colak, C. Effect of preoperative iron deficiency in liver transplant recipients on length of intensive care unit stay. *Transplant. Proc.* **2013**, *45*, 978–981. [CrossRef]
8. Nakanishi, T.; Hasuike, Y.; Otaki, Y.; Kida, A.; Nonoguchi, H.; Kuragano, T. Hepcidin: Another culprit for complications in patients with chronic kidney disease? *Nephrol. Dial. Transplant.* **2011**, *26*, 3092–3100. [CrossRef]
9. Anastasiou, O.E.; Kälsch, J.; Hakmouni, M.; Kucukoglu, O.; Heider, D.; Korth, J.; Manka, P.; Sowa, J.P.; Bechmann, L.; Saner, F.H.; et al. Low transferrin and high ferritin concentrations are associated with worse outcome in acute liver failure. *Liver Int.* **2017**, *37*, 1032–1041. [CrossRef]
10. Bardou-Jacquet, E.; Philip, J.; Lorho, R.; Ropert, M.; Latournerie, M.; Houssel-Debry, P.; Guyader, D.; Loréal, O.; Boudjema, K.; Brissot, P. Liver transplantation normalizes serum hepcidin level and cures iron metabolism alterations in HFE hemochromatosis. *Hepatology* **2014**, *59*, 839–847. [CrossRef]
11. Sarnak, M.J.; Coronado, B.E.; Greene, T.; Wang, S.R.; Kusek, J.W.; Beck, G.J.; Levey, A.S. Cardiovascular disease risk factors in chronic renal insufficiency. *Clin. Nephrol.* **2002**, *57*, 327–335. [CrossRef] [PubMed]
12. Kasiske, B.L. Cardiovascular disease after renal transplantation. *Semin. Nephrol.* **2000**, *20*, 176–187. [CrossRef]
13. Djamali, A.; Samaniego, M.; Muth, B.; Muehrer, R.; Hofmann, R.M.; Pirsch, J.; Howard, A.; Mourad, G.; Becker, B.N. Medical care of kidney transplant recipients after the first posttransplant year. *Clin. J. Am. Soc. Nephrol.* **2006**, *1*, 623–640. [CrossRef] [PubMed]
14. Imoagene-Oyedeji, A.E.; Rosas, S.E.; Doyle, A.M.; Goral, S.; Bloom, R. Posttransplantation anemia at 12 months in kidney recipients treated with mycophenolate mofetil: Risk factors and implications for mortality. *Clin. J. Am. Soc. Nephrol.* **2006**, *17*, 3240–3247. [CrossRef] [PubMed]
15. Molnar, M.Z.; Czira, M.; Ambrus, C.; Szeifert, L.; Szentkiralyi, A.; Beko, G.; Rosivall, L.; Remport, A.; Novak, M.; Mucsi, I. Anemia isassociated with mortality in kidney-transplanted patients: Aprospective cohort study. *Am. J. Transplant.* **2007**, *7*, 818–824. [CrossRef]
16. Zhang, J. A Reaction of Aspirin with Ferrous Gluconate. *AAPS Pharmscitech* **2015**, *16*, 1495–1499. [CrossRef] [PubMed]
17. Schwarz, K.B.; Arey, B.J.; Tolman, K.; Mahanty, S. Iron chelation as a possible mechanism for aspirin-induced malondialdehyde production by mouse liver microsomes and mitochondria. *J. Clin. Investig.* **1988**, *81*, 165–170. [CrossRef]
18. Li, W.Y.; Li, F.M.; Zhou, Y.F.; Wen, Z.M.; Ma, J.; Ya, K.; Qian, Z.M. Aspirin down regulates hepcidin by inhibiting NF-κB and IL6/JAK2/STAT3 pathways in BV-2 microglial cells treated with lipopolysaccharide. *Int. J. Mol. Sci.* **2016**, *17*, 1921. [CrossRef] [PubMed]

19. Xu, Y.X.; Du, F.; Jiang, L.R.; Gong, J.; Zhou, Y.-F.; Luo, Q.Q.; Qian, Z.M.; Ke, Y. Effects of aspirin on expression of iron transport and storage proteins in BV-2 microglial cells. *Neurochem. Int.* **2015**, *91*, 72–77. [CrossRef]

20. Oberle, S.; Polte, T.; Abate, A.; Podhaisky, H.P.; Schröder, H. Aspirin increases ferritin synthesis in endothelial cells: A novel antioxidant pathway. *Circ. Res.* **1998**, *82*, 1016–1020. [CrossRef]

21. Chang, C.C.; Chiu, P.F.; Chen, H.L.; Chang, T.L.; Chang, Y.J.; Huang, C.H. Simvastatin downregulates the expression of hepcidin and erythropoietin in HepG2 cells. *Hemodial. Int.* **2013**, *17*, 116–121. [CrossRef]

22. Chiu, P.F.; Ko, S.Y.; Chang, C.C. Vitamin C affects the expression of hepcidin and erythropoietin receptor in HepG2 cells. *J. Ren. Nutr.* **2012**, *22*, 373–376. [CrossRef]

23. Diekmann, F.; Rovira, J.; Diaz-Ricart, M.; Arellano, E.M.; Vodenik, B.; Jou, J.M.; Vives-Corrons, J.L.; Escolar, G.; Campistol, J.M. mTOR inhibition and erythropoiesis: Microcytosis or anaemia? *Nephrol. Dial. Transpl.* **2012**, *27*, 537–541. [CrossRef] [PubMed]

24. Bayeva, M.; Khechaduri, A.; Puig, S.; Chang, H.-C.; Patial, S.; Blackshear, P.J.; Ardehali, H. mTOR regulates cellular iron homeostasis through tristetraprolin. *Cell Metab.* **2012**, *16*, 645–657. [CrossRef]

25. Kolouchova, G.; Brcakova, E.; Hirsova, P.; Cermanova, J.; Fuksa, L.; Mokry, J.; Nachtigal, P.; Lastuvkova, H.; Micuda, S. Modification of hepatic iron metabolism induced by pravastatin during obstructive cholestasis in rats. *Life Sci.* **2011**, *89*, 717–724. [CrossRef]

26. Müller, A.C.; Dairam, A.; Limson, J.L.; Daya, S. Mechanisms by which acyclovir reduces the oxidative neurotoxicity and biosynthesis of quinolinic acid. *Life Sci.* **2007**, *80*, 918–925. [CrossRef] [PubMed]

27. Kramer, J.H.; Spurney, C.F.; Iantorno, M.; Tziros, C.; Chmielinska, J.J.; Mak, I.T.; Weglicki, W.B. d-Propranolol protects against oxidative stress and progressive cardiac dysfunction in iron overloaded rats. *Can. J. Physiol. Pharmacol.* **2012**, *90*, 1257–1268. [CrossRef] [PubMed]

28. McDonald, M.A.; Gustafsson, F.; Almasood, A.; Barth, D.; Ross, H.J. Sirolimus is associated with impaired hematopoiesis in heart transplant patients? A retrospective analysis. *Transplant. Proc.* **2010**, *42*, 2693–2696. [CrossRef]

29. Sofroniadou, S.; Kassimatis, T.; Goldsmith, D. Anaemia, microcytosis and sirolimus—Is iron the missing link? *Nephrol. Dial. Transplant.* **2010**, *25*, 1667–1675. [CrossRef]

30. Maiorano, A.; Stallone, G.; Schena, A.; Infante, B.; Pontrelli, P.; Schena, F.P.; Grandaliano, G. Sirolimus interferes with iron homeostasis in renal transplant recipients. *Transplantation* **2006**, *82*, 908–912. [CrossRef]

31. Thaunat, O.; Beaumont, C.; Chatenoud, L.; Lechaton, S.; Mamzer-Bruneel, M.-F.; Varet, B.; Kreis, H.; Morelon, E. Anemia after late introduction of sirolimus may correlate with biochemical evidence of a chronic inflammatory state. *Transplantation* **2005**, *80*, 1212–1219. [CrossRef]

32. Sánchez Fructuoso, A.; Calvo, N.; Moreno, M.A.; Giorgi, M.; Barrientos, A. Study of anemia after late introduction of everolimus in the immunosuppressive treatment of renal transplant patients. *Transplant. Proc.* **2007**, *39*, 2242–2244. [CrossRef]

33. Masetti, M.; Rompianesi, G.; Montalti, R.; Romano, A.; Spaggiari, M.; Ballarin, R.; Guerrini, G.; Gerunda, G. Effects of everolimus monotherapy on hematological parameters and iron homeostasis in de novo liver transplant recipients: Preliminary results. *Transplant. Proc.* **2008**, *40*, 1947–1949. [CrossRef]

34. Kanellopoulou, T. Autoimmune hemolytic anemia in solid organ transplantation-The role of immunosuppression. *Clin. Transplant.* **2017**, *31*, e13031. [CrossRef] [PubMed]

35. Friend, P.; Russ, G.; Oberbauer, R.; Murgia, M.G.; Tufveson, G.; Chapman, J.; Blancho, G.; Mota, A.; Grandaliano, G.; Campistol, J.M.; et al. Rapamune Maintenance Regimen (RMR) Study Group. Incidence of anemia in sirolimus-treated renal transplant recipients: The importance of preserving renal function. *Transpl. Int.* **2007**, *20*, 754–760. [CrossRef] [PubMed]

36. Patil, M.R.; Choudhury, A.R.; Chohwanglim, M.; Divyaveer, S.; Mahajan, C.; Pandey, R. Post renal transplant pure red cell aplasia-is tacrolimus a culprit? *Clin. Kidney J.* **2016**, *9*, 603–605. [CrossRef]

37. Al-Uzri, A.; Yorgin, P.D.; Kling, P.J. Anemia in children after transplantation: Etiology and the effect of immunosuppressive therapy on erythropoiesis. *Pediatr. Transplant.* **2003**, *7*, 253–264. [CrossRef] [PubMed]

38. Vlahakos, D.V.; Canzanello, V.J.; Madaio, M.P.; Madias, N.E. Enalapril-associated anemia in renal transplant recipients treated for hypertension. *Am. J. Kidney Dis.* **1991**, *17*, 199–205. [CrossRef]

39. Graafland, A.D.; Doorenbos, C.J.; van Saase, J.C. Enalapril-induced anemia in two kidney transplant recipients. *Transpl. Int.* **1992**, *5*, 51–53. [CrossRef]

40. Frieder, J.; Mouabbi, J.A.; Zein, R.; Hadid, T. Autoimmune hemolytic anemia associated with trimethoprim-sulfamethoxazole use. *Am. J. Health Syst. Pharm.* **2017**, *74*, 894–897. [CrossRef]

41. Arndt, P.A.; Garratty, G.; Wolf, C.F.; Rivera, M. Haemolytic anaemia and renal failure associated with antibodies to trimethoprim and sulfamethoxazole. *Transfus. Med.* **2011**, *21*, 194–198. [CrossRef] [PubMed]

42. Arndt, P.A.; Leger, R.M.; Garratty, G. Serology of antibodies to second-and third-generation cephalosporins associated with immune hemolytic anemia and/or positive direct antiglobulin tests. *Transfusion* **1999**, *39*, 1239–1246. [CrossRef]

43. Chisholm-Burns, M.A.; Patanwala, A.E.; Spivey, C.A. Aseptic meningitis, hemolytic anemia, hepatitis, and orthostatic hypotension in a patient treated with trimethoprim-sulfamethoxazole. *Am. J. Health Syst. Pharm.* **2010**, *67*, 123–127. [CrossRef] [PubMed]

44. Habas, E.; Khammaj, A.; Rayani, A. Hematologic side effects of azathioprine and mycophenolate in kidney transplantation. *Transplant. Proc.* **2011**, *43*, 504–506. [CrossRef] [PubMed]

45. Maheshwari, A.; Mishra, R.; Thuluvath, P.J. Post-liver-transplant anemia: Etiology and management. *Liver Transplant.* **2004**, *10*, 165–173. [CrossRef]

46. Willms, H.; Gutjahr, K.; Juergens, U.R.; Hammerschmidt, S.; Gessner, C.; Hoheisel, G.; Wirtz, H.; Gillissen, A. Die idiopathische pulmonale Hämosiderose. *Med. Klin.* **2007**, *102*, 445–450. [CrossRef]
47. Fernandes, J.L.; Sampaio, E.F.; Fertrin, K.; Coelho, O.R.; Loggetto, S.; Piga, A.; Verissimo, M.; Saad, S.T. Amlodipine reduces cardiac iron overload in patients with thalassemia major: A pilot trial. *Am. J. Med.* **2013**, *126*, 834–837. [CrossRef]
48. Hammerman-Rozenberg, R.; Jacobs, J.M.; Azoulay, D.; Stessman, J. Aspirin prophylaxis and the prevalence of anaemia. *Age Ageing* **2006**, *35*, 514–517. [CrossRef] [PubMed]
49. Berset, C.; Trachsel, H.; Altmann, M. The TOR (target of rapamycin) signal transduction pathway regulates the stability of translation initiation factor eIF4G in the yeast Saccharomyces cerevisiae. *Proc. Natl. Acad. Sci. USA* **1998**, *95*, 4264–4269. [CrossRef]
50. Fleming, D.J.; Jacques, P.F.; Massaro, J.M.; D'Agostino, R.B.S.; Wilson, P.W.; Wood, R.J. Aspirin intake and the use of serum ferritin as a measure of iron status. *Am. J. Clin. Nutr.* **2001**, *74*, 219–226. [CrossRef]
51. Liu, J.M.; Hankinson, S.E.; Stampfer, M.J.; Rifai, N.; Willett, W.C.; Ma, J. Body iron stores and their determinants in healthy postmenopausal US women. *Am. J. Clin. Nutr.* **2003**, *78*, 1160–1167. [CrossRef] [PubMed]
52. Wang, X.; Uzu, T.; Isshiki, K.; Kanasaki, M.; Hirata, K.; Soumura, M.; Nakazawa, J.; Kashiwagi, A.; Takaya, K.; Isono, M.; et al. Clinical Dialysis Meeting. Iron status and the use of non-steroidal anti-inflammatory drugs in hemodialysis patients. *Ther. Apher. Dial.* **2007**, *11*, 215–219. [CrossRef]
53. Arabul, M.; Gullulu, M.; Yilmaz, Y.; Akdag, I.; Kahvecioglu, S.; Eren, M.A.; Dilek, K. Effect of fluvastatin on serum prohepcidin levels in patients with end-stage renal disease. *Clin. Biochem.* **2008**, *41*, 1055–1058. [CrossRef]
54. Li, X.-Y.; Chang, J.-P.; Su, Z.-W.; Li, J.-H.; Peng, B.-S.; Zhu, S.-L.; Cai, A.-J.; Zhang, J.; Jiang, Y. How does short-term low-dose simvastatin influence serum prohepcidin levels in patients with end-stage renal disease? A pilot study. *Ther. Apher. Dial.* **2010**, *14*, 308–314. [CrossRef]
55. Kim, Y.; Park, B.; Ryu, C.; Park, S.; Kang, S.; Kim, Y.; Song, I.; Shon, J. Allopurinol-induced aplastic anemia in a patient with chronic kidney disease. *Clin. Nephrol.* **2009**, *71*, 203–206. [CrossRef]
56. Liu, Q.; Li, S.; Quan, H.; Li, J. Vitamin B12 status in metformin treated patients: Systematic review. *PLoS ONE* **2014**, *9*, e100379. [CrossRef] [PubMed]
57. Rusu, I.G.; Suharoschi, R.; Vodnar, D.C.; Pop, C.R.; Socaci, S.A.; Vulturar, R.; Istrati, M.; Moroșan, I.; Fărcaș, A.C.; Kerezsi, A.D.; et al. Iron Supplementation Influence on the Gut Microbiota and Probiotic Intake Effect in Iron Deficiency-A Literature-Based Review. *Nutrients* **2020**, *12*, 1993. [CrossRef] [PubMed]
58. Ponka, P.; Tenenbein, M.; Eaton, J.W. Iron. Chapter 41. In *Handbook on the Toxicology of Metals*; Nordberg, G., Fowler, B., Nordberg, M., Friberg, L., Eds.; Elsevier: New York, NY, USA, 2015; pp. 879–902.
59. Nemeth, E.; Tuttle, M.S.; Powelson, J.; Vaughn, M.B.; Donovan, A.; Ward, D.M.; Ganz, T.; Kaplan, J. Hepcidin regulates cellular iron efflux by binding to ferroportin and inducing its internalization. *Science* **2004**, *306*, 2090–2093. [CrossRef]
60. Ganz, T. Hepcidin and iron regulation, 10 years later. *Blood* **2011**, *117*, 4425–4433. [CrossRef] [PubMed]
61. Nicolas, G.; Bennoun, M.; Porteu, A.; Mativet, S.; Beaumont, C.; Grandchamp, B.; Sirito, M.; Sawadogo, M.; Kahn, A.; Vaulont, S. Severe iron deficiency anemia in transgenic mice expressing liver hepcidin. *Proc. Natl. Acad. Sci. USA* **2002**, *99*, 4596–4601. [CrossRef]
62. Przybylowski, P.; Wasilewski, G.; Golabek, K.; Bachorzewska-Gajewska, H.; Dobrzycki, S.; Koc-Zorawska, E.; Malyszko, J. Absolute and Functional Iron Deficiency Is a Common Finding in Patients With Heart Failure and After Heart Transplantation. *Transplant. Proc.* **2016**, *48*, 173–176. [CrossRef]
63. Malyszko, J.; Malyszko, J.S.; Mysliwiec, M. Anemia prevalence and possible role of hepcidin in its pathogenesis in kidney allograft recipients. *Transplant. Proc.* **2009**, *4*, 3056–3059. [CrossRef] [PubMed]
64. Malyszko, J.; Malyszko, J.S.; Pawlak, K.; Mysliwiec, M. Hepcidin, iron status, and renal function in chronic renal failure, kidney transplantation, and hemodialysis. *Am. J. Hematol.* **2006**, *81*, 832–837. [CrossRef]
65. Silvestri, L.; Pagani, A.; Nai, A.; De Dominico, I.; Kaplan, J.; Camaschella, C. The serine protease matriptase-2 (TMPRSS6) inhibits hepcidin activation by cleaving membrane hemojuvelin. *Cell Metab.* **2008**, *8*, 502. [CrossRef] [PubMed]
66. Anderson, G.J.; Frazer, D.M.; McLaren, G.D. Iron absorption and metabolism. *Curr. Opin. Gastroenterol.* **2009**, *25*, 129–135. [CrossRef] [PubMed]
67. Langer, A.L.; Ginzburg, Y.Z. Role of hepcidin-ferroportin axis in the pathophysiology, diagnosis, and treatment of anemia of chronic inflammation. *Hemodial. Int.* **2017**, *21*, 37–46. [CrossRef]
68. Donovan, A.; Lima, C.A.; Pinkus, J.L.; Pinkus, G.S.; Zon, L.I.; Robine, S.; Andrews, N.C. The iron exporter ferroportin/Slc40a1 is essential for iron homeostasis. *Cell Metab.* **2005**, *1*, 191–200. [CrossRef]
69. Roy, C.N.; Enns, C.A. Iron homeostasis: New tales from the crypt. *Blood* **2000**, *96*, 4020–4027. [CrossRef] [PubMed]
70. Cassat, J.E.; Skaar, E.P. Iron in infection and immunity. *Cell Host Microbe* **2013**, *13*, 509–519. [CrossRef]
71. Beard, J.L. Iron biology in immune function, muscle metabolism and neuronal functioning. *J. Nutr.* **2001**, *131*, 568–579. [CrossRef] [PubMed]
72. Bacon, B.R.; O'Neill, R.; Park, C.H. Iron-induced peroxidative injury to isolated rat hepatic mitochondria. *J. Free Radic. Biol. Med.* **1986**, *2*, 339–347. [CrossRef]
73. Freinbichler, W.; Tipton, K.F.; Corte, L.D.; Linert, W. Mechanistic aspects of the Fenton reaction under conditions approximated to the extracellular fluid. *J. Inorg. Biochem.* **2009**, *103*, 28–34. [CrossRef]

74. Hassan, T.H.; Badr, M.A.; Karam, N.A.; Zkaria, M.; El Saadany, H.F.; Rahman, D.M.A.; Shahbah, D.A.; Al Morshedy, S.M.; Fathy, M.; Esh, A.M.H.; et al. Impact of iron deficiency anemia on the function of the immune system in children. *Medicine* **2016**, *95*, e5395. [CrossRef] [PubMed]

75. Mleczko-Sanecka, K.; Roche, F.; da Silva, A.R.; Call, D.; D'Alessio, F.; Ragab, A.; Lapinski, P.E.; Ummanni, R.; Korf, U.; Oakes, C.; et al. Unbiased RNAi screen for hepcidin regulators links hepcidin suppression to proliferative Ras/RAF and nutrient-dependent mTOR signaling. *Blood* **2014**, *123*, 1574–1585. [CrossRef]

76. Saha, K.; Mukhopadhyay, D.; Roy, S.; Raychaudhuri, G.; Chatterjee, M.; Mitra, P.; Das, I. Impact of iron deficiency anemia on cell-mediated and humoral immunity in children: A case control study. *J. Nat. Sci. Biol. Med.* **2014**, *5*, 158–163. [CrossRef]

77. Attia, M.A.; Essa, S.A.; Nosair, N.A.; Amin, A.M.; El-Agamy, O.A. Effect of iron deficiency anemia and its treatment on cell mediated immunity. *Indian J. Hematol. Blood Transfus.* **2009**, *25*, 70–77. [CrossRef] [PubMed]

78. Kim, C.M.; Park, Y.J.; Shin, S.H. A widespread deferoxamine-mediated iron-uptake system in Vibrio vulnificus. *J. Infect. Dis.* **2007**, *196*, 1537–1545. [CrossRef]

79. Cassat, J.E.; Skaar, E.P. Metal ion acquisition in Staphylococcus aureus: Overcoming nutritional immunity. *Semin. Immunopathol.* **2012**, *34*, 215–235. [CrossRef]

80. Adamiak, P.; Calmettes, C.; Moraes, T.F.; Schryvers, A.B. Patterns of structural and sequence variation within isotype lineages of the Neisseria meningitidis transferrin receptor system. *Microbiologyopen* **2015**, *4*, 491–504. [CrossRef]

81. Nevitt, T. War-Fe-re: Iron at the core of fungal virulence and host immunity. *Biometals* **2011**, *24*, 547–558. [CrossRef]

82. Kim, H.C.; Park, S.B.; Han, S.Y.; Whang, E.A. Anemia following renal transplantation. *Transplant. Proc.* **2003**, *35*, 302–303. [CrossRef] [PubMed]

83. Moffatt, S.; Yaegashi, N.; Tada, K.; Tanaka, N.; Sugamura, K. Human parvovirus B19 nonstructural (NS1) protein induces apo-ptosis in erythroid lineage cells. *J. Virol.* **1998**, *72*, 3018–3028. [CrossRef] [PubMed]

84. Baranski, B.; Armstrong, G.; Truman, J.T.; Quinnan, G.V., Jr.; Straus, S.E.; Young, N.S. Epstein-Barr virus in the bone marrow of pa-tients with aplastic anemia. *Ann. Intern. Med.* **1988**, *109*, 695–704. [CrossRef] [PubMed]

85. McMillan, M.A.; Muirhead, C.S.; Lucie, N.P.; Briggs, J.D.; Junor, B.J. Autoimmune haemolytic anaemia relatedto cyclosporin with ABO-compatible kidney donor andrecipient. *Nephrol. Dial. Transplant.* **1991**, *6*, 57–59. [CrossRef] [PubMed]

86. Sehgal, S.N. Rapamune (Sirolimus, rapamycin): An overview and mechanism of action. *Ther. Drug. Monit.* **1995**, *17*, 660–665. [CrossRef] [PubMed]

87. Kaffes, A.; Cullen, J.; Mitchell, H.; Katelaris, P.H. Effect of Helicobacter pylori infection and low-dose aspirin use on iron stores in the elderly. *J. Gastroenterol. Hepatol.* **2003**, *18*, 1024–1028. [CrossRef] [PubMed]

88. Golčić, M.; Petković, M. Changes in metabolic profile, iron and ferritin levels during the treatment of metastatic renal cancer-A new potential biomarker? *Med. Hypotheses* **2016**, *94*, 148–150. [CrossRef]

89. Przybylowski, P.; Malyszko, J.S.; Macdougall, I.C.; Malyszko, J. Iron metabolism, hepcidin, and anemia in orthotopic heart transplantation recipients treated with mammalian target of rapamycin. *Transpl. Proc.* **2013**, *45*, 387. [CrossRef] [PubMed]

90. Mix, T.; Waqar, K.; Samina, K.; Robin, R.; Richard, R.; Brian, P.; Annamaria, K. Anemia: A Continuing Problem Following Kidney Transplantation. *Am. J. Transplant.* **2003**, *3*, 1426–1433. [CrossRef] [PubMed]

91. Clarke, R.; Evans, J.G.; Schneede, J.; Nexo, E.; Bates, C.; Fletcher, A.; Prentice, A.; Johnston, C.; Ueland, P.M.; Refsum, H.; et al. Vitamin B12 and folate deficiency in later life. *Age Ageing* **2004**, *33*, 34–41. [CrossRef]

92. Stebbins, R.; Scott, J.; Herbert, V. Drug-induced megaloblastic anemias. *Semin. Hematol.* **1973**, *10*, 235–251.

93. Rao, K.V. eChapter 24. Drug-Induced Hematologic Disorders. In *Pharmacotherapy: A Pathophysiologic Approach, 9e*; DiPiro, J.T., Talbert, R.L., Yee, G.C., Matzke, G.R., Wells, B.G., Posey, L., Eds.; McGraw-Hill: New York, NY, USA, 2020.

94. Jenkins, G.C.; Hughes, D.T.; Hall, P.C. A haematological study of patients receiving long-term treatment with trimethoprim and sulphonamide. *J. Clin. Pathol.* **1970**, *23*, 392–396. [CrossRef]

95. Wormser, G.P.; Keusch, G.T.; Heel, R.C. Co-trimoxazole(Trimethoprim-sulfamethoxazole) an updated review of its antibacterial activity and clinical efficacy. *Drugs* **1982**, *24*, 459–518. [CrossRef] [PubMed]

96. Huang, X.-J.; Jiang, Q.; Chen, H.; Xu, L.; Liu, D.; Chen, Y.; Han, W.; Zhang, Y.; Liu, K.; Lu, D. Low-dose methotrexate for the treatment of graft-versus-host disease after allogeneic hematopoietic stem cell transplantation. *Bone Marrow Transplant.* **2005**, *36*, 343–348. [CrossRef]

97. Brown, P.M.; Pratt, A.G.; Isaacs, J.D. Mechanism of Action of Methotrexate in Rheumatoid Arthritis, and the Search for Biomarkers. *Nat. Rev. Rheumatol.* **2016**, *12*, 731–742. [CrossRef] [PubMed]

98. Blain, H.; Hamdan, K.A.; Blain, A.; Jeandel, C. Aplastic anemia induced by phenytoin: A geriatric case with severe folic acid deficiency. *J. Am. Geriatr. Soc.* **2002**, *50*, 396–397. [CrossRef]

99. Dessypris, E.N. Pure red cell aplasia. In *Hematology Basic Principles and Practice*, 3rd ed.; Hoffman, R., Benz, E.J., Jr., Shattil, S.J., Eds.; Churchill Livingstone: New York, NY, USA, 2000; pp. 342–354.

100. Thompson, D.F.; Gales, M.A. Drug-induced pure red cell aplasia. *Pharmacotherapy* **1996**, *16*, 1002–1008.

101. Moffatt, B.A.; Ashihara, H. Purine and pyrimidine nucleotide synthesis and metabolism. *Arab. Book* **2002**, *1*, e0018. [CrossRef] [PubMed]

102. Chiang, T.T.; Hung, C.T.; Wang, W.M.; Lee, J.T.; Yang, F.C. Recreational nitrous oxide abuse-induced vitamin B12 deficiency in a patient presenting with hyperpigmenta-tion of the skin. *Case Rep. Dermatol.* **2013**, *5*, 186–191. [CrossRef]

103. Gewirtz, A.M.; Hoffman, R. Current considerations of the etiology of aplastic anemia. *Crit. Rev. Oncol. Hematol.* **1985**, *4*, 1–30. [CrossRef]
104. Malkin, D.; Koren, G.; Saunders, E.F. Drug-induced aplastic anemia: Pathogenesis and clinical aspects. *Am. J. Pediatr. Hematol. Oncol.* **1990**, *12*, 402–410. [CrossRef]
105. Ajmal, A.; Gessert, C.E.; Johnson, B.P.; Renier, C.M.; Palcher, J.A. Effect of angiotensin converting enzyme inhibitors and angiotensin receptor blockers on hemoglobin levels. *BMC Res. Notes* **2013**, *6*, 443. [CrossRef]
106. Dailey, J.W.; Reith, M.E.; Steidley, K.R.; Milbrandt, J.C.; Jobe, P.C. Carbamazepine-Induced Release of Serotonin from Rat Hippocampus In Vitro. *Epilepsia* **1998**, *39*, 1054–1063. [CrossRef] [PubMed]
107. Hart, R.G.; Easton, J.D. Carbamazepine and hematological monitoring. *Ann. Neurol.* **1982**, *11*, 309–312. [CrossRef] [PubMed]
108. Garratty, G. Immune hemolytic anemia associated with drug therapy. *Blood Rev.* **2010**, *24*, 143–150. [CrossRef]
109. Paterson, D.L.; Bonomo, R.A. Extended-spectrum beta-lactamases: A clinical update. *Clin. Microbiol. Rev.* **2005**, *18*, 657–686. [CrossRef]
110. Young, N.S.; Maciejewski, J. The pathophysi-ology of acquired aplastic anemia. *N. Engl. J. Med.* **1997**, *336*, 1365–1372. [CrossRef]
111. Nkhoma, E.T.; Poole, C.; Vannappagari, V.; Hall, S.A.; Beutler, E. The global prevalence of glucose-6-phosphate dehydrogenase deficiency: A systematic review and meta-analysis. *Blood. Cells Mol. Dis.* **2009**, *42*, 267–278. [CrossRef] [PubMed]

 *nutrients*

Article

Nitrosative and Oxidative Stress, Reduced Antioxidant Capacity, and Fiber Type Switch in Iron-Deficient COPD Patients: Analysis of Muscle and Systemic Compartments

Maria Pérez-Peiró [1,2], Mariela Alvarado Miranda [1,2,3], Clara Martín-Ontiyuelo [2,4], Diego A. Rodríguez-Chiaradía [1,2] and Esther Barreiro [1,2,*]

[1] Muscle Wasting and Cachexia in Chronic Respiratory Diseases and Lung Cancer Research Group, Pulmonology Department, Department of Medicine and Life Sciences (MELIS), Hospital del Mar, Medical Research Institute (IMIM), Parc de Salut Mar, Universitat Pompeu Fabra (UPF), Barcelona Biomedical Research Park (PRBB), 08003 Barcelona, Spain

[2] Centro de Investigación en Red de Enfermedades Respiratorias (CIBERES), Instituto de Salud Carlos III (ISCIII), 08003 Barcelona, Spain

[3] Department of Pulmonary Medicine, Hospital Universitari Sant Joan, 43204 Reus, Tarragona, Spain

[4] Department of Pulmonary Medicine, Hospital Clínic-Institut d'Investigacions Biomèdiques August Pi i Sunyer (IDIBAPS), University of Barcelona, 08036 Barcelona, Spain

* Correspondence: ebarreiro@imim.es; Tel.: +34-93-316-0385; Fax: +34-93-316-0410

Abstract: We hypothesized that a rise in the levels of oxidative/nitrosative stress markers and a decline in antioxidants might take place in systemic and muscle compartments of chronic obstructive pulmonary disease (COPD) patients with non-anemic iron deficiency. In COPD patients with/without iron depletion ($n = 20$/group), markers of oxidative/nitrosative stress and antioxidants were determined in blood and vastus lateralis (biopsies, muscle fiber phenotype). Iron metabolism, exercise, and limb muscle strength were assessed in all patients. In iron-deficient COPD compared to non-iron deficient patients, oxidative (lipofuscin) and nitrosative stress levels were greater in muscle and blood compartments and proportions of fast-twitch fibers, whereas levels of mitochondrial superoxide dismutase (SOD) and Trolox equivalent antioxidant capacity (TEAC) decreased. In severe COPD, nitrosative stress and reduced antioxidant capacity were demonstrated in vastus lateralis and systemic compartments of iron-deficient patients. The slow- to fast-twitch muscle fiber switch towards a less resistant phenotype was significantly more prominent in muscles of these patients. Iron deficiency is associated with a specific pattern of nitrosative and oxidative stress and reduced antioxidant capacity in severe COPD irrespective of quadriceps muscle function. In clinical settings, parameters of iron metabolism and content should be routinely quantify given its implications in redox balance and exercise tolerance.

Keywords: COPD; non-anemic iron deficiency; nitrosative stress; antioxidant systems; lipofuscin inclusions; muscle fiber type switch; muscle and systemic compartments

Citation: Pérez-Peiró, M.; Alvarado Miranda, M.; Martín-Ontiyuelo, C.; Rodríguez-Chiaradía, D.A.; Barreiro, E. Nitrosative and Oxidative Stress, Reduced Antioxidant Capacity, and Fiber Type Switch in Iron-Deficient COPD Patients: Analysis of Muscle and Systemic Compartments. *Nutrients* **2023**, *15*, 1454. https://doi.org/10.3390/nu15061454

Academic Editor: Dietmar Enko

Received: 14 February 2023
Revised: 7 March 2023
Accepted: 15 March 2023
Published: 17 March 2023

1. Introduction

Patients with chronic respiratory and cardiac disorders, such as chronic obstructive pulmonary disease (COPD), experience systemic manifestations and comorbidities that affect different organs other than the lungs and airways, particularly in patients with a more advanced disease [1,2]. Skeletal muscle weakness and nutritional abnormalities are encountered among the most prominent manifestations in COPD due to their clinical implications in disease prognosis, including overall survival and impaired quality of life [1–3].

The transition metal iron (Fe) is involved in several key cellular processes in mammals. In humans, iron is mostly found in hemoglobin and myoglobin proteins that are responsible for oxygen transport in blood and oxygen storage in muscles, respectively. Furthermore, iron is also present in the active site of redox enzymes that are involved in important chemical reactions, such as respiration, oxidation, and reduction within the cells [4–6]. A minimum iron uptake from the diet is necessary in order to keep the required homeostatic levels in humans [4–6].

However, in the elderly and in patients with chronic heart and renal failure, the levels of iron may be severely lowered [7–9]. In patients with COPD, iron depletion has also been documented in previous investigations [10–14]. In a recent study [15], a potential crosstalk between systemic and muscle compartments was observed in COPD patients with iron deficiency. Moreover, lung function and exercise capacity were associated with several markers involved in the regulation of iron metabolism in the same cohort of patients [15]. Hence, in iron content depletion is common among patients with chronic diseases, including COPD [10–12,16,17].

Oxidative stress defined as the imbalance between prooxidants and antioxidants in favor of the former has been shown to induce damage to cells and tissues. Increased levels of oxidative and nitrosative stress have been consistently demonstrated in the vastus lateralis muscle of patients with COPD [18–21]. Enhanced proteolysis of major structural proteins is a relevant deleterious effect of oxidative stress within the myofibers, leading to muscle atrophy and even sarcopenia in patients with COPD [18–23]. Other mechanisms, such as apoptosis and autophagy, may also be signaled by high levels of reactive oxygen species (ROS) within the skeletal muscle fibers [21,24–26]. ROS are generated by different sources in the cells and skeletal muscle fibers. Moreover, Fenton reactions between iron and hydrogen peroxide also lead to the formation of the powerful ROS hydroxyl radicals, with a great potential to oxidize other molecules in cells [27]. Whether variations in the levels of iron content may also alter the levels of oxidative stress in the muscle fibers of patients with COPD without anemia warrants further attention. The analysis of whether changes in iron levels may also influence the muscle phenotype in COPD also remains to be answered.

Iron deficiency anemia has been reliably associated with a rise in oxidative stress levels in the systemic compartment and the myocardium [28–32]. The levels of antioxidants were also significantly reduced in patients with iron-deficiency anemia [28,32]. Whether iron depletion without anemia may also entail an increase in oxidative stress levels in patients with COPD needs to be thoroughly investigated. On this basis, we hypothesized that a rise in the levels of oxidative/nitrosative stress markers along with a decline in antioxidants may take place in the systemic and muscle compartments of COPD patients with non-anemic iron deficiency. The study objectives were that in non-anemic COPD patients with and without iron depletion, markers of oxidative and nitrosative stress and antioxidants were determined in blood and vastus lateralis muscle specimens along with characterization of the muscle fiber phenotype. Additionally, all the patients were clinically evaluated, including iron metabolism parameters, and both exercise capacity and limb muscle strength were also determined. For the purpose of the investigation, two different experimental groups were recruited: COPD patients with non-anemic iron deficiency and COPD patients with normal iron content as the control group.

2. Materials and Methods

2.1. Study Population

This was a cross-sectional study, in which forty severe COPD patients were recruited. The Global Strategy of Management of COPD patients (GOLD) was used to diagnose and classify the patients according to disease severity [33]. Clinical parameters and dyspnea score (modified medical research council, mMRC) were obtained from all the participants [34]. All patients were recruited during the years 2018–2021 in the Pulmonology Department at Hospital del Mar (Barcelona, Spain). The following criteria were used to define iron deficiency: hemoglobin > 12 g/dL in women and >13 g/dL in men, ferritin < 100 ng/mL, or ferritin 100–299 ng/mL with a transferrin saturation <20% [16,35]. Iron deficiency without anemia was detected in half of the patients. Hence, on the basis of the iron status, patients were subdivided into two different groups: iron deficiency and non-iron deficiency patients (N = 20/group, respectively, 12 male patients/group).

The current investigation was approved by the Ethics Committee on Human Investigation at Hospital de Mar (Hospital del Mar-IMIM, Barcelona, project number # 2017/7691/I). Ethical standards on human experimentation from our institution, the World Medical Association guidelines (Seventh revision of the Declaration of Helsinki, Fortaleza, Brazil, 2013) [36], and the guidelines established by the International Committee of Medical Journal Editors (ICMJE) were followed. All the participants signed the informed written consent.

2.2. Exclusion Criteria

The following exclusion criteria were established: (1) acute exacerbations in the last three months; (2) other chronic diseases with lung and airways implications (e.g., long-term oxygen therapy, bronchiectasis, asthma); (3) cardiovascular disorders; (4) musculoskeletal alterations; (5) neurological, metabolic, kidney, chronic liver disease, or uncontrolled psychiatric disorders; (6) obesity (body mass index > 30 Kg/m^2); (7) pharmacological treatment with drugs known to alter muscle structure and/or function including oral corticosteroids; (8) active oncologic disease; and (9) history of potentially bleeding conditions.

2.3. Clinical Assessment

In all the patients, lung function was measured through conventional spirometry and references values for a Mediterranean population were taken [37–39]. Anthropometric evaluation included body mass index (BMI) and fat-free mass index (FFMI), which was measured using bioelectrical impedance in all the participants [40,41]. The six-minute walk test was used to determine the exercise capacity in all the patients [42]. The six-minute walk test was conducted along a 30-m indoor flat corridor with no obstacles. Patients were encouraged every minute and were allowed to rest if needed to resume the exercise as soon as they could walk again. The test lasted for six minutes in all the patients independently of whether they had to stop to rest.

Upper (handgrip) and lower (quadriceps) limb muscle function was evaluated in all the participants. As such, the maximum voluntary contraction of the flexor muscles was measured using a specific hand dynamometer (Jamar 030J1, Chicago, IL, USA). Reference values from Luna-Heredia et al. [43] were used. Quadriceps muscle strength was quantified through the determination of the isometric maximum voluntary contraction (QMVC) following standard procedures [44]. Briefly, a fixed handheld dynamometer [MicroFet 2^{TM}, Hoggan Scientific, Salt Lake City, UT, USA] was placed on the tibia and the QMVC was obtained through the exerted compression force. In both tests, three reproducible measurements (<5% variability among them) were obtained in each patient. The highest value out of three maneuvers was selected as the handgrip or QMVC measurements. Reference values from Seymour et al. [45] were used.

2.4. Blood Samples

After an overnight fasting period, blood samples were drawn from the arm vein of each patient. Serum specimens were obtained through the centrifugation of blood samples collected into Vacuette$^®$ serum tubes (with clot activator) at 1600× g for 15 min. Serum samples were immediately stored in the −80 °C freezers until further use.

2.5. Muscle Biopsies

As previously described, specimens from the vastus lateralis were obtained using the open biopsy technique from all the study patients [15,19–21,23]. Muscle samples were immediately frozen in liquid nitrogen and then stored at −80 °C (temperature under alarm control) for the molecular biology experiments. The second half of the muscle specimen was immersed in a series of alcohol baths to be thereafter embedded in paraffin. Paraffin-embedded samples were used for the purpose of the structural analyses carried out on the muscle specimens from all the patients.

2.6. Biological Analyses

Systemic iron status. Conventional analytical parameters were analyzed in the systemic compartment in all the patients: ferritin, transferrin saturation total iron, transferrin, soluble transferrin receptor, hepcidin, hematocrit, hemoglobin, mean corpuscular (erythrocyte) volume (MCV), mean corpuscular hemoglobin (MCH), and mean corpuscular hemoglobin concentration (MCHC).

Hepcidin-25. Levels of hepcidin-25 concentration were quantified in serum samples using the Human hepcidin (Hepc) ELISA kit (Biorbyt, Cambridgeshire, UK) following the manufacturer's instructions and previously described methodologies [14]. In each well of the hepcidin antibody pre-coated microplate, 50 µL 5-fold diluted serum samples or standard and 50 µL HRP-conjugate were poured. Samples were then incubated at 37 °C for 1 h and were washed three times. Moreover, 50 µL of substrate A and 50 µL of substrate B were added and incubated at 37 °C for 15 min. The enzymatic reaction was stopped by adding 50 µL stop solution. Absorbance of each sample was read at a 450 nm wavelength in a plate-reader (Infinite M Nano, Tecan Group Ltd., Zürich, Switzerland). A standard curve was always generated with each assay run. Intra-assay coefficients of variation (CV) for all the samples ranged from 0.1% to 11.0%. The intra-assay CV for all the analyzed markers was calculated as a result of dividing the standard deviation of each replicate value by its mean, which was then multiplied by 100. Standard curves data and intra-assay CVs of all kits used in this study are shown in the Supplementary Materials.

Protein tyrosine nitration. Levels of nitrated proteins were analyzed using the Elabscience® 3-NT(3-Nytrotyrosine) ELISA kit (Elabscience, Houston, TX, USA) in the serum samples of all the patients following standard procedures and previous investigations [14,46]. All reagents were equilibrated at room temperature before the beginning of the ELISA procedures. Thus, 50 µL 7-fold diluted serum samples and standards were added in the corresponding wells of the 3-nitrotyrosine antibody pre-coated microplates. Samples were incubated at 37 °C with the biotinylated antibody for 45 min. After three consecutive washes, HRP conjugate working solution was added to each well, and samples were then incubated at 37 °C for 30 min. Finally, after five additional washes, samples were incubated at 37 °C with the substrate reagent for 15 more minutes. Following this incubation, enzyme substrate reaction was stopped by adding the stop solution. Optical densities in each well were determined by reading the absorbance of the samples at a 450 nm wavelength in a plate-reader (Infinite M Nano, Tecan Group Ltd., Zürich, Switzerland). A standard curve was always generated with each assay run. Intra-assay coefficients of variation for all the samples ranged from 0.27% to 9.81%.

Reactive carbonyls in proteins. The Protein Carbonyl Content Assay Kit (Sigma-Aldrich, St. Louis, MO, USA) was used to assess the levels of reactive carbonyls in serum samples of the study patients following standard procedures published in previous reports [47,48]. The protein concentrations of serum samples in all study patients were determined using the Pierce BCA Protein Assay Kit (Thermo Scientific, Rockford, IL, USA). Samples were diluted with purified water to a protein concentration of 5 mg/mL. All samples and standards were run in duplicate. Samples were incubated at room temperature for 15 min with 10 µL of the 10% Streptozocin solution per 100 µL of sample to degrade the remaining nucleic acids of samples. Following this incubation, samples were centrifuged at 13,000× *g* for 5 min and then the supernatant was transferred to a new tube. Afterwards, 100 µL of DNPH solution was added to each sample and gently vortexed. After an incubation of 10 min at room temperature, 30 µL of the 87% TCA solution was poured to each sample and gently vortexed. Samples were incubated with TCA solution on ice for 5 min and then centrifugated at 13,000× *g* for 2 min. Supernatant was careful removed to not disturbed the pellet and 500 µL of ice-cold acetone was added to each pellet and placed in a sonication bath for 30 s. Subsequently, samples were incubated at −20 °C for 5 min and then centrifuged at 13,000× *g* for 2 min. Acetone were carefully removed from pellet and then 200 µL of 6 M Guanidine solution was poured to pellet and sonicate briefly. Successively, pellets were spined briefly to remove any insolubilized material. Then, 100 µL of each sample was transferred to the 96-well plate to analyze the levels of serum protein carbonylation and 5 µL of the remaining sample was used to determine the protein concentration of each sample. Absorbances were read in each well at a 450 nm wavelength in a plate-reader (Infinite M Nano, Tecan Group Ltd., Zürich, Switzerland) to detect serum levels of protein carbonyls. A standard curve was always generated with each assay run. Intra-assay coefficients of variation for all the samples ranged from 0.24% to 6.43%.

Levels of malondialdehyde-protein adducts. The OxiSelect™ MDA Adduct Competitive ELISA Kit (Cell Biolabs, Inc., San Diego, CA, USA) was used to quantify the levels of MDA-protein adducts in serum samples of the study patients following standard procedures as previously reported [14,49]. Fifty µL serum and MDA-BSA standards were mixed with the MDA conjugate preabsorbed ELISA plate. Samples were then incubated at room temperature for 10 min. Subsequently, the primary antibody was added and incubated at room temperature for one hour. After three washes, samples were incubated with the HRP conjugated secondary antibody at room temperature for an hour. Subsequently, the plate was washed three more times and the substrate solution was incubated at room temperature for 20 min. The enzyme reaction was stopped by adding 100 µL stop solution into each well. Absorbances were read in each well at a 450 nm wavelength in a plate-reader (Infinite M Nano, Tecan Group Ltd., Zürich, Switzerland). A standard curve was always generated with each assay run. Intra-assay coefficients of variation for all the samples ranged from 0.13% to 9.74%.

Activity of superoxide Dismutase (SOD). The Superoxide Dismutase Assay Kit (Cayman chemical, Ann Arbor, MI, USA) was used to quantify the levels of SOD activity in serum samples of all the patients following standard procedures that were reported previously [14,21,49]. Reagents were equilibrated to room temperature before beginning the assay. Serum samples were diluted 1:5 with sample buffer before assaying for SOD activity. Subsequently, 10 µL diluted samples, standards, and 200 µL of the diluted Radical Detector were added in the designated wells on the plates. To initiate the enzymatic reaction, 20 µL xanthine oxidase was added to the wells. Samples were incubated at room temperature for 30 min and the absorbances were read at 440 nm in a plate-reader (Infinite M Nano, Tecan Group Ltd., Zürich, Switzerland). A standard curve was always generated with each assay run. Intra-assay coefficients of variation for all the samples ranged from 0.12% to 8.92%.

Catalase activity. The Catalase Assay Kit (Cayman chemical, Ann Arbor, MN, USA) was used to analyze catalase activity in the serum samples of all the patients following standard procedures that had been previously reported [14,21,49]. All reagents were equilibrated to room temperature before beginning the assay. Briefly, 20 µL samples, standards, and positive control were diluted in 100 µL assay buffer, which were added to the corresponding wells. Subsequently, 30 µL methanol was poured into each well and the reaction took place by adding 20 µL hydrogen peroxide. Following a 20-minute incubation on a shaker at room temperature, 30 µL potassium hydroxide was added to terminate the reaction. Subsequently, 30 µL catalase purpald (chromogen) was incubated with the samples for ten minutes. Finally, 10 µL catalase potassium periodate was incubated with the samples at room temperature for 5 min and the absorbances were read at a 540 nm wavelength in a plate-reader (Infinite M Nano, Tecan Group Ltd., Zürich, Switzerland). A standard curve was always generated with each assay run. Intra-assay coefficients of variation for all the samples ranged from 0.42% to 9.25%.

Levels of reduced glutathione (GSH). The Human Reduced Glutathione (GSH) ELISA Kit (MyBioSource, San Diego, CA, USA) was used to determine GSH levels in serum samples of all the patients following standard procedures that were previously reported [14,46]. All reagents and samples needed to be equilibrated to room temperature (18–25 °C) before starting the procedure. Briefly, 50 µL samples and standards were added to each well and incubated with horseradish (HRP)-conjugate reagent at 37 °C for 60 min. After four washes, 50 µL of chromogen solution A and 50 µL of chromogen solution B were added to each well. Samples were then incubated at 37 °C in the dark for 15 min. Finally, 50 µL of the stop solution was incubated for five minutes, and the absorbance in each sample was read at a 450 nm wavelength in a plate-reader (Infinite M Nano, Tecan Group Ltd., Zürich, Switzerland). A standard curve was always generated with each assay run. Intra-assay coefficients for all the samples ranged from 0.06% to 7.78%.

Levels of Trolox Equivalent Antioxidant Capacity (TEAC). The OxiSelectTM TEAC (TEAC Assay Kit (ABTS) (Cell Biolabs, Inc., San Diego, CA, USA) was used to analyze the total antioxidant capacity of the serum samples in all the patients following standard procedures previously reported [14,50]. Briefly, 25 µL of 20-fold diluted serum samples were added to each well. Subsequently, 150 µL of the diluted 2,2′-azino-bis (3-ethylbenzothiazoline-6-sulfonic acid) (ABTS) reagent was mixed vigorously with the samples to be incubated for five minutes. Finally, the absorbances were read at a 405 nm wavelength in a plate-reader (Infinite M Nano, Tecan Group Ltd., Zürich, Switzerland). A standard curve was always generated with each assay run. Antioxidant activity was determined by comparison with the Trolox standards. Intra-assay coefficients of variation for all the samples ranged from 0.07% to 5.28%.

2.7. Muscle Redox Markers

Tissue homogenization and protein quantification. Protein homogenates were obtained from all the frozen muscle specimens following standard procedures [21,51]. Briefly, frozen specimens of vastus lateralis were mechanically homogenized in specific lysis buffer containing 50 mM 4-(2-hydroxyethyl)-1-piperazineethanesulfonic acid (HEPES), 150 mM NaCl, 100 nM NaF, 10 mM Na pyrophosphate, 5 mM EDTA, 0.5%Triton-X, 2 µg/mL leupeptin, 100 µg/mL phenylmethylsulfonyl fluoride (PMSF), 2 µg/mL aprotinin, and 10 µg/mL pepstatin A using a tissue homogenizer. Samples were kept on ice during the entire procedure. After homogenization, samples were centrifugated at 3600 rpm, at 4 °C for 30 min. Then, pellets were discarded, and supernatants were aliquoted to further protein quantification. The Bradford methodologies were used to analyze protein concentration [21,51].

Immunoblotting procedures. Protein samples (5–20 µg, according to antigen and antibody) were diluted 1:1 with 2X Laemmli buffer (Bio-Rad Laboratories, Inc., Hercules, CA, USA) and 10% of 2-mercaptoethanol (Bio-Rad Laboratories, Inc., Hercules, CA, USA). The samples were then boiled at 95 °C for five minutes to be exposed to electrophoretic separation according to their molecular weight. After the electrophoresis, the proteins were transferred onto polyvinylidene difluoride (PVDF) membranes (Merck KGaA, Darmstadt, Germany) to be blocked with bovine serum albumin (BSA) (NZYTech, Lisbon, Portugal) or 5% nonfat milk, depending on the antibody. Immediately afterwards, the membranes were incubated with the corresponding primary antibodies at 4 °C (cold room) overnight. The following antibodies were used in order to determine the target biomarkers in the study: reactive carbonyls in proteins (protein carbonyl assay kit, Abcam, Cambridge, UK), MDA-protein adducts (anti-MDA-protein adducts antibody, Academy Bio-Chemical Company, Houston, TX, USA), protein tyrosine nitration as measured by levels of 3-nitrotyrosine (anti-3-nitrotyrosine antibody, Thermo Fisher Scientific, Waltham, MA, USA), SOD-1 (anti-SOD-1 antibody, Santa Cruz Biotechnology, Dallas, TX, USA), SOD-2 (anti-SOD-2 antibody, Santa Cruz Biotechnology), catalase (anti-catalase antibody, Merck KGaA, Darmstadt, Germany), and the loading control glyceraldehyde-3-phosphate dehydrogenase (GAPDH, anti-GAPDH antibody, Santa Cruz Biotechnology). The corresponding antigens were detected using HRP-conjugated secondary antibodies (Jackson ImmunoResearch Inc., West Grove, PA, USA) and a chemiluminescence kit (Thermo Scientific, Rockford, IL, USA). The Alliance Q9 Advanced (Uvitec, Cambridge, UK) equipment was used to identify each of the antigens in the PVDF membranes. The forty samples analyzed in the study were always exposed to identical conditions during the entire immunoblotting procedures, including chemiluminescence detection and image capturing. The optical densities of the specific protein bands were quantified using the ImageJ software (National Institute of Health, available at http://rsb.info.nih.gov/ij/, accessed on 1 November 2022). The optical densities were normalized with those of the glycolytic enzyme GAPDH in all the immunoblots. Negative control experiments in which primary antibodies were omitted were also performed in the study. For technical-methodological reasons, nineteen and twenty muscle specimens were analyzed in COPD patients without and with iron deficiency, respectively.

2.8. Muscle Phenotype and Damage

Muscle fiber composition and morphometry. The proteins MyHC-I and -II isoforms were identified to analyze muscle fiber phenotype using immunohistochemistry [40,41,51]. Following a deparaffination process with xylene and a graded ethanol series (rehydration), samples were subject to antigen retrieval. As such, samples were incubated with 1 mM EDTA buffer with 0.05% Tween20 (pH 8.0) at 95 °C for forty minutes, to be cooled down for another thirty minutes. Endogenous peroxidase activity was blocked with 6% H_2O_2. Incubation with the corresponding primary antibodies, anti-MyHCI and anti-MyHCII antibodies (Abcam, Cambridge, UK), respectively, at room temperature for 40 min followed. Sections were than washed three times with PBS. Subsequently, afterwards, slides were incubated with horseradish peroxidase (HRP) Polymer-antiMouse/Rabbit IgG at room temperature for thirty minutes (Neobiotech, Seoul, Republic of Korea). After three more washes with PBS, slides were then incubated with 3,3′-Diaminobenzidine (DAB) (Neobiotech, Seoul, Republic of Korea) solution until the appropriate color (brown) was reached. Subsequently, sections were rinsed with tap water to be counterstained with hematoxylin. Finally, all the muscle sections were dehydrated in a battery of alcohol and xylene solutions to be mounted in dibutylphthalate polystyrene xylene (DPX) media (Sigma-Aldrich, St. Louis, MO, USA). A negative control section (primary antibody omission) was always prepared in each sample. Images were captured using a light microscope (x20 objective, Olympus, Series BX50F3, Olympus Optical Co., Hamburg, Germany) to be analyzed. In all the muscle sections, the following items were assessed: cross-sectional area, mean least diameter, and the proportions of type I, type II, and those of the hybrid muscle fibers. The morphometric analysis was performed using ImageJ software (National Institute of Health, available at http://rsb.info.nih.gov/ij/, accessed on 1 November 2022). In each muscle cross-section, at least 100 fibers were measured in both groups of patients. The proportions of fiber types were also counted in each muscle preparation from both groups of patients. For technical-methodological reasons, fiber morphometry and composition analyses were assessed in eighteen and sixteen muscle specimens from in COPD patients without and with iron deficiency, respectively.

Muscle structural abnormalities. Structural abnormalities were assessed on the 3-μm paraffin-embedded muscle sections following standard procedures [51,52]. Images of the stained sections were captured using a light microscope ($\times 40$ objective, Olympus, Series BX50F3, Olympus Optical Co., Hamburg, Germany). The images were superimposed on a 63-square grid composed of a 7×9 rectangular pattern to analyze the abnormal and normal muscle fractions through the evaluation of several markers: (1) normal muscle, (2) internal nucleus, (3) inflammatory cell, (4) lipofuscin, (5) abnormal viable fiber, (6) inflamed/necrotic fiber, (7) blood vessel, and (0) no count. The normal muscle fraction was calculated as the percentage of all the points that fell into the first category relative to the total number of points counted on the viable fields, whereas the abnormal fraction was defined as the percentage of points that fell between categories 2 and 6 relative to the total number of counted points. Categories 0 and 7 were not considered in the counting.

Terminal deoxynucleotidyl transferase-mediated dUTP nick-end labeling (TUNEL) assay. The TUNEL assay (ApopTag Peroxidase In Situ Apoptosis Detection Kit; Merck Millipore) was used to identify the apoptotic nuclei [53,54]. In brief, DNA strand breaks that are generated during nuclear activation can be identified by labeling the 3′-OH terminal groups. The labeling of the 3′-OH groups with modified nucleotides is carried out by an enzymatic reaction catalyzed by the terminal deoxynucleotidyl transferase (TdT) enzyme. Muscle sections were fixed, permeabilized, and immediately incubated with the TUNEL Working Strength TdT Enzyme and the anti-Digoxigenin Conjugate. TdT catalyzed the adding of digoxigenin-dNTP at 3′-OH terminal groups in single- and double-stranded DNA. After several washes, the digoxigenin-nucleotide that were bound to DNA fragments were detected using an anti-digoxigenin antibody conjugated with peroxidase, which resulted in a brown color upon reaction. Methyl green counterstaining was additionally performed to distinguish the negatively stained nuclei. Negative control experiments, in which the TdT enzyme was not added, were also performed. Only the nuclei located within the muscle fiber boundary were counted in the study. The final count of positive and total nuclei was carried out by two trained observers (correlation coefficient 95%). Apoptotic nuclei were expressed as the percentage of the TUNEL-positive nuclei to the total number of counted nuclei in each muscle preparation [53,54].

2.9. Statistical Analysis

The normality of all variables was analyzed using the Shapiro –Wilk test and histograms. The normality of the study variables determined the test used to assess potential differences between the two groups: the independent-sample Student's *t* test (parametric) and the Mann–Whitney U test (nonparametric). The Chi-squared test was used to analyze differences for the categorical variables between the two groups. Potential associations were assessed using the Pearson's correlation coefficient. Such correlations were explored within each group of patients individually and as whole. Clinical characteristics of the patients are presented in tables, while results from the biological analyses are depicted in graphs. The results are expressed as mean and standard deviation for each variable in each study group. In the tables, results are presented as mean and standard deviation in numbers, while individual data points are shown in the graphs. Moreover, the fiber type distributions in both patient groups are represented in histograms of the percentages of slow- and fast-twitch muscle fibers in a specific figure.

For the calculation of the sample size, the hepcidin was used as the target variable. In a two-sided test, to accept an α-risk of 0.05 and a β-risk of 0.2 (80% power), at least 16 patients in each group were required to detect a minimum difference of 200 ng/mL in hepcidin levels between the two groups. Statistical significance was established at $p \leq 0.05$. All the statistical analyses were performed using the software SPSS 23.0 (SPSS Inc., Chicago, IL, USA).

3. Results

3.1. Clinical Characteristics of the Study Patients

Anthropometry, smoking history, lung function, and GOLD classification were similar in both groups of patients (Table 1). In iron-deficiency patients, serum levels of ferritin, transferrin saturation, and hepcidin were lower, while those of soluble transferrin receptor and transferrin were higher than in the non-iron deficiency patients (Table 1). The six-minute walk test distance was reduced (absolute and reference values) in iron-deficiency COPD patients (Table 2). Muscle strength of the upper and lower limb muscles did not significantly differ between the two study groups (Table 2).

Table 1. General clinical characteristics of the study patients.

	COPD Patients		
	Non-Iron Deficiency	Iron Deficiency	*p*-Value
	N = 20	N = 20	
Anthropometry			
Age (years)	67.3 ± 8.0	65.9 ± 8.1	0.602
Males/Females	12/8	12/8	1.000
Body weight (Kg)	59.72 ± 11.33	63.97 ± 13.07	0.280
BMI (Kg/m^2)	22.73 ± 3.64	24.08 ± 4.28	0.301
FFMI (Kg/m^2)	15.65 ± 2.30	14.72 ± 2.15	0.202
Smoking history			
Smoking status (active/ex-smoker)	9/11	12/8	0.515
Packs-year	53.2 ± 33.4	44.9 ± 24.1	0.406
Lung Function			
FEV$_1$ (L)	1.20 ± 0.37	1.23 ± 0.41	0.759
FEV$_1$ (% predicted)	47.26 ± 13.50	43.42 ± 11.13	0.345
FVC (L)	2.82 ± 0.50	2.76 ± 0.76	0.753
FVC (% predicted)	85.05 ± 14.52	76.74 ± 12.24	0.064
FEV$_1$/FVC	45.19 ± 11.73	46.62 ± 11.44	0.706
GOLD classification			
1, (%)	0	0	
2, (%)	45	25	0.224
3, (%)	45	60	
4, (%)	10	15	
A, (%)	55	45	
B, (%)	35	45	0.699
C, (%)	5	5	
D, (%)	5	5	
Iron status			
Hemoglobin (g/dL)	15.23 ± 1.46	14.99 ± 1.64	0.641
Hematocrit (%)	45.40 ± 4.81	44.97 ± 4.45	0.778
MCV (fL)	93.84 ± 3.56	91.61 ± 6.84	0.214
MCH (pg)	31.52 ± 1.39	30.53 ± 2.78	0.176
MCHC (g/dL)	33.59 ± 1.26	33.31 ± 1.01	0.447
Ferritin (ng/mL)	212.16 ± 60.77	68.52 ± 32.62	0.000
Transferrin saturation (%)	31.45 ± 8.36	24.78 ± 7.02	0.016
Transferrin (g/dL)	237.26 ± 31.65	259.11 ± 28.71	0.032
Soluble transferrin receptor (mg/L)	2.18 ± 0.49	2.99 ± 0.80	0.001
Serum iron (µg/dL)	106.54 ± 26.84	91.44 ± 27.35	0.099
Hepcidin (ng/mL)	411.30 ± 113.92	87.78 ± 68.25	0.000

Data are presented as mean ± SD. Abbreviations: COPD, chronic obstructive pulmonary disease; BMI, body mass index; FFMI, fat-free mass index; N, number of patients; FEV$_1$, forced expiratory volume in one second; FVC, forced vital capacity; MCV, mean corpuscular (erythrocyte) volume; MCH, mean corpuscular hemoglobin; MCHC, mean corpuscular hemoglobin concentration.

Table 2. Exercise and muscle function assessment of the study patients.

	COPD Patients		
	Non-Iron Deficiency	Iron Deficiency	*p*-Value
	N = 20	N = 20	
Six-minute walk test			
Distance (m)	481.67 ± 59.45	435.12 ± 57.82	0.025
Distance (% predicted)	98.61 ± 17.21	84.47 ± 14.56	0.013
Upper limb muscle strength			
D-HGS (Kg)	26.50 ± 6.98	27.78 ± 8.16	0.617
D-HGS (% predicted)	89.91 ± 15.48	91.20 ± 19.02	0.887
ND-HGS (Kg)	24.16 ± 7.36	24.56 ± 8.89	0.825
ND-HGS (% predicted)	91.44 ± 22.32	87.68 ± 19.57	0.595
Lower limb muscle strength			
D-QMVC (Kg)	22.12 ± 6.74	22.58 ± 4.35	0.831
D-QMVC (% predicted)	62.23 ± 22.00	63.81 ± 11.46	0.784
ND-QMVC (Kg)	21.44 ± 5.85	22.00 ± 4.86	0.825
ND-QMVC (% predicted)	60.79 ± 22.07	62.03 ± 12.32	0.858

Data are presented as mean ± SD. Abbreviations: COPD, chronic obstructive pulmonary disease; D, dominant; ND, non-dominant; HGS, handgrip strength; QMVC, quadriceps maximum voluntary contraction.

3.2. Pro-Oxidant Markers in Vastus Lateralis and Blood of COPD Patients

Protein tyrosine nitration levels were greater in both muscle and blood compartments of iron-deficiency than in the non-iron deficiency COPD patients (Figures 1 and 2A). No significant differences were seen in either protein carbonylation or MDA-protein adducts in any of the compartments between the two patient groups (Figures 1 and 2B,C).

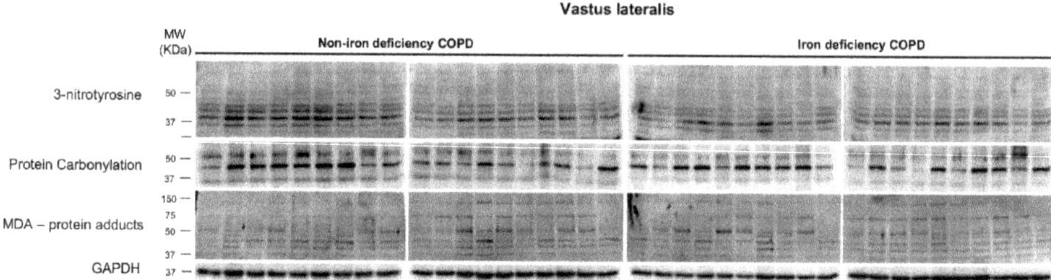

Figure 1. Representative immunoblots of 3-nitrotyrosine, protein carbonylation, MDA-protein adducts, and GAPDH proteins in the vastus lateralis of all COPD patients. Definition of abbreviations: MDA, malonaldehyde; GAPDH, glyceraldehyde-3-phosphate dehydrogenase; MW, molecular weight; kDa, kilodalton.

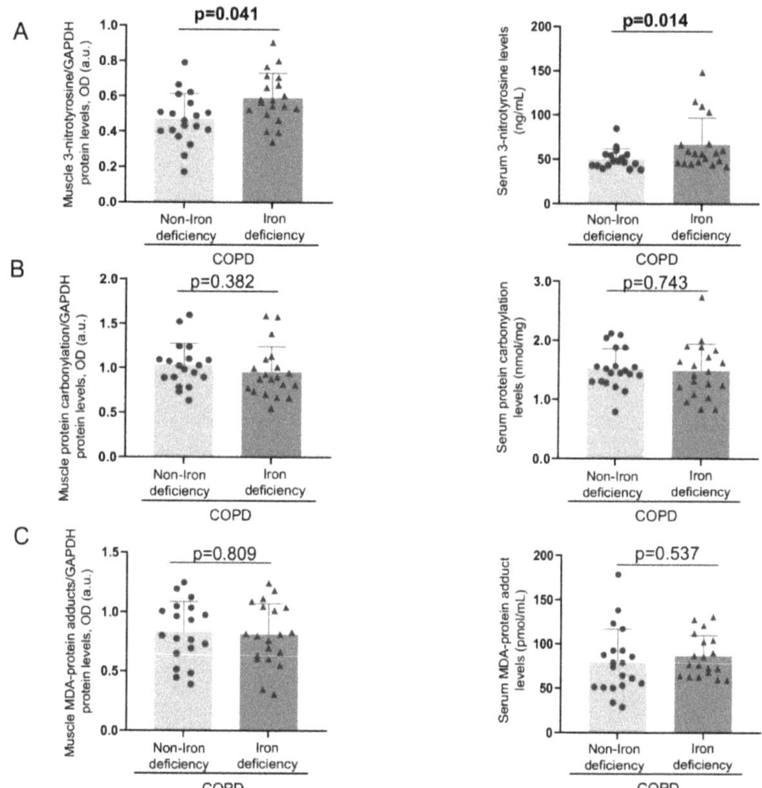

Figure 2. Mean values and standard deviation of 3-nitrotyrosine (**A**), protein carbonylation (**B**), and MDA-protein adducts levels (**C**) in vastus lateralis (left-hand side panels, N = 19 non-iron deficient COPD, N = 20 iron-deficient COPD) and serum (right-hand side panels, N = 20 in each group). Definition of abbreviations: MDA, malonaldehyde; GAPDH, glyceraldehyde-3-phosphate dehydrogenase; MW, molecular weight; kDa, kilodalton. Muscle protein content was measured by optical densities in arbitrary units (OD, a.u.).

3.3. Antioxidants Markers in Vastus Lateralis and Blood of COPD Patients

Protein levels of mitochondrial SOD were significantly lower in the vastus lateralis of iron-deficiency than in the non-iron deficiency patients (Figures 3 and 4A). No significant differences in levels of muscle SOD-1 protein content or serum SOD activity were detected between the two study groups (Figures 3 and 4B,C). Muscle protein content or serum activity levels of catalase did not significantly differ between the two COPD groups of patients (Figure 4D,E). Total antioxidant activity measured as TEAC levels significantly decreased in the iron- deficiency group compared to the non-iron deficiency patients (Figure 5A). Serum antioxidant GSH levels did not significantly differ between the two study groups (Figure 5B).

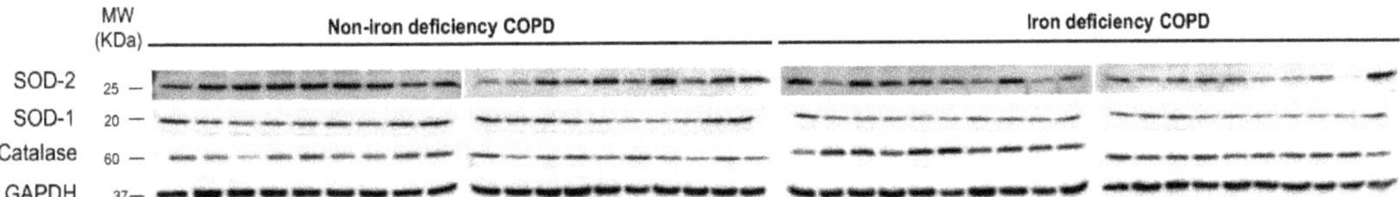

Figure 3. Representative immunoblots of SOD-2, SOD-1, catalase, and GAPDH proteins in the vastus lateralis of all COPD patients. Definition of abbreviations: SOD, superoxide dismutase; GAPDH, glyceraldehyde-3-phosphate dehydrogenase; MW, molecular weight; kDa, kilodalton.

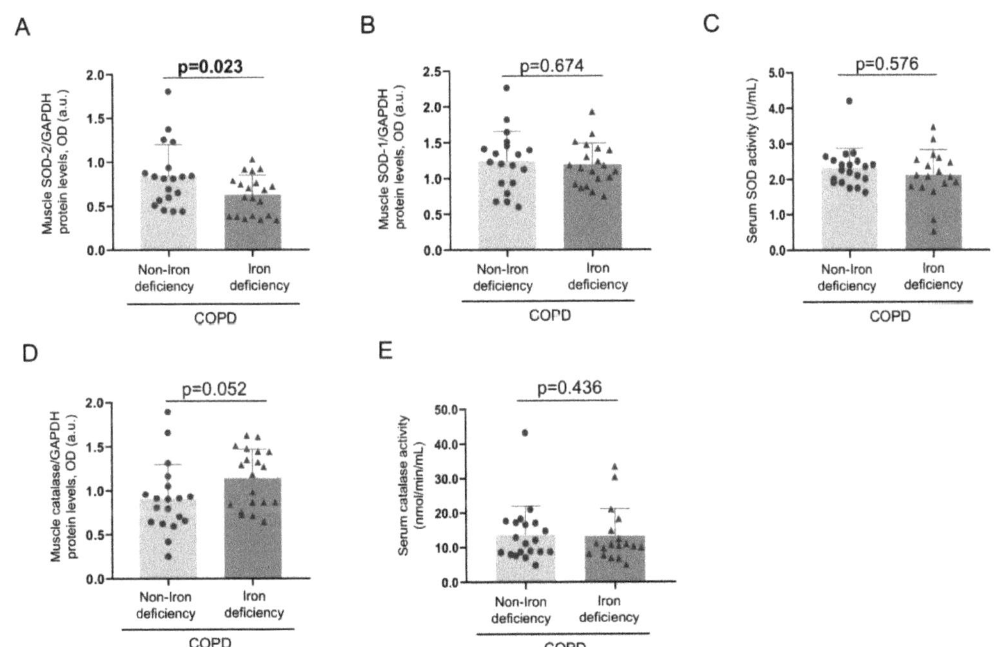

Figure 4. Mean values and standard deviation of muscle protein content of SOD-2 (**A**), SOD-1 (**B**), and serum SOD activity (**C**). Mean values and standard deviation of muscle protein content of catalase (**D**) and serum catalase activity (**E**). Muscle protein content was assessed in 19 non-iron deficiency and 20 iron-deficient COPD patients, while the activity of the target enzymes was analyzed in the serum of all the study patients (n = 20/group). Definition of abbreviations: SOD, superoxide dismutase; GAPDH, glyceraldehyde-3-phosphate dehydrogenase.

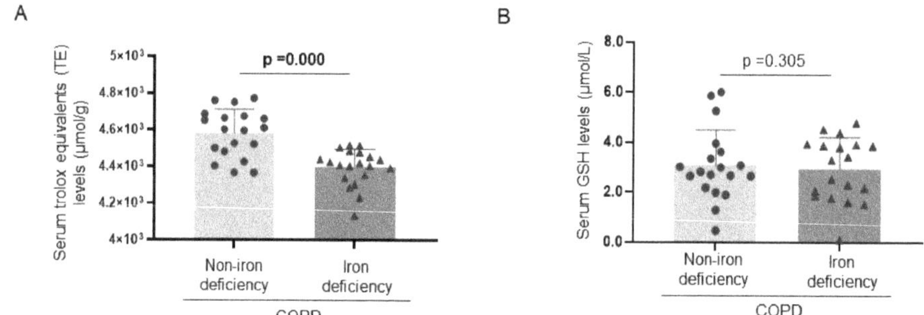

Figure 5. Mean values and standard deviation of serum TEAC levels (**A**) and serum GSH levels (**B**) (N = 20 in each group). Definition of abbreviations: GSH, reduced glutathione.

3.4. Muscle Structural Features of the Study Patients

The proportions of fast-twitch fibers increased in the vastus lateralis of the iron-deficiency compared to the non-iron deficiency patients, whereas the percentages of the slow-twitch fibers significantly declined in the former group (Table 3, Figure 6A,B). The cross-sectional area of the analyzed muscle fibers did not significantly differ between the two study groups (Table 3, Figure 6A). The proportions or size of the hybrid fibers did not significant differ between the two study groups. The proportions of the slow-twitch fibers positively correlated with muscle SOD-2 protein content among all COPD patients (Figure 7A). This correlation was lost when non-iron deficiency COPD patients were analyzed independently, whereas in the iron-deficiency patients, this correlation was almost maintained (r = 0.481 and p = 0.059). The proportions of fast-twitch fibers negatively correlated with muscle SOD-2 protein levels among all COPD patients and in the iron-deficiency group (Figure 7B). Such a correlation was lost in the non-iron deficiency group of patients (Figure 7B). A significant increase in lipofuscin levels was detected in the vastus lateralis of the iron-deficiency COPD compared to the non-iron deficiency patients (Table 3, Figure 8). No significant differences were observed in the markers of muscle damage: total abnormal fraction, internal nuclei count, inflammatory cells, abnormal cells, necrotic cells, or apoptotic nuclei between the study groups.

Table 3. Structural characteristics of vastus lateralis of the study patients.

	COPD Patients		
	Non-Iron Deficiency	**Iron Deficiency**	***p*-Value**
	N = 18	**N = 16**	
Muscle fiber type proportions			
Type I fibers (%)	27.36 ± 7.70	20.04 ± 7.84	0.008
Type II fibers (%)	66.52 ± 10.11	76.85 ± 8.97	0.003
Hybrid fibers (%)	6.12 ± 7.01	3.11 ± 2.83	0.109
Cross-sectional fiber type areas			
Type I fibers (μm^2)	2661.72 ± 717.11	2837.05 ± 889.59	0.539
Type II fibers (μm^2)	1886.94 ± 645.03	1970.29 ± 781.53	0.744
Hybrid fibers (μm^2)	1973.64 ± 987.46	2179.66 ± 1161.20	0.639
Muscle structural abnormalities	N = 20	N = 20	
Total abnormal fraction (%)	1.38 ± 0.81	1.70 ± 0.76	0.215
Internal nuclei count (%)	0.90 ± 0.42	1.12 ± 0.74	0.237
Inflammatory cells (%)	0.08 ± 0.08	0.10 ± 0.08	0.538
Lipofuscin (%)	0.01 ± 0.03	0.07 ± 0.10	0.014
Abnormal cells (%)	0.11 ± 0.15	0.15 ± 0.16	0.321
Necrotic cells (%)	0.30 ± 0.61	0.23 ± 0.40	0.698
Apoptotic nuclei (%)	53.48 ± 7.70	57.43 ± 8.81	0.139

Data are presented as mean ± SD. Abbreviations: COPD, chronic obstructive pulmonary disease; CSA, cross-sectional fiber area.

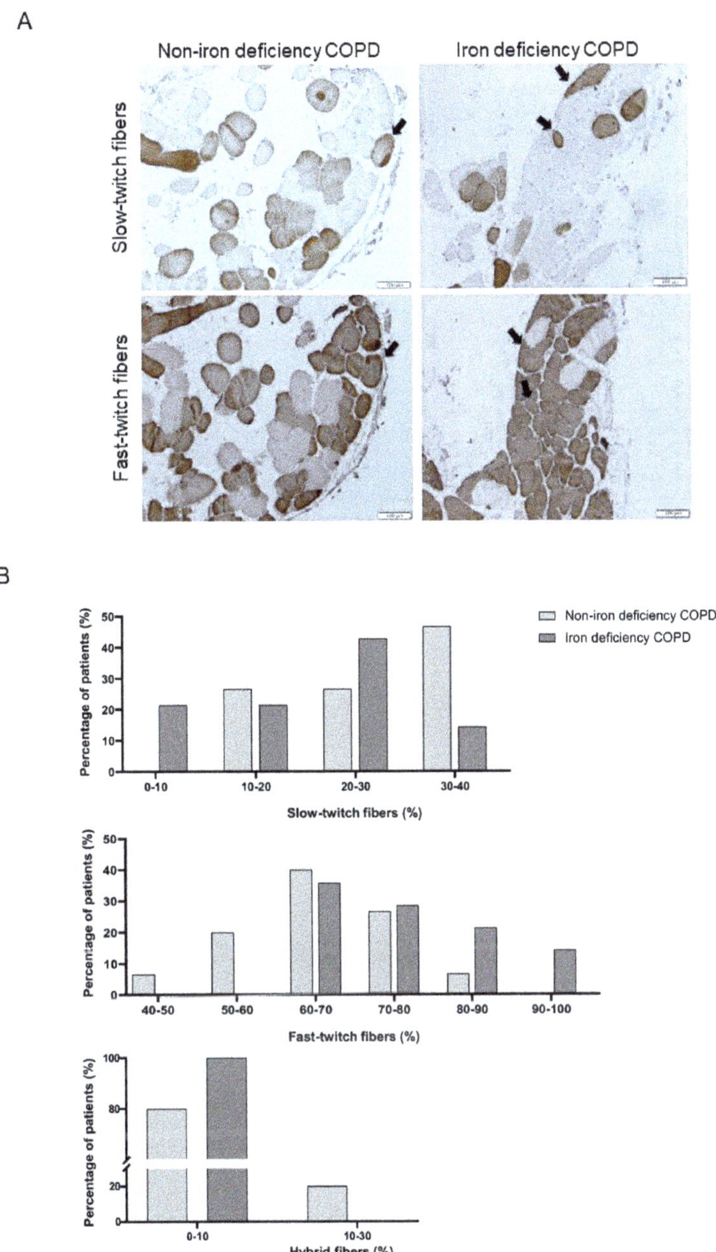

Figure 6. Representative images of vastus lateralis muscle fibers. Myofibers with anti-MyHC type I antibody staining (brown color) are shown in the top panel and anti-MyHC type II antibody staining in the bottom panel. Hybrid fibers (arrows) are seen in both panels. Scale bar = 100 μm, 20× magnification (**A**). Histograms of the percentages of patients (ordinate axis) of the fiber type distributions (abscissa axis by ranges) of slow-twitch (top panel), fast-twitch (middle panel), and hybrid (bottom panel) muscle fibers in the two study groups of patients (N = 18 non-iron deficient COPD, N = 16 iron-deficient COPD) (**B**).

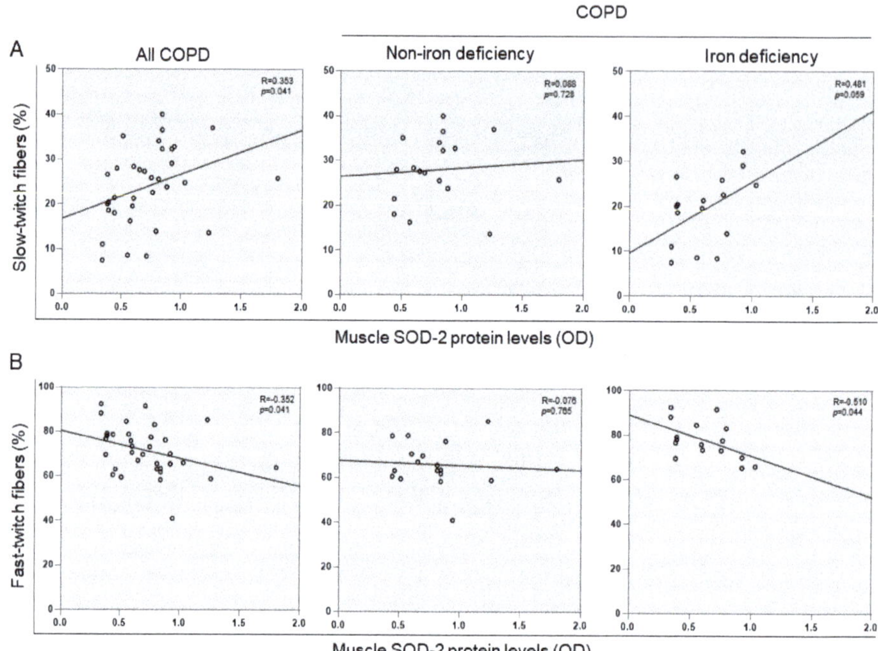

Figure 7. Scatter plots representation of correlations between muscle SOD-2 protein levels and the percentage of slow-twitch fibers (**A**) and fast-twitch fibers (**B**) in all COPD patients (left panels, N = 34), non-iron deficiency (middle panels, N = 18), and iron deficiency (right panels, N = 16) COPD patients. Definition of abbreviations: COPD, chronic obstructive pulmonary disease.

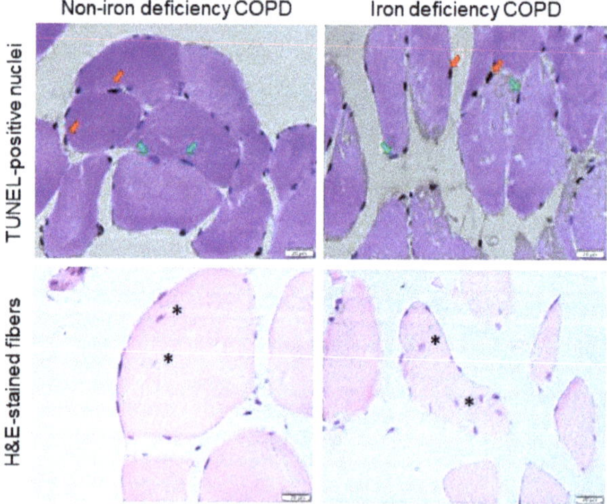

Figure 8. Representative images of TUNEL-positive nuclei (**top panel**) and muscle morphology within the vastus lateralis (**bottom panel**). Examples of TUNEL-positively stained nuclei (brown, red arrows), TUNEL-negative nuclei (purple, blue arrows), and internal nuclei (asterisks) are indicated in the panels. Definition of abbreviations: COPD, chronic obstructive pulmonary disease; TUNEL, Terminal deoxynucleotidyl transferase dUTP nick end labeling; H&E, hematoxylin and eosin. Scale bar = 20 µm, 40× magnification.

4. Discussion

The study hypothesis has been confirmed in this study. Oxidative/nitrosative stress levels were greater in the muscle and blood compartments of COPD patients with iron deficiency. Importantly, levels of the powerful antioxidants SOD and TEAC were reduced in the vastus lateralis and the systemic compartment of the patients with iron depletion compared to those with normal iron content. The population of COPD patients recruited in the present study has been well characterized as determined by the lower levels observed in the markers of iron metabolism among patients with iron deficiency, whereas in patients with preserved iron content, the levels of those biomarkers were within normal ranges. The predominance of fast-twitch fibers within the vastus lateralis of the iron-depleted patients (almost 80%) compared to the control COPD patients was also a novel finding in the present investigation. All these results are discussed below.

The assessment of the systemic and muscle compartments from the same patients was a relevant approach in the study. Consistently, levels of protein tyrosine nitration as measured by the 3-nitrotyrosine marker were significantly greater in the lower limb muscles and systemic compartments of COPD patients with iron deficiency as compared to patients with normal iron content. These are relevant findings that point towards the existence of a crosstalk between the two compartments as also previously reported in a study [15] in which a similar approach was used. The confirmation of the crosstalk between these two specific compartments has clinical implications, since therapeutic strategies will be able to target the two compartments simultaneously.

Other relevant findings in the investigation were the significant decrease observed in the powerful antioxidants SOD and TEAC. SOD scavenges superoxide anion to produce hydrogen peroxide, a substrate for catalase enzyme activity. An excess of superoxide anion production leads to the generation of the powerful reactive nitrogen species (RNS) peroxynitrite resulting from its reaction with nitric oxide (NO). The reaction of peroxynitrite with aromatic amino acids (i.e., tyrosine) results in the formation of protein tyrosine nitration events in tissues [55,56]. Indeed, protein tyrosine nitration is the most relevant marker of nitrosative stress in vivo with a potential to inactivate enzymes or prevent phosphorylation of tyrosine kinase substrates [57]. The decrease in mitochondrial SOD observed in the vastus lateralis and blood compartment of the iron-deficiency patients may account for the increased levels of protein tyrosine nitration observed in the same compartments in these patients. These are very consistent findings from a biochemical standpoint. The mechanisms whereby iron depletion influences mitochondrial SOD content in both muscle and the systemic compartments warrant further attention in future investigations.

The biomarker TEAC measures the antioxidant capacity based on the compound Trolox (a water-soluble analog of vitamin E). Trolox equivalency is calculated on the basis of the ABTS (2,2′-azino-bis [3-ethylbenzothiazoline-6-sulfonic acid]) decolorization assay [58,59]. In the systemic compartment of the iron-depleted patients, TEAC levels were reduced compared to patients with normal iron content. As far as we are concerned, this is the first report in which TEAC levels were shown to be decreased in COPD patients with iron deficiency without anemia. Previous investigations also demonstrated a significant decrease in TEAC levels in the blood of patients with anemia of different etiology [60,61]. These findings are also in line with the decrease seen in the levels of mitochondrial SOD in the muscles of the patients with iron deficiency. Collectively, these results suggest that in stable COPD patients, iron deficiency is associated with a depleted antioxidant capacity in both muscle and systemic compartments.

The compound lipofuscin is a marker of unsaturated fatty acid oxidation, particularly of damaged membranes, mitochondria, and lysosomes. Additionally, lipofuscin may also contain sugars and metals such as iron, mercury, copper, and zinc [62]. Oxidized proteins may also be identified in the lipofuscin aggregates [63]. In general, lipofuscin inclusions are considered to be a marker of lipid peroxidation in tissues. They tend to accumulate in postmitotic cells as a marker of aging [64]. In the vastus lateralis of patients with severe COPD, lipofuscin aggregates were also significantly greater than in control subjects [65].

Consistently, in the current study, lipofuscin inclusions were significantly greater in muscles of the iron-deficient patients than in those of patients with normal iron content, suggesting that lipid peroxidation was probably more prominent among patients with iron deficiency. Whether the lipofuscin aggregates may contain higher levels of iron in these patients needs to be explored in future investigations.

A switch from the slow- to fast-twitch muscle phenotype has been consistently demonstrated in vastus lateralis patients with COPD, particularly in those with a severe disease [51,66,67]. Several factors, such as reduced physical activity, deconditioning, and malnourishment, play a crucial role in the development of a less fatigue-resistant phenotype of the lower limb muscles in COPD patients and in other chronic diseases [68–72]. In the present investigation, the switch to a less resistant phenotype was significantly more pronounced in patients with iron deficiency than in patients with normal iron content. This finding was independent of the muscle function, as no differences were seen in quadriceps muscle function between the two study groups. Furthermore, a significant positive correlation was observed between muscle levels of mitochondrial SOD and the proportions of slow-twitch fibers in all the patients as a whole and in the iron-deficient group. Interestingly, inverse relationships were also seen between protein levels of SOD2 and those of the fast-twitch muscle fibers among all the COPD patients as a group and in those with iron depletion. These are all relevant findings that suggest that mitochondrial SOD expression follows a specific fiber type distribution in the lower limb muscles of COPD patients and iron may be involved in the underlying mechanism. In fact, a decrease in the proportions of oxidative fibers was also reported in mice with iron deficiency [73], suggesting that the element iron is deeply involved in the maintenance of an adequate proportion of oxidative fibers within the muscles.

5. Study Limitations

A potential limitation in the study relies on its descriptive nature, which is based on the design of a cross-sectional investigation. Treatment with iron for several months may partly revert the effects seen in the systemic and muscle compartments of the patients. This would have required the design of a longitudinal study in which patients should be evaluated at different time-points, with the difficulties at the time of obtaining the biological specimens. Despite these potential limitations, we believe that the current investigation opens a new avenue of research in COPD patients with iron depletion. In addition, the model used in the study has enabled us to analyze the systemic and muscle compartments in the same population of COPD patients.

6. Conclusions

In severe COPD, nitrosative and oxidative stress along with a reduction in the antioxidant capacity were demonstrated in the vastus lateralis and systemic compartments of patients with iron deficiency without anemia. The slow- to fast-twitch muscle fiber switch towards a less resistant phenotype was also significantly more prominent in the muscles of these patients, suggesting that iron content may be involved in this phenomenon. Iron deficiency is associated with a specific pattern of oxidative and nitrosative stress and reduced antioxidant capacity in severe COPD irrespective of muscle function of the lower limbs. In the clinics, parameters of iron metabolism and content should be routinely quantify given its implications in redox balance and exercise tolerance. These results should also be taken into consideration in the design of therapeutic strategies in severe COPD, including pulmonary rehabilitation programs.

Supplementary Materials: The following supporting information can be downloaded at: https://www.mdpi.com/article/10.3390/nu15061454/s1; Figure S1: Standard curve Elabscience®3-NT(3-Nytrotyrosine) ELISA kit; Figure S2: Standard curve— Protein Carbonyl Content Assay Kit; Figure S3: Standard curve— OxiSelectTM MDA Adduct Competitive ELISA Kit; Figure S4: Standard curve—Superoxide Dismutase Assay Kit; Figure S5: Standard curve—Catalase Assay kit; Figure S6: Standard curve—The OxiSelectTM TEAC (TEAC Assay Kit (ABTS); Figure S7: Standard curve—The human Reduced Glutathione (GSH) ELISA Kit; Figure S8: Standard curve—The Human hepcidin (Hepc) ELISA kit; Figure S9: Intra-assay coefficients of variation.

Author Contributions: Study conception and design: E.B., D.A.R.-C. and M.P.-P.; Patients recruitment: M.A.M. and C.M.-O.; Assessment of functional parameters: M.A.M. and M.P.-P.; Laboratory experiments: M.P.-P.; Statistical analyses and data interpretation: M.P.-P. and E.B.; manuscript drafting and intellectual input: M.P.-P., D.A.R.-C. and E.B.; manuscript writing final version: E.B. All authors have read and agreed to the published version of the manuscript.

Funding: María Pérez-Peiró was a recipient of a predoctoral fellowship from FIS Contratos Predoctorales de Formación en Investigación en Salud del Instituto de Salud Carlos III (ISC-III, FI19/00001) (Co-funded by European Social Fund "Investing in your future"). Mariela Alvarado was a recipient of a Contrato de Intensificación from FIS (ISC-III, INT19/00002). The research conducted in the study has been funded by Project FIS 21/00215 funded by ISC-III and co-funded by the European Union, CIBER—Consorcio Centro de Investigación Biomédica en Red—(CIBERES), Instituto de Salud Carlos III, Spanish Ministry of Science and Innovation and European Union, Spanish Respiratory Society (SEPAR), grant SEPAR 2020, and an unrestricted grant from Vifor Pharma 2018. Vifor Pharma did not have any role in the study or statistical analysis.

Institutional Review Board Statement: The research followed the guidelines of the World Medical Association for Research in Humans (Seventh revision of the Declaration of Helsinki, Fortaleza, Brazil, 2013) and the guidelines established by the International Committee of Medical Journal Editors (ICMJE) The editorial group MDPI strictly follows the guidelines recommended by ICMJE. The study was approved in 17 January 2018 by the local Ethics Committee at Hospital del Mar (CEIm Parc de Salut Mar, registration # 2017/7691/I).

Informed Consent Statement: All patients signed the informed written consent to participate in the study.

Data Availability Statement: The datasets generated and analyzed during the current study are available from the corresponding author on reasonable request.

Conflicts of Interest: The authors have none to disclose regarding this study.

References

1. Machado, A.; Marques, A.; Burtin, C. Extra-pulmonary manifestations of COPD and the role of pulmonary rehabilitation: A symptom-centered approach. *Expert Rev. Respir. Med.* **2020**, *15*, 131–142. [CrossRef]
2. Kaluźniak-Szymanowska, A.; Krzymińska-Siemaszko, R.; Deskur-śmielecka, E.; Lewandowicz, M.; Kaczmarek, B.; Wieczorowska-Tobis, K. Malnutrition, Sarcopenia, and Malnutrition-Sarcopenia Syndrome in Older Adults with COPD. *Nutrients* **2021**, *14*, 44. [CrossRef] [PubMed]
3. Jaitovich, A.; Barreiro, E. Skeletal muscle dysfunction in chronic obstructive pulmonary disease what we know and can do for our patients. *Am. J. Respir. Crit. Care Med.* **2018**, *198*, 175–186. [CrossRef] [PubMed]
4. Winter, W.E.; Bazydlo, L.A.L.; Harris, N.S. The molecular biology of human iron metabolism. *Lab. Med.* **2014**, *45*, 92–102. [CrossRef] [PubMed]
5. Abbaspour, N.; Hurrell, R.; Kelishadi, R. Review on iron and its importance for human health. *J. Res. Med. Sci.* **2014**, *19*, 164.
6. Paul, B.T.; Manz, D.H.; Torti, F.M.; Torti, S.V. Mitochondria and Iron: Current Questions. *Expert Rev. Hematol.* **2017**, *10*, 65. [CrossRef]
7. Lukaszyk, E.; Lukaszyk, M.; Koc-Zorawska, E.; Bodzenta-Lukaszyk, A.; Malyszko, J. GDF-15, iron, and inflammation in early chronic kidney disease among elderly patients. *Int. Urol. Nephrol.* **2016**, *48*, 839–844. [CrossRef]
8. González-D'Gregorio, J.; Miñana, G.; Núñez, J.; Núñez, E.; Ruiz, V.; García-Blas, S.; Bonanad, C.; Mollar, A.; Valero, E.; Amiguet, M.; et al. Iron deficiency and long-term mortality in elderly patients with acute coronary syndrome. *Biomark. Med.* **2018**, *12*, 987–999. [CrossRef]
9. Chopra, V.K.; Anker, S.D. Anaemia, iron deficiency and heart failure in 2020: Facts and numbers. *ESC Heart Fail.* **2020**, *7*, 2007–2011. [CrossRef]

10. Cloonan, S.M.; Mumby, S.; Adcock, I.M.; Choi, A.M.K.; Chung, K.F.; Quinlan, G.J. The iron-y of iron overload and iron deficiency in chronic obstructive pulmonary disease. *Am. J. Respir. Crit. Care Med.* **2017**, *196*, 1103–1112. [CrossRef]

11. Silverberg, D.S.; Mor, R.; Weu, M.T.; Schwartz, D.; Schwartz, I.F.; Chernin, G. Anemia and iron deficiency in COPD patients: Prevalence and the effects of correction of the anemia with erythropoiesis stimulating agents and intravenous iron. *BMC Pulm. Med.* **2014**, *14*, 24. [CrossRef] [PubMed]

12. Rathi, V.; Ish, P.; Singh, G.; Tiwari, M.; Goel, N.; Gaur, S.N. Iron deficiency in non-anemic chronic obstructive pulmonary disease in a predominantly male population: An ignored entity. *Monaldi Arch. Chest Dis./Arch. Monaldi Mal. Torace* **2020**, *90*, 38–42. [CrossRef]

13. Martín-Ontiyuelo, C.; Rodó-Pin, A.; Echeverría-Esnal, D.; Admetlló, M.; Duran-Jordà, X.; Alvarado, M.; Gea, J.; Barreiro, E.; Rodríguez-Chiaradía, D.A. Intravenous Iron Replacement Improves Exercise Tolerance in COPD: A Single-Blind Randomized Trial. *Arch. Bronconeumol.* **2021**, *58*, 689–698. [CrossRef]

14. Pérez-Peiró, M.; Martín-Ontiyuelo, C.; Rodó-Pi, A.; Piccari, L.; Admetlló, M.; Durán, X.; Rodríguez-Chiaradía, D.A.; Barreiro, E. Iron Replacement and Redox Balance in Non-Anemic and Mildly Anemic Iron Deficiency COPD Patients: Insights from a Clinical Trial. *Biomedicines* **2021**, *9*, 1191. [CrossRef] [PubMed]

15. Pérez-Peiró, M.; Alvarado, M.; Martín-Ontiyuelo, C.; Duran, X.; Rodríguez-Chiaradía, D.A.; Barreiro, E. Iron Depletion in Systemic and Muscle Compartments Defines a Specific Phenotype of Severe COPD in Female and Male Patients: Implications in Exercise Tolerance. *Nutrients* **2022**, *14*, 3929. [CrossRef] [PubMed]

16. Barberan-Garcia, A.; Rodríguez, D.A.; Blanco, I.; Gea, J.; Torralba, Y.; Arbillaga-Etxarri, A.; Barberà, J.A.; Vilaró, J.; Roca, J.; Orozco-Levi, M. Non-anaemic iron deficiency impairs response to pulmonary rehabilitation in COPD. *Respirology* **2015**, *20*, 1089–1095. [CrossRef] [PubMed]

17. Nickol, A.H.; Frise, M.C.; Cheng, H.Y.; McGahey, A.; McFadyen, B.M.; Harris-Wright, T.; Bart, N.K.; Curtis, M.K.; Khandwala, S.; O'Neill, D.P.; et al. A cross-sectional study of the prevalence and associations of iron deficiency in a cohort of patients with chronic obstructive pulmonary disease. *BMJ Open* **2015**, *5*, e007911. [CrossRef]

18. Barreiro, E.; Gea, J. Molecular and biological pathways of skeletal muscle dysfunction in chronic obstructive pulmonary disease. *Chron. Respir. Dis.* **2016**, *13*, 297–311. [CrossRef]

19. Barreiro, E.; Gea, J.; Corominas, J.M.; Hussain, S.N.A. Nitric oxide synthases and protein oxidation in the quadriceps femoris of patients with chronic obstructive pulmonary disease. *Am. J. Respir. Cell Mol. Biol.* **2003**, *29*, 771–778. [CrossRef]

20. Barreiro, E.; Peinado, V.I.; Galdiz, J.B.; Ferrer, E.; Marin-Corral, J.; Sánchez, F.; Gea, J.; Barberà, J.A. Cigarette smoke-induced oxidative stress: A role in chronic obstructive pulmonary disease skeletal muscle dysfunction. *Am. J. Respir. Crit. Care Med.* **2010**, *182*, 477–488. [CrossRef]

21. Puig-Vilanova, E.; Rodriguez, D.A.; Lloreta, J.; Ausin, P.; Pascual-Guardia, S.; Broquetas, J.; Roca, J.; Gea, J.; Barreiro, E. Oxidative stress, redox signaling pathways, and autophagy in cachectic muscles of male patients with advanced COPD and lung cancer. *Free. Radic. Biol. Med.* **2015**, *79*, 91–108. [CrossRef] [PubMed]

22. Marin-Corral, J.; Fontes, C.C.; Pascual-Guardia, S.; Sanchez, F.; Olivan, M.; Argilés, J.M.; Busquets, S.; López-Soriano, F.J.; Barreiro, E. Redox balance and carbonylated proteins in limb and heart muscles of cachectic rats. *Antioxid. Redox Signal.* **2010**, *12*, 365–380. [CrossRef] [PubMed]

23. Fermoselle, C.; Rabinovich, R.; Ausín, P.; Puig-Vilanova, E.; Coronell, C.; Sanchez, F.; Roca, J.; Gea, J.; Barreiro, E. Does oxidative stress modulate limb muscle atrophy in severe COPD patients? *Eur. Respir. J.* **2012**, *40*, 851–862. [CrossRef]

24. Wu, P.; Wang, A.; Cheng, J.; Chen, L.; Pan, Y.; Li, H.; Zhang, Q.; Zhang, J.; Chu, W.; Zhang, J. Effects of Starvation on Antioxidant-Related Signaling Molecules, Oxidative Stress, and Autophagy in Juvenile Chinese Perch Skeletal Muscle. *Mar. Biotechnol.* **2020**, *22*, 81–93. [CrossRef]

25. O'Leary, M.F.N.; Hood, D.A. Denervation-induced oxidative stress and autophagy signaling in muscle. *Autophagy* **2009**, *5*, 230–231. [CrossRef] [PubMed]

26. Sadeghi, A.; Shabani, M.; Alizadeh, S.; Meshkani, R. Interplay between oxidative stress and autophagy function and its role in inflammatory cytokine expression induced by palmitate in skeletal muscle cells. *Cytokine* **2020**, *125*, 154835. [CrossRef] [PubMed]

27. Galaris, D.; Barbouti, A.; Pantopoulos, K. Iron homeostasis and oxidative stress: An intimate relationship. *Biochim. Biophys. Acta Mol. Cell Res.* **2019**, *1866*, 118535. [CrossRef]

28. Yoo, J.H.; Maeng, H.Y.; Sun, Y.K.; Kim, Y.A.; Park, D.W.; Tae, S.P.; Seung, T.L.; Choi, J.R. Oxidative status in iron-deficiency anemia. *J. Clin. Lab. Anal.* **2009**, *23*, 319–323. [CrossRef]

29. Zhang, H.; Jamieson, K.L.; Grenier, J.; Nikhanj, A.; Tang, Z.; Wang, F.; Wang, S.; Seidman, J.G.; Seidman, C.E.; Thompson, R.; et al. Myocardial Iron Deficiency and Mitochondrial Dysfunction in Advanced Heart Failure in Humans. *J. Am. Heart Assoc.* **2022**, *11*, e022853. [CrossRef]

30. Choi, J.; Pai, S.; Kim, S.; Ito, M.; Park, C.; Cha, Y. Iron deficiency anemia increases nitric oxide production in healthy adolescents. *Ann. Hematol.* **2002**, *81*, 1–6. [CrossRef]

31. Inserte, J.; Barrabés, J.A.; Aluja, D.; Otaegui, I.; Bañeras, J.; Castellote, L.; Sánchez, A.; Rodríguez-Palomares, J.F.; Pineda, V.; Miró-Casas, E.; et al. Implications of Iron Deficiency in STEMI Patients and in a Murine Model of Myocardial Infarction. *JACC Basic Transl. Sci.* **2021**, *6*, 567–580. [CrossRef]

32. Sharif Usman, S.; Dahiru, M.; Abdullahi, B.; Abdullahi, S.B.; Maigari, U.M.; Ibrahim Uba, A. Status of malondialdehyde, catalase and superoxide dismutase levels/activities in schoolchildren with iron deficiency and iron-deficiency anemia of Kashere and its environs in Gombe State, Nigeria. *Heliyon* **2019**, *5*, e02214. [CrossRef] [PubMed]
33. Venkatesan, P. GOLD COPD report: 2023 update. *Lancet Respir. Med.* **2023**, *11*, 18. [CrossRef]
34. Venkatesan, P. GOLD report: 2022 update. *Lancet Respir. Med.* **2022**, *10*, e20. [CrossRef]
35. Anker, S.D.; Comin Colet, J.; Filippatos, G.; Willenheimer, R.; Dickstein, K.; Drexler, H.; Lüscher, T.F.; Bart, B.; Banasiak, W.; Niegowska, J.; et al. Ferric carboxymaltose in patients with heart failure and iron deficiency. *N. Engl. J. Med.* **2009**, *361*, 2436–2448. [CrossRef]
36. Shrestha, B.; Dunn, L. The Declaration of Helsinki on Medical Research involving Human Subjects: A Review of Seventh Revision. *J. Nepal. Health Res. Counc.* **2020**, *17*, 548–552. [CrossRef] [PubMed]
37. Roca, J.; Vargas, C.; Cano, I.; Selivanov, V.; Barreiro, E.; Maier, D.; Falciani, F.; Wagner, P.; Cascante, M.; Garcia-Aymerich, J.; et al. Chronic Obstructive Pulmonary Disease heterogeneity: Challenges for health risk assessment, stratification and management. *J. Transl. Med.* **2014**, *12*, S3. [CrossRef] [PubMed]
38. Roca, J.; Burgos, F.; Barberà, J.A.; Sunyer, J.; Rodriguez-Roisin, R.; Castellsagué, J.; Sanchis, J.; Antóo, J.M.; Casan, P.; Clausen, J.L. Prediction equations for plethysmographic lung volumes. *Respir. Med.* **1998**, *92*, 454–460. [CrossRef] [PubMed]
39. Roca, J.; Burgos, F.; Sunyer, J.; Saez, M.; Chinn, S.; Anto, J.; Rodriguez-Roisin, R.; Quanjer, P.; Nowak, D.; Burney, P. References values for forced spirometry. Group of the European Community Respiratory Health Survey. *Eur. Respir. J.* **1998**, *11*, 1354–1362. [CrossRef]
40. Barreiro, E.; Salazar-Degracia, A.; Sancho-Muñoz, A.; Gea, J. Endoplasmic reticulum stress and unfolded protein response profile in quadriceps of sarcopenic patients with respiratory diseases. *J. Cell Physiol.* **2019**, *234*, 11315–11329. [CrossRef]
41. Puig-Vilanova, E.; Martínez-Llorens, J.; Ausin, P.; Roca, J.; Gea, J.; Barreiro, E. Quadriceps muscle weakness and atrophy are associated with a differential epigenetic profile in advanced COPD. *Clin. Sci.* **2015**, *128*, 905–921. [CrossRef]
42. Rodriguez, D.A.; Kalko, S.; Puig-Vilanova, E.; Perez-Olabarría, M.; Falciani, F.; Gea, J.; Cascante, M.; Barreiro, E.; Roca, J. Muscle and blood redox status after exercise training in severe COPD patients. *Free Radic. Biol. Med.* **2012**, *52*, 88–94. [CrossRef]
43. Luna-Heredia, E.; Martín-Peña, G.; Ruiz-Galiana, J. Handgrip dynamometry in healthy adults. *Clin. Nutr.* **2005**, *24*, 250–258. [CrossRef]
44. Bui, K.L.; Mathur, S.; Dechman, G.; Maltais, F.; Camp, P.; Saey, D. Fixed Handheld Dynamometry Provides Reliable and Valid Values for Quadriceps Isometric Strength in People With Chronic Obstructive Pulmonary Disease: A Multicenter Study. *Phys. Ther.* **2019**, *99*, 1255–1267. [CrossRef] [PubMed]
45. Seymour, J.M.; Spruit, M.A.; Hopkinson, N.S.; Natanek, S.A.; Man, W.D.C.; Jackson, A.; Gosker, H.R.; Schols, A.M.W.J.; Moxham, J.; Polkey, M.I.; et al. The prevalence of quadriceps weakness in COPD and the relationship with disease severity. *Eur. Respir. J.* **2010**, *36*, 81–88. [CrossRef] [PubMed]
46. Mateu-Jiménez, M.; Sánchez-Font, A.; Rodríguez-Fuster, A.; Aguiló, R.; Pijuan, L.; Fermoselle, C.; Gea, J.; Curull, V.; Barreiro, E. Redox imbalance in lung cancer of patients with underlying chronic respiratory conditions. *Mol. Med.* **2016**, *22*, 85–98. [CrossRef] [PubMed]
47. Thiele, R.H.; Osuru, H.P.; Paila, U.; Ikeda, K.; Zuo, Z. Impact of inflammation on brain subcellular energetics in anesthetized rats. *BMC Neurosci.* **2019**, *20*, 34. [CrossRef]
48. González-Ruiz, R.; Peregrino-Uriarte, A.B.; Valenzuela-Soto, E.M.; Cinco-Moroyoqui, F.J.; Martínez-Téllez, M.A.; Yepiz-Plascencia, G. Mitochondrial manganese superoxide dismutase knock-down increases oxidative stress and caspase-3 activity in the white shrimp Litopenaeus vannamei exposed to high temperature, hypoxia, and reoxygenation. *Comp. Biochem. Physiol. Part A Mol. Integr. Physiol.* **2021**, *252*, 110826. [CrossRef]
49. Barreiro, E.; Fermoselle, C.; Mateu-Jimenez, M.; Sanchez-Font, A.; Pijuan, L.; Gea, J.; Curull, V. Oxidative stress and inflammation in the normal airways and blood of patients with lung cancer and COPD. *Free Radic. Biol. Med.* **2013**, *65*, 859–871. [CrossRef]
50. Qin, L.; Guitart, M.; Admetlló, M.; Esteban-Cucó, S.; Maiques, J.M.; Xia, Y.; Zha, J.; Carbullanca, S.; Duran, X.; Wang, X.; et al. Do Redox Balance and Inflammatory Events Take Place in Mild Bronchiectasis? A Hint to Clinical Implications. *J. Clin. Med.* **2021**, *10*, 4534. [CrossRef]
51. Sancho-Muñoz, A.; Guitart, M.; Rodríguez, D.A.; Gea, J.; Martínez-Llorens, J.; Barreiro, E. Deficient muscle regeneration potential in sarcopenic COPD patients: Role of satellite cells. *J. Cell Physiol.* **2021**, *236*, 3083–3098. [CrossRef]
52. Pérez-Peiró, M.; Duran, X.; Yélamos, J.; Barreiro, E. Attenuation of Muscle Damage, Structural Abnormalities, and Physical Activity in Respiratory and Limb Muscles following Treatment with Rucaparib in Lung Cancer Cachexia Mice. *Cancers* **2022**, *14*, 2894. [CrossRef]
53. Barreiro, E.; Ferrer, D.; Sanchez, F.; Minguella, J.; Marin-Corral, J.; Martinez-Llorens, J.; Lloreta, J.; Gea, J. Inflammatory cells and apoptosis in respiratory and limb muscles of patients with COPD. *J. Appl. Physiol.* **2011**, *111*, 808–817. [CrossRef]
54. Salazar-Degracia, A.; Blanco, D.; Vilà-Ubach, M.; Biurrun, G.; Solórzano, C.O.; Montuenga, L.M.; Barreiro, E. Phenotypic and metabolic features of mouse diaphragm and gastrocnemius muscles in chronic lung carcinogenesis: Influence of underlying emphysema. *J. Transl. Med.* **2016**, *14*, 244. [CrossRef]
55. Ahmad, R.; Hussain, A.; Ahsan, H. Peroxynitrite: Cellular pathology and implications in autoimmunity. *J. Immunoass. Immunochem.* **2019**, *40*, 123–138. [CrossRef] [PubMed]

56. Bartesaghi, S.; Radi, R. Fundamentals on the biochemistry of peroxynitrite and protein tyrosine nitration. *Redox Biol.* **2018**, *14*, 618–625. [CrossRef]

57. Ferrer-Sueta, G.; Campolo, N.; Trujillo, M.; Bartesaghi, S.; Carballal, S.; Romero, N.; Alvarez, B.; Radi, R. Biochemistry of Peroxynitrite and Protein Tyrosine Nitration. *Chem. Rev.* **2018**, *118*, 1338–1408. [CrossRef]

58. Re, R.; Pellegrini, N.; Proteggente, A.; Pannala, A.; Yang, M.; Rice-Evans, C. Antioxidant activity applying an improved ABTS radical cation decolorization assay. *Free Radic. Biol. Med.* **1999**, *26*, 1231–1237. [CrossRef] [PubMed]

59. Sánchez-Moreno, C. Review: Methods Used to Evaluate the Free Radical Scavenging Activity in Foods and Biological Systems. *Food Sci. Technol. Int.* **2016**, *8*, 121–137. [CrossRef]

60. Shimauti, E.L.T.; Silva, D.G.H.; de Almeida, E.A.; Zamaro, P.J.A.; Belini Junior, E.; Bonini-Domingos, C.R. Serum melatonin level and oxidative stress in sickle cell anemia. *Blood Cells Mol. Dis.* **2010**, *45*, 297–301. [CrossRef]

61. Belini, E.; Da Silva, D.G.H.; De Souza Torres, L.; De Almeida, E.A.; Cancado, R.D.; Chiattone, C.; Bonini-Domingos, C.R. Oxidative stress and antioxidant capacity in sickle cell anaemia patients receiving different treatments and medications for different periods of time. *Ann. Hematol.* **2012**, *91*, 479–489. [CrossRef]

62. Lu, J.Q.; Monaco, C.M.F.; Hawke, T.J.; Yan, C.; Tarnopolsky, M.A. Increased intra-mitochondrial lipofuscin aggregates with spherical dense body formation in mitochondrial myopathy. *J. Neurol. Sci.* **2020**, *413*, 116816. [CrossRef]

63. Double, K.L.; Dedov, V.N.; Fedorow, H.; Kettle, E.; Halliday, G.M.; Garner, B.; Brunk, U.T. The comparative biology of neurome-lanin and lipofuscin in the human brain. *Cell Mol. Life Sci.* **2008**, *65*, 1669–1682. [CrossRef] [PubMed]

64. Sohal, R.S.; Brunk, U.T. Lipofuscin as an indicator of oxidative stress and aging. *Adv. Exp. Med. Biol.* **1989**, *266*, 17–29. [CrossRef]

65. Allaire, J.; Maltais, F.; Leblanc, P.; Simard, P.M.; Whittom, F.; Doyon, J.F.; Simard, C.; Jobin, J. Lipofuscin accumulation in the vastus lateralis muscle in patients with chronic obstructive pulmonary disease. *Muscle Nerve* **2002**, *25*, 383–389. [CrossRef]

66. Gosker, H.R.; Zeegers, M.P.; Wouters, E.F.M.; Schols, A.M.W.J. Muscle fibre type shifting in the vastus lateralis of patients with COPD is associated with disease severity: A systematic review and meta-analysis. *Thorax* **2007**, *62*, 944–949. [CrossRef]

67. Gosker, H.R.; Mariëlle Engelen, P.K.J.; Van Mameren, H.; Van Dijk, P.J.; Van Der Vusse, G.J.; Wouters, E.F.M.; Annemie Schols, M.W.J. Muscle fiber type IIX atrophy is involved in the loss of fat-free mass in chronic obstructive pulmonary disease. *Am. J. Clin. Nutr.* **2002**, *76*, 113–119. [CrossRef] [PubMed]

68. Shrikrishna, D.; Patel, M.; Tanner, R.J.; Seymour, J.M.; Connolly, B.A.; Puthucheary, Z.A.; Walsh, S.L.F.; Bloch, S.A.; Sidhu, P.S.; Hart, N.; et al. Quadriceps wasting and physical inactivity in patients with COPD. *Eur. Respir. J.* **2012**, *40*, 1115–1122. [CrossRef] [PubMed]

69. Serres, I.; Gautier, V.; Varray, A.; Préfaut, C. Impaired skeletal muscle endurance related to physical inactivity and altered lung function in COPD patients. *Chest* **1998**, *113*, 900–905. [CrossRef] [PubMed]

70. Plotkin, D.L.; Roberts, M.D.; Haun, C.T.; Schoenfeld, B.J. Muscle Fiber Type Transitions with Exercise Training: Shifting Perspectives. *Sports* **2021**, *9*, 127. [CrossRef]

71. Thompson, L. V Skeletal Muscle Adaptations with Age, Inactivity, and Therapeutic Exercise. *J. Orthop. Sports Phys. Ther.* **2002**, *32*, 44–57. [CrossRef] [PubMed]

72. Gea, J.; Sancho-Muñoz, A.; Chalela, R. Nutritional status and muscle dysfunction in chronic respiratory diseases: Stable phase versus acute exacerbations. *J. Thorac. Dis.* **2018**, *10*, S1332. [CrossRef] [PubMed]

73. Ohira, Y.; Gill, S.L. Effects of Dietary Iron Deficiency on Muscle Fiber Characteristics and Whole-Body Distribution of Hemoglobin in Mice. *J. Nutr.* **1983**, *113*, 1811–1818. [CrossRef] [PubMed]

MDPI

St. Alban-Anlage 66

4052 Basel

Switzerland

www.mdpi.com

Nutrients Editorial Office

E-mail: nutrients@mdpi.com

www.mdpi.com/journal/nutrients

* 9 7 8 3 0 3 6 5 9 5 9 0 0 *